Minnesota's Geology

to Dad

from Thomas &
Jeun

12/1982

The University of Minnesota Press
gratefully acknowledges the assistance for publication
provided by David and Sherrill Fesler.

Minnesota's Geology

Richard W. Ojakangas

Charles L. Matsch

Drawings, charts, and graphs by Dan Beedy

University of Minnesota Press

Minneapolis

Library of Congress Cataloging in Publication Data

Ojakangas, Richard W.
    Minnesota's geology.

    Bibliography: p.
    Includes index.
    1. Geology—Minnesota. I. Matsch, Charles L.
II. Title.
QE127.038   557.76   81-14709
ISBN 0-8166-0950-0   AACR2
ISBN 0-8166-0953-5 (pbk.)

DEDICATION

To the memory of George M. Schwartz and George A. Thiel,
two Minnesota geologists who were pioneers in the art
of communicating geological information to everyone.

Contents

Acknowledgments

We wish to thank the many people whose contributions were vital to the completion of this book. Two deserving of special accolades are Dan Beedy, who drew all of the artwork, including the many maps, and Ken Moran, who produced excellent black and white photographs from our color transparencies.

The many persons and organizations who kindly supplied illustrations are acknowledged in the figure captions. Elizabeth Knight of the Minnesota Historical Society was especially helpful. Typing was expertly done by Jackie Gollinger, Mary Ankarlo, and Mary Nash. We also thank our colleagues at the University of Minnesota, Duluth, and our families for their support throughout this project. Finally, we thank the countless students and friends who provided a constant stimulus by asking, "When will that book be done?"

Preface

This book about Minnesota's geology was written in an effort to share with all interested persons the excitement of scientific discovery and the fascinating story held within the rocks and landscape of an area we both have come to know as field geologists and as continuing students of earth history. Just as the earth itself is ever changing, so is the depth and breadth of our understanding of geological processes and our perception of its history. Therefore, it is not surprising that this book contains a great deal of geologic data and history not available for inclusion in its predecessor, *Minnesota's Rocks and Waters*, masterfully written by the late George M. Schwartz and the late George A. Thiel and published in 1954. Because many things have not changed, and because the personal experiences of individuals always influence what they write about, that book will continue to stand as an important and useful guide to Minnesota's geology. Our book is intended as a companion volume that brings to the public the views of current geologists, reflecting the fact that geological knowledge has grown by leaps and bounds during the last three decades.

Although at first glance the language of geology appears to be formidable to the point of deflecting the reader to more conventional prose, the slight effort required to master the basics will surely be rewarded by the understanding of so important a part of the natural world. Make the effort, and share with us the excitement of exploring the rich geological heritage of this great state. Widespread mountains, explosive volcanoes, relentless glaciers, and long-extinct life forms of both land and sea have been a part of Minnesota's ancient history. It is a story for us all, young and old.

Our total experience as geologists, although adequate to give us perspective and to contribute specific details about some of the geology, by no means was sufficient to cover all these geological riches. We acknowledge with gratitude the contributions of the many earth scientists whose observations and ideas permeate the entire book. Also, our personal involvement with geology has been enriched not only by the geologic investigations of others but by our long association with curious, energetic, and altogether stimulating students as well. For them, we have special thanks.

PART I

Background

In order for you to better appreciate the geologic events that have shaped Minnesota, some introduction of general geologic concepts seems necessary. This we have presented in the first chapter. If you have already had some background in geology and this chapter seems to be a repetition of material, we apologize. It is just that we want everyone to be able to understand and appreciate the science that we two geologists find to be both career and avocation.

Chapter 2 indicates how Minnesota's geologic past has been studied over the last century and how Minnesota fits into the grander geologic picture.

A part of Minnesota's history is revealed in these rocky bluffs along the
Mississippi River and Lake Pepin in southeastern Minnesota.

1

The Briefing

The Minnesota River Valley is a trench that originates on the western border of Minnesota, slices southeastward to Mankato, and then bends sharply to the northeast, where it joins the Mississippi River Valley at St. Paul near Fort Snelling. This capacious valley, up to 60 meters (m) deep and in places almost 5 kilometers (km) wide, was excavated about 12,000 years ago when the rising waters of a large, newly formed lake, fed by melting glacial ice, overtopped a natural earth dam near what is now Browns Valley. The ensuing flood not only dropped the level of the lake significantly but scarred irreparably the land surface across which it raced. Successive floods, each as energetic as the first, pulsed down the watercourse established by the initial burst. The turbulent waters eroded deeply into the soil and underlying bedrock before the glacial lake was finally drained dry. Eventually, the Minnesota River established itself as the feeble successor to the mighty outlet of that glacial lake. Although an important event in its own right as part of the geologic history of Minnesota, the excavation of that great trench across the midsection of the state also brought into view through a window 400 km long a geologic past spanning more than 3,500 million years.

Low outcrops of dark red rock on the valley floor near Montevideo, blasted through to make way for U.S. Highway 212, sparkle to the passing motorist as the sun is reflected back from the crystal faces of minerals that are among the oldest materials ever to be identified in the crust of the earth (Figure 1-1). From quarries at Morton huge blocks of rock equally as old are raised by cranes to be loaded and shipped throughout the United States as architectural stone. The importance of these rocks to an understanding of the early history of the earth is so great that geologists from all over the world come to western Minnesota to study them in their natural surroundings.

The minerals and structures in these primitive rocks reflect an environment of formation much different from that in which they now lie as outcrops. They indicate recurring melting, crystallization, and deformation at temperatures and pressures known to exist only tens of kilometers beneath the earth's surface. Elevation of the rocks to their present position in the crust and their exposure to today's environment occurred as the result of a remarkable series of geologic events, from mountain building to the erosion of great volumes of overlying bedrock.

These oldest rocks, so solid and fresh looking in the Morton quarries, at other places along the sides of the river valley have been decomposed into soft, white clay, a result of prolonged exposure to subtropical weather conditions more than 70 million years ago. The contemporaneous effects of erosion by rivers and deposition in the floodplains, lakes, and seas of a long-vanished landscape are preserved in layered sequences of sand and mud at Redwood Falls, New Ulm, and many other places in the western part of the valley. Plant fragments, seashells, shark teeth, and even crocodile bones document the complexities of life in that distant past.

Northeastward from Mankato, the sides of the valley are buttressed with the sediments of even earlier seas. Layers of sandstone, limestone, and shale deposited about 450 million years ago thicken and interfinger toward the Twin Cities. Not only is the history of repeated marine invasions preserved in fine detail here, but the evolutionary course of early life is shown by the fossil record.

Away from the valley, the surrounding uplands are generally devoid of bedrock exposures. Instead, the land surface is underlain by a thick blanket of stony sediments that represents still another drastic change in the environment, a great ice age that furnished a cover of boulders for most of

Figure 1-1. This outcrop of rock near Montevideo in western Minnesota exposes some of the earth's oldest crustal material. Determined to be more than 3,600 million years old, the rock, called the Montevideo Gneiss, is still a billion years younger than the earth itself.

the state. That period of cold climate and glacial activity ended just 12,000 years ago.

Elsewhere in Minnesota, rocks and other earth materials existing in natural exposures or encountered by drill bits in search of water or mineral deposits reflect even more complex events, from volcanic eruptions to the splitting apart of the continent itself. To explore the rock record and properly read the history it contains, it is necessary to understand the processes by which earth materials are formed and destroyed; to recognize the existence, map the boundaries, and determine the duration of ancient environments; and to establish the relative chronological order of a complicated series of geologic events. Finally, the most difficult question of all must be answered: when did these events in Minnesota's geologic past take place within the framework of real time?

EARTH MATERIALS

Rocks

Rocks make up the solid foundation underlying the surface of the earth. They are so varied in composition, texture, and color that the casual observer is almost immediately discouraged from trying to learn about their origin and significance. Too bad. Lost is the excitement of exploration and discovery that is every prospector's reward. Lost also is the opportunity to read geological history firsthand, for it is the rocks that contain the record of the earth's past. Rocks originate in all of the diverse environments that exist within the planetary system. The number of the possible

combinations of temperature, pressure, and chemical compositions within the earth and upon its surface makes the construction of a framework for understanding the origin or rocks appear to be an awesome task. However, the number of combinations is not limitless.

Igneous Rock

The black, fine-textured rocks that rim Lake Superior's northern shore at Duluth contrast sharply with the coarse-grained, multicolored outcrops exposed in the quarries around St. Cloud. Yet they have one thing in common: both rock types are collections of minerals that formed during the cooling of a hot liquid material called *magma*. They are both igneous rock. These two varieties of igneous rock are the most common constituents of the cold, brittle outer layer of the earth, its crust. The rocks of the North Shore, oddly enough, are the common constituents of ocean basins, whereas those exposed at St. Cloud are typically present in most continental rock masses.

The fine-grained texture of the North Shore rocks resulted from rapid cooling of molten material spilled out onto the earth's surface as lava flows. The minerals they contain reflect the chemical composition of the mother magma, in this case a mixture rich in iron, magnesium, aluminum, silicon, and oxygen. These two characteristics, its texture and its mineral composition, form the basis for giving the rock a specific name—basalt. St. Cloud's granite, on the other hand, cooled beneath the surface and it cooled so slowly that larger crystals formed. The composition of the granite-forming magma was very different from that which solidified into basalt. Granite is richer in potassium, sodium, silicon, and oxygen. Granitic magma, if it reaches the earth's surface as lava, will cool to form a fine-textured rock called rhyolite. The coarse-grained equivalent of basalt is gabbro, a rock common to northeastern Minnesota (Figure 1-2; Plate 6).

Many other varieties of igneous rock exist in the earth's crust, and they have been given names, depending upon variations in their mineral composition and texture. Table 1-1 is a summary of the more common igneous rocks and their characteristics. The general term *intrusive* applies to igneous rock that crystallized beneath the surface of the earth. *Extrusive* igneous rock is rock that formed in the surface environment from hot material ejected during volcanic activity.

Sedimentary Rock

Sedimentary rock, such as the sandstone, shale, and limestone that line the bluffs of the Mississippi River Valley south of St. Anthony Falls in Minneapolis, is the reconstituted debris that results from the destruction of earth materials in the surface environment (Figure 1-3). Bedrock exposed to weathering and erosion either breaks up to form sediments or decomposes chemically. Eventually the rock and mineral fragments, along with dissolved material, are swept away by the moving fluids that bathe the entire surface of the earth, especially running water. From the continents, most of the debris is washed into the ocean basins. There, waves winnow out finer material, leaving behind sandy beaches, and finer muds settle in the quiet depths. Under the proper circumstances the dissolved material is precipitated in the form of chemical sediments. Compaction and cementation turn the sediment into rock. Sedimentary rocks are given names depending upon their composition and the size of particles they contain, or if they are chemical precipitates, upon their chemical composition alone (Table 1-2).

Metamorphic Rock

A third group of rocks results from natural changes that take place when earth materials are subjected to heat, pressure, and chemically active fluids. These metamorphic rocks owe their particular character to the composition of the parent material and to the intensity of the agents of metamorphism. Thus, limestone recrystallizes into marble, shale changes to slate or to schist, and granite may be reconsti-

tuted into gneiss. Metamorphism generally incites the growth of new minerals from the ingredients of the parent rocks, and these, in turn, record the temperature and pressure conditions of their formation.

One characteristic of some metamorphic rocks is the growth of platy mineral crystals, especially mica, all in the same parallel direction. This alignment of flat crystals gives

Figure 1-2. The most abundant rocks in the earth's outer shell are granite (lower left) and basalt (upper right). Granite and its fine-textured chemical equivalent, rhyolite (upper left), make up the bulk of the continents. Basalt and its coarse-textured relative, gabbro (lower right), generally make up the rocks beneath the ocean basins. The granite is from the St. Cloud area, the rhyolite is from Grand Marais, the basalt is from the North Shore of Lake Superior, and the gabbro is from Duluth. See Plate 6.

Table 1-1. Characteristics of the Common Igneous Rocks

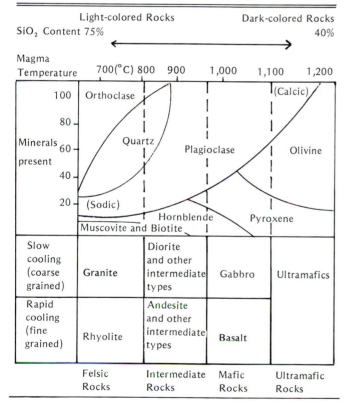

Note: The two major igneous rock types, granite and basalt, are shown by bold print. The other rock types (and hundreds of others could be included) are volumetrically minor. Minerals present in each rock can be determined by reading upward from the rock name. For example, granite can contain muscovite, biotite, hornblende, sodic plagioclase, quartz, and orthoclase, but the essential minerals are orthoclase and quartz, which occupy most of the upper part of the box.

Figure 1-3. This exposure of dolomite in Goodhue County displays horizontal layering, or *bedding*, the most distinctive characteristic of sedimentary rocks. (Courtesy of David Stone.)

Figure 1-4. Some metamorphic rocks, such as this specimen, gneiss, have a layered aspect called foliation. Foliated rocks result when minerals, especially micas, grow in the same planar orientation. The hand lens provides a scale.

Table 1-2. Classification and Identification of the Common Sedimentary Rocks

CLASTIC SEDIMENTARY ROCKS

Grain Size	Unconsolidated Sediment	Sedimentary Rock
>2 mm	Gravel	Conglomerate (rounded grains); Breccia (angular grains)
2 mm to $\frac{1}{16}$ mm	Sand	Sandstone: quartz sandstone (composed of nearly 100% quartz); arkose (contains abundant feldspar); graywacke (poorly sorted mixture of quartz, feldspar, and rock fragments in a fine-grained matrix)
$\frac{1}{16}$ mm to $\frac{1}{256}$ mm	Silt	Siltstone
$<\frac{1}{256}$ mm	Clay	Shale

CHEMICAL SEDIMENTARY ROCKS

Dominant Mineral	Occurrence	Rock Name
Calcite ($CaCO_3$)	Small crystals; contains fossils	Crystalline limestone; fossiliferous limestone
Dolomite $[CaMg(CO_3)_2]$	Small crystals	Dolostone
Quartz (SiO_2)	Smooth, gray or black	Chert
Halite (NaCl)		Rock salt
Gypsum ($CaSo_4 \cdot 2H_2O$)		Rock gypsum

the rock a layered aspect called *foliation* (Figure 1-4). Slate, schist, and gneiss all display this mineral layering, and these rocks tend to break along the planes of mica concentration. Quartzite, which is metamorphosed sandstone, and marble, which is recrystallized limestone, are coarse and granular. Both lack foliation. Table1-3 is a summary of the common types of metamorphic rock.

Minerals

Minerals are the natural chemical compounds of which rocks are composed. Minerals are combinations of the various elements, the basic kinds of matter that reside in the planetary system. Seven or eight of the elements are so abundant that their chemical characteristics control the way things are. Two of these, oxygen and silicon, are the most important constituents of the earth's crust, making up about 93% of it by volume. The combination of silicon (Si) and oxygen (O) produces the mineral quartz, which has the chemical formula SiO_2. This glassy-looking mineral is an important constituent of granite, and because of its durability in the surface environment it is an important constituent of most sediments. Chemical bonding of silicon and oxygen with the other elements, notably aluminum (Al), iron (Fe), calcium (Ca), potassium (K), sodium (Na), and magnesium (Mg), produces most of the other common minerals, the silicates, that are found in the earth's crust.

Only a few rocks do not contain the silicate minerals as important constituents. Limestone is a chemical sediment that has the mineral calcite, or calcium carbonate ($CaCO_3$), as its major ingredient. Dolostone is composed of dolomite, or calcium magnesium carbonate $[CaMg(CO_3)_2]$.

Because granite and basalt are the most common rocks in the crust, their constituents dominate the mineral kingdom. Learning to identify just half a dozen minerals, such as quartz, several different feldspars, mica, hornblende, augite, and calcite, enables one to recognize the most commonly encountered earth materials (Figure 1-5; Plate 9). The real challenge comes from a close encounter with some of the less abundant and more exotic of the approximately 2,800 other combinations of elements found in nature. No wonder that mineral collecting is such a popular and stimulating hobby. Prospecting for economically valuable mineral deposits is a primary task of geologists.

Minerals have characteristic physical and chemical properties determined by the nature of the elements they contain and by the way those elements interact as particles of matter bonded together in the solid state. Generally, the size and electrical properties of individual atoms decree a particular internal arrangement for each different mineral. Such a condition of internal order is often reflected in the external form as a symmetrical array of smooth crystal faces. The geometry of individual crystals or *crystal form*, is one of the clues useful in the identification of minerals. Even when such exterior form is not present, however, the internal order of the elements within a mineral fragment can be detected by X-ray examination.

In many minerals, the strength of bonding between adjacent atoms is not constant in all directions. A plane of weakness, along which a mineral will preferentially break when subjected to stress, is called a *cleavage* plane. The number and orientation of such cleavages are important physical properties in the identification of certain mineral species. Mica, for example, has one cleavage, and it is easily broken into thin, shiny sheets. Feldspar, the most common mineral in crustal rocks, breaks along two planes almost at right angles. Galena, a compound of lead and sulfur, cleaves into cubes because it has three directions of weak bonding. Quartz, one of the toughest minerals, is equally strong in all directions. Rather than displaying smooth cleavage faces, quartz fragments are bounded by irregular surfaces.

The strength of its internal atomic bonding also determines how resistant a mineral is to abrasion. The relative *hardness* among a group of different minerals is established by a simple test. The hardest will scratch all the others. The weakest will make its mark on none. Geologists have long used a standardized scale on which talc (assigned a hardness of 1) is the softest mineral and diamond (assigned a hardness of 10) is the hardest mineral.

Luster is the term used to describe the way light is reflected from a mineral's surface. Descriptive words range

Figure 1-5. These minerals are the most important constituents of the earth's crust. Quartz, the feldspars, augite, hornblende, and the micas form into igneous and metamorphic rocks. Calcite precipitates to form limestone. The feldspars weather to clay, which is the dominant mineral in sedimentary rocks. All, in fact, can be found as fragments in sedimentary rocks. Clockwise, from upper left, the minerals are orthoclase, plagioclase, quartz, muscovite mica, calcite, clay, biotite, mica, olivine, augite, and hornblende. See Plate 9.

Table 1-3. Identification of the Common Metamorphic Rocks

Name	Structure	Texture	Mineralogy	Location in Minnesota and Other Remarks
FOLIATED				
Gneiss	Coarsely to finely banded or streaked	Coarse to fine grained	Quartz, feldspar, micas, and hornblende common;	Minnesota River Valley. Common building stone; bands usually of different composition
Schist	Finely foliated or schistose; breaks along foliation	Coarse to fine grained	Micas predominant; talc, chlorite, hornblende, garnet, quartz, etc.	South of Little Falls; northern Minnesota. Staurolite crystals can be collected near Little Falls; many platy or elongate minerals
Slate	Finely foliated; mineral grains not visible; has slaty cleavage	Extremely fine grained	Variable; mineral grains not visible to the naked eye	Thomson, Carlton; Vermilion district. Very smooth rock cleavages, used for roofing, etc.
MASSIVE				
Quartzite	Massive, occasionally gneissic or schistose	Coarse to fine grained; equigranular	Predominantly quartz, may have micas, feldspars, etc.	Iron Range; southwestern Minnesota. Very glassy luster; breaks through grains
Marble	Massive, occasionally gneissic or schistose	Coarse to fine grained; usually equigranular	Calcite, dolomite, sometimes with serpentine or other minerals	Effervesces in dilute HCL; cleavages visible
Granofels	Massive	Coarse to fine grained; usually equigranular	Variable; usually nonplaty minerals (quartz, feldspar, garnet, pyroxene, calcite, etc.)	General name for any massive metamorphic rock

7

from "glassy" to "metallic." The brilliant luster of diamond is distinctively different from the pearly reflection of talc or the resinous appearance of sphalerite. *Color*, too, is very helpful in mineral identification. Unfortunately, some of the common minerals display such a wide range of color that other characteristics must be resorted to. A collection of all the different hues of quartz could be used to construct a crystal rainbow. Milky, purple (amethyst), yellow (citrine), rose, and smoky quartz are the most common variations from the uncolored, transparent rock crystal form of quartz. A more consistent and useful property for mineral identification than surface color is the color of the powder produced by abrasion, called *streak*.

The right combination of hardness, luster, and color in certain mineral species results in the durability and beauty that makes them appealing as decorations and possessions. Precious and semiprecious gemstones, such as diamond, emerald, sapphire, ruby, and aquamarine, are valued for their beauty and their rarity. Minnesota's rocks, unfortunately, do not contain gemstones of any great value. However, many attractive varieties of quartz and other silicate minerals such as amethyst, agate, and thomsonite give mineral collectors ample rewards for their searches.

Rock Bodies

The conditions of formation of the three general rock types and their mineral constituents result in characteristic regional geometries for the rock bodies that eventually are formed (Figure 1-6). Recall that igneous rocks are intrusive when magma is injected beneath the earth's surface and that magma flowing out onto the earth from beneath cools into extrusive lava flows. A *dike* is an intrusive rock body that cuts across the grain of other rocks. A *sill* follows the general trend of the rock layers it intrudes. A large body of intrusive rock, especially granite, measuring tens of kilometers across is called a *batholith*.

Sediments accumulate to significant thicknesses wherever there are profound low spots on the earth's surface. A pile of sedimentary rock, therefore, is generally a wedge or prism of material whose dimension is controlled by the geometry of the depression it fills. From large ocean basins of irregular shape to the smallest lakes, depressions eventually acquire a distinctive sequence of horizontally layered sedimentary material. Other environments of deposition of sedimentary material are river valleys, glaciated highlands, and windswept deserts.

Metamorphic rocks are generally associated with mountain belts. They crop out in places where erosion has exhumed the more ancient crust from deep burial. Because they are formed under extreme heat and pressure, these rocks can also be found along the margins of large igneous bodies, where heat from the intruding magma has cooked the surrounding (or *country*) rock.

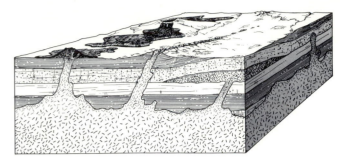

Figure 1-6. This block diagram shows the major forms of extrusive and intrusive igneous rock bodies, Slower cooling of magma beneath the surface of the earth generally results in a coarser texture (i.e., in larger crystals).

Soils

The term *soil* means different things to different people. To the engineer, soil is any unconsolidated sediment, in contrast to hard bedrock, the rock closest to the earth's surface at any given place. To the agronomist, it is a complicated physical, chemical, and biological system that provides the necessary nutrients for plant growth. Geologists know it as a rind of rotten rock resulting from the exposure of bedrock to the surface environment. At any rate, fresh bedrock makes up only a small percentage of the surface area of Minnesota because a variety of processes are constantly at work to break it down and decompose it. A good term (because it is less ambiguous than *soil*) for the resulting loose aggregate of rock and mineral fragments that eventually blankets most bedrock is *regolith*. Most of Minnesota's regolith has been moved from its original bedrock source and redeposited as sediment by glaciers, rivers, and wind.

An excavation into the regolith generally reveals a layered sequence of colors and of materials. Regolith consists of a gray to black organically rich horizon (i.e., layer) at the surface, then a horizon of brown or reddish brown fine-textured material, and finally a horizon of relatively fresh sediment or bedrock. This *soil profile* results from chemical changes induced by the climate and by biological activity (Figure 1-7). The surface horizon (black dirt) is rich in decayed organic material called humus. Below that is a zone of redeposited chemicals, especially iron oxides, and fine particles of clay washed down by percolating water. The parent material can be anything from bouldery glacial sediment to fine silt on a river floodplain to bedrock.

Variations in vegetation, topographic position, and length of time a material is exposed to soil-forming processes result in great differences in soil profiles. Recognizing and mapping the extent of different soil types is an ongoing task of the Soil Conservation Service, a branch of the United States Department of Agriculture. Soils in Minnesota are generally developed on parent material associated with past glacial activity and the melting of sediment-laden glacial ice. Most of Minnesota's soils have formed within the last 15,000 years, and some are much younger. The relative youth of the regolith and the freshness of the parent material are two reasons that the soils of Minnesota are so fertile. Not enough time has transpired to remove the nutrients from their mineral sources in the relatively young parent material.

Fossils

The life of past ages in the history of earth is known from the discovery and study of fossils, the remains of organisms buried and preserved, mainly in sediments and sedimentary rocks (Figure 1-8). Fossils consist not only of hard body parts such as bone and shell but also of tracks, trails, feces, and burrows. Paleontology is the science of this fossil record. Geology encompasses inquiry into the history of both the physical and the biological changes recorded in the rocks.

An important conclusion from a systematic study of fossils is that life forms have changed significantly through time. From very simple and limited types of living things in the older rocks, there appears to be a progressive development toward complexity and diversity. Another fascinating aspect of the fossil record is the indication that organisms periodically became extinct, never to reappear in a later geologic period. Knowledge of the evolution of species and their eventual extinction helps geologists in their attempts to place rock units in a time frame and to identify rock bodies of the same age from place to place. In 1962, a young Minnesota geologist, G. F. Webers, while exploring a mountain range in Antarctica, discovered a layer of fossil-rich limestone in a steep cliff. He recognized immediately that the fossils were similar to those he had seen in the bluffs of the St. Croix River south of Taylors Falls, and he concluded, correctly, that those two rock bodies were formed at about the same time. Correlation of other pairs of rock bodies based on their fossil content allows the reconstruction of the geography of the earth at various times in the past.

Another important use of fossils is the reconstruction of ancient environments. Communities of organisms today are adjusted to particular living conditions. Each assemblage is a dynamic system, mutually dependent and adjusted to a variety of complex influences. The tidal zone along the coast of Maine is populated by a distinctly different group of plants and animals than is the shoreline of Lake Superior. Knowledge of the ecological details of modern environments allows the identification of past environments from the characteristics of the rocks and fossils they contain, because it can be assumed that natural processes evident today occurred in the same way in the past. This assumption, known as the *principle of uniformitarianism*, has played a valuable role in the development of the science of geology since the latter part of the eighteenth century. Simply stated, the principle is that the present is the key to the past. Obviously, present processes include both those that are slow and those that are catastrophic.

Figure 1-7. The distinctive layering in this exposure of soil is called a soil profile. Decayed organic matter gives the uppermost layer a black color. Other layers are the result of chemical and physical changes collectively called soil-forming processes.

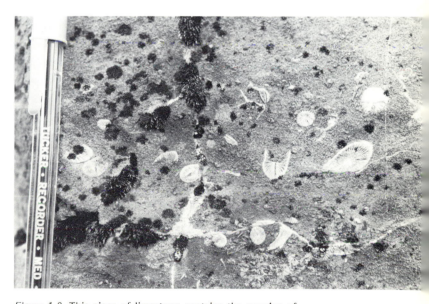

Figure 1-8. This piece of limestone contains the remains of organisms that lived in a warm, shallow sea about 350 million years ago. Such fossils are important records of past life, and they are useful in determining the ages of the rocks that contain them.

Figure 1-9. The relationships displayed in this exposure of crosscutting igneous dikes indicates that the lightest-colored rock is the youngest.

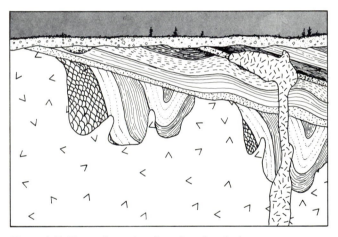

Figure 1-10. Try to determine the order of geologic events portrayed in this generalized cross section of northeastern Minnesota. You will see this sequence again in Chapters 3, 4, and 5. By that time the answers should be evident.

GEOLOGIC TIME

Long before the real ages of rocks could be determined, geologists were able to order many geological events in a relative way from oldest to youngest by making use of two simple but important commonsense principles. The *principle of superposition* states that, in a pile of undisturbed sedimentary rocks, the oldest resides on the bottom and the youngest on the top. The same principle applies to lava flows. One special requirement for the application of this principle to rock bodies that have been broken and folded is recognizing the original top and bottom of the pile. Another useful relationship concerns rock bodies such as

igneous dikes and mineral-filled veins that cut across other formations (Figure 1-9). Such a rock body is always younger than the rocks it crosscuts, according to the *principle of crosscutting relationships.*

Applying these principles to a generalized cross section of the rocks in northeastern Minnesota (Figure 1-10) gives the following sequence of events, listed from the oldest to the youngest: (1) eruption of lavas and sedimentation; (2) folding, granite intrusion, and erosion; (3) deposition of sediments; (4) folding and erosion; (5) eruption of flows; (6) intrusion of gabbro and tilling; (7) deposition of glacial debris. Thus, the broad outlines of a geologic history are developed from a careful investigation of the rocks and their relationships to each other.

Methods to determine the real age of earth materials are based mainly on the measurement of certain radioactive elements and the products of their eventual decay. Granite in the Minnesota River Valley cooled and crystallized into a mass of minerals, primarily feldspar, quartz, and mica. One minor mineral found in that rock is zircon, an extremely durable combination of zirconium, silicon, and oxygen. But also trapped within the crystal structure are small amounts of the element uranium in several forms, or *isotopes.* Uranium is radioactive; that is, its atoms emit from their nuclei particles of mass and wave energy. In a series of such emissions, an atom of uranium changes, or decays, into lead, which is stable. The rate at which each isotope of uranium decays into the so-called daughter products is known. Any given amount of the isotope uranium-238, for example, will be reduced by one-half in about 4,500 million years, which coincidentally is the approximate age of the earth. This *half-life* of 4,500 million years for uranium predicts that, since the earth's beginning, uranium-238 should be less abundant by one-half and lead should be more abundant. To return to the granite in the Minnesota River Valley, careful measurements of the relative amounts of the uranium-238 and the lead-206, its final daughter product, trapped in the zircons since the rock formed allows the calculation of the age of the rock, that is, of how long decay has been turning uranium into lead. Measurements show that this process has been going on for about 3,600 million years.

Other radioactive elements such as rubidium, thorium, potassium, and carbon-14 (an unstable form of carbon) are the bases for measuring the ages of other minerals. The constancy of their half-lives makes these elements radioactive clocks. Each radioactive element has a different half-life, some of very short duration. Carbon-14 decays to half its original volume in 5,730 years.

Not all rocks can be age dated. Some may not contain the required radioactive elements; others may not have been properly sealed since their formation. Leaky systems that allow parent and daughter elements to come and go are worthless for age determination. Igneous rocks are the most valuable for age determinations, because their minerals can crystallize into effective traps for radioactive elements and the daughter products they spawn. Sedimentary rocks can rarely be dated directly, and by their very nature they are mixtures of rock and mineral fragments derived from bedrock of different ages. Determining the age of a granite pebble contained in an outcrop of sandstone tells nothing about when the pebble finally came to rest in the sandstone after erosion from its original place of formation and hence nothing about the age of the sandstone itself. Metamorphic rocks commonly have experienced such extreme heat and pressure that the original minerals have been destroyed and then reconstituted. Age determinations of such rocks generally indicate the time of metamorphism, but not the antiquity of the original materials before metamorphism.

Carbon is an important constituent of all living things. Carbon-14 is manufactured in the atmosphere from nitrogen atoms struck by high-energy particles of matter, and, during the life cycle, organisms are contaminated with a small amount of carbon-14. The ratio of radioactive carbon to stable carbon is the same for all living organisms. After death, the radioactive carbon decays slowly back to nitrogen, with no replacement occurring. Thus, measuring the amount of carbon-14 left in organic material at any time fixes the number of years that have passed since the death of the organism. Because of the short half-life of carbon-14 and the fact that it is produced in very small quantities, this dating technique is useful only for objects less than about 50,000 years old.

Even before the real age of the earth was known, geologists recognized the enormous span of time it must have taken for all the rock-forming events and geologic structures open to view on the earth's surface to occur. In fact, early estimates relating to the age of the earth were based on calculations of the rate at which sedimentary rocks accumulated. As the earth's crust was systematically explored, geologic time was subdivided into segments represented by particular bodies of rock. Fossil content, deformation, and buried erosion surfaces were all used to provide important time boundaries in the subdivision. The largest segments of time are called eras; those, in turn, are divided into geologic periods. An epoch is a time segment even smaller than a period. The eras and periods are all formally named, but an epoch is generally designated as an early, middle, or late segment of a period. For example, the first part of the Cambrian Period, the Early Cambrian Epoch, is not represented by rocks in Minnesota, nor is the next segment of that period, the Middle Cambrian. However, events in the Late Cambrian Epoch left a significant rock record, rich in fossils, indicating a major shift in the position of the ocean boundaries in North America at that time. All of the earth materials formed during a geologic period of time make up a geologic *system* of rocks, and rock units within the system are further divided into lower, middle, and upper parts. Thus, the term "Lower Cambrian rocks" includes all of the

rocks formed during the Early Cambrian Epoch. Figure 1–11 is the geologic time scale currently in use in North America.

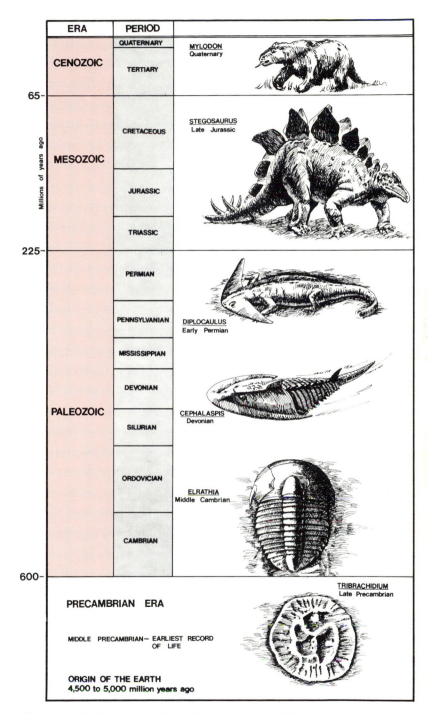

Figure 1-11. Geologic time is divided into eras and subdivided into periods. The animals shown were selected to represent the development of life forms through time.

EARTH STRUCTURES

From geologic studies over the entire surface of the earth it is evident that the earth's history has been punctuated by periods of major igneous activity and mountain building followed by long periods of erosion and the accumulation of sediments. A period of crustal upheaval and magma emplacement is called an *orogeny*. The forces accompanying such an event are great enough to squeeze, break, and lift large segments of the crust, compressing sedimentary layers into *folds*. Upfolds are called *anticlines*, and downfolds are called *synclines*. If the stress is great enough, the rocks are broken along *faults*, and enormous blocks thus produced may be moved thousands of meters vertically or horizontally. Eventually, these mountain-building forces stall, and erosion wears the highlands down. Subsequent burial of the old land surface by deposition produces an *unconformity*. Unconformities in the geologic record generally represent gaps in the earth's history, because important pages have been removed by erosion. Nowhere is there a complete rock record spanning all of geologic time, because the crust has experienced many upheavals accompanied by erosion.

Weaker vertical movement of the earth's crust is called *epeirogeny*; such minor shifting is constantly going on in the form of gentle uplift and subsidence in response to gravity. The effect of this movement can be dramatic in areas of low relief with elevations near sea level; in such environments, marine transgressions and regressions result.

PLATE TECTONICS

On an even larger scale, the earth's crust consists of three major structures—continents, island arcs, and ocean basins. Ocean basins alone cover more than 70% of the surface of the earth. One obvious difference between these major features is that ocean basins are low spots and continents stand high. More important to understanding their origin is the fact that oceans, beneath a thin layer of sediment, are floored by basaltic rocks whereas continents are composed mainly of granitic material. Granite is less dense than basalt (i.e., it contains less mass per unit of volume). In a way, then, the continents are floating in a "sea" of basalt.

Oceanographic surveys were accelerated during World War II, and worldwide studies of the ocean basins since then have revealed some startling facts about their nature. A widely held view that the ocean depths are essentially flat, featureless surfaces was shattered with the discovery of thousands of submarine volcanoes. Then an oceanic mountain system that crisscrosses the entire planet like the stitching on a baseball was charted. Island arcs, such as the

Aleutians and Japan, were found to be bordered by deep trenches. Eventually, the true picture developed: submerged beneath the oceans' waters is a topographic surface even more varied than the landscape of the continents.

That discovery, along with observations about the geological nature of the various features, led to a theory about the origin of ocean basins based on the notion of sea-floor spreading. This widely accepted theory, the *plate tectonic theory*, identifies the oceanic ridges as places where basaltic magma rises to the surface and cools to form new ocean crust. Elsewhere, along certain continental margins, ocean crust sinks back into the *mantle* (the 2,900-km-thick zone beneath the thin crust), where it remelts into magma at the same rate that crust is generated at the ridge systems. Some of this magma rises to the surface as volcanoes, forming island arcs. The development of this enormous recycling system has broken the brittle outer part of the earth into large segments, called plates, that move about relative to each other driven by forces still not well understood. North and South America, for example, are embedded in one thick plate like huge logs frozen into great slabs of lake ice. Continents drift along with these lithospheric plates at modest rates averaging a few centimeters each year.

According to the plate tectonic theory, the present Atlantic Ocean Basin began to form just 200 million years ago, when a supercontinent, called Pangaea, developed a series of cracks or rifts, along which magma began to rise. Slowly, South America separated from Africa and North America separated from Europe and North Africa as new ocean floor developed along the Mid-Atlantic Ridge. Along the leading edge of the westward-moving plate, older ocean crust was consumed as it sank beneath the plate boundary. Magma generation and other instabilities along those boundaries resulted in volcanic and even mountain-building activity. And, of course, the continents themselves drifted across planetary climatic boundaries, resulting in slow but profound changes in the surface environment of each through time.

Minnesota's midcontinental position was thus attained only as a result of the opening of the Atlantic Ocean. The state continues to drift westward embedded in the middle of the North American raft, a relatively stable position far from the stresses generated at the plate boundaries. That stability is reflected in the low incidence of earthquakes, an absence of volcanic activity, and no mountain-building activity. Vertical movements in the form of broad warping, both up and down, have been the only signs of unsteady behavior for hundreds of millions of years.

Minnesota's geologic history is contained in its rocks and landscape. Understanding geologic processes as they occur today and knowing the formations they produce allow the correct interpretation of ancient sequences of rocks, because natural laws have not changed. Fossils provide a valuable record of the development of life forms on the planet. The presence of similar fossil assemblages in widely separated rock formations allows the correlation of these formations to the same period in geologic time. Geologic events can generally be placed in a correct chronological sequence by applying the principles of superposition and crosscutting relationships. Real ages are determinable where certain radioactive elements have been trapped in rock-forming minerals. Although the oldest rocks yet discovered are about 3,600 million years of age, the earth is probably much older, in the range of 4,500 to 5,000 million years. Finally, igneous, sedimentary, and metamorphic rocks form in distinctive environments. Studying rock bodies, their structures, and the minerals and fossils they contain allows the reconstruction of the events that produced them. The application of all this knowledge to Minnesota's rocks and landscape has produced the detailed discussion that makes up the remainder of this book.

Tiled beds of the Middle Precambrian Thomson Formation in Jay Cooke State Park, in northeastern Minnesota, display a long history of sedimentation, metamorphism, uplift, and erosion.

2

Minnesota's Place in Geologic History

Minnesota is situated near the low-relief center of the North American continent, about equidistant from North America's two big mountain belts, the Rockies and the Appalachians. One might think it unfortunate that Minnesota has no mountains. But Minnesota had its day. Twice—2,700 million years ago and again about 1,800 million years ago—sizable mountains were part of the Minnesota and Lake Superior scene. Spectacular volcanism was commonplace at those times and 1,100 million years ago as well. But 2,700 million years is a tremendously long expanse of time for erosion to have been at work. The 225-million-year-old Appalachians and the 60-million-year-old Rockies are still around only because they are so young.

Minnesota sits astride two of North America's largest physiographic or topographic provinces, the Laurentian Upland and the Interior Lowland (Figure 2-1). Both provinces —the Laurentian Upland with its ancient Precambrian igneous, metamorphic, and sedimentary rocks and the Interior Lowland with its much younger Paleozoic sedimentary rocks —are low lying and quite featureless. Why? For the last 1,000 million years or so, what is now the state of Minnesota has been stable and quiet, with no mountain building, no volcanism, probably only local faulting and earthquake activity, some gentle and broad upwarping and downwarping of the underlying crust, and slow advances and withdrawals of the seas. Certainly the most exciting midcontinent events since the last volcanism were the repeated advances and retreats of ice sheets 1,000 m thick that originated in northern Canada. During the last couple of million years, this process of advance and retreat occurred four times. Geologically speaking, the last ice sheet arrived a couple of days ago and melted back just yesterday. Although no people were here to witness the mountains, volcanoes, and seas, it is quite possible that paleo-Indians saw the last glacier retreat to northern latitudes.

The ancient Precambrian volcanism and sedimentation was on a grand scale and resulted in many thousands of meters of layered rocks. The subsequent mountain building tilted and folded these thick sequences. As granites rose into these rocks, adjacent parts of the tilted sequences sank into the crust. Even though the long erosion has removed the tops of the mountains, the mountain roots remain. These roots are detailed evidence of early, violent events. The history is here in the rocks of northern Minnesota and the Minnesota River Valley. It is the interpretation of these rock pages of nature's most ancient history books that both delights and frustrates the earth scientist.

North America, like every other continent, has a nucleus or *shield* of old Precambrian rocks. Most of North America's history—from 3,600 million years ago until 600 million years ago—is documented within the shield. Note that North America's shield, called the Canadian Shield (Figure 2-2), essentially coincides with the Laurentian Upland of Figure 2-1. The southwestern portion of the shield is fairly well exposed in the Lake Superior region, where more than 3,000 million years of ancient earth history are recorded. The Lake Superior region is one of the best studied parts of any of the world's Precambrian shields.

Minnesota's legacy from its long stability and unceasing erosion is an essentially featureless, low-relief surface. Running water slowly but surely carved the face of the state.

Onto an early flat surface, Paleozoic seas advanced repeatedly, only to retreat as the crust was gently raised by internal earth forces. Yet evidence of the presence of several seas remains in southeastern Minnesota. Horizontal beds of sandstone, shale, and limestone (Figure 2-3), most containing fossils of now long-extinct invertebrates, offer the unmistakable clues to these ancient invasions. After several marine advances and retreats, Minnesota was left high and dry for at least 200 million years.

Figure 2-1. Map of North America showing the major physiographic provinces. Province 1 is the Laurentian Upland; province 2 is the Interior Plains and Lowlands; province 3 is the Appalachians; province 4 is the Western Mountains, including the Rockies and the Coast Ranges; province 5 is the Coastal Plains. Note Minnesota's location in two provinces, the Laurentian Upland and the Interior Lowland. (After Judson, Deffeyes, and Hargraves, 1976.)

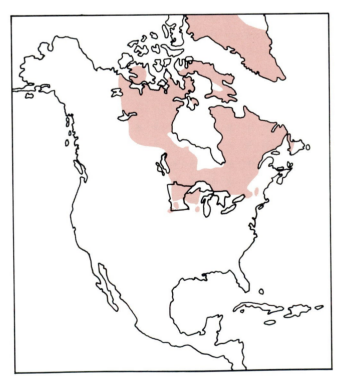

Figure 2-2. Map of North America showing the Canadian Shield. Note that the shield encompasses the Lake Superior region and most of Minnesota.

During Cretaceous time, about 100 million years ago, the last great advance of the seas into the interior of North America and into Minnesota took place. In this sea swam various invertebrates, sharks, and reptiles. After tens of millions of years, the sea withdrew for the last time, and erosion once again went to work.

Finally, "yesterday's" glacial action helped to remove the weathered rock detritus and further leveled parts of the state by filling in low areas with glacial deposits. Today, Minnesota is a *peneplain* ("almost a plain"), the end result of the long cycles of erosion. One need only fly over Minnesota to appreciate the fact that its topography is like that of a pancake (Figure 2-4).

As long as Minnesota remains above sea level, as it is today, the same slow but steady erosion that has been wearing down the surface will continue to wear it down. Eventually, the thin veneer of horizontal Paleozoic and Cretaceous sedimentary rock, at most 700 m thick, will also be gone.

Some rock outcrops in Minnesota are splendid, for the glacial abrasion removed all weathered and rotten rock and even polished the rock it left behind. These glaciated rock exposures, especially on wave-washed lakeshores where lichen cannot grow, are among the most revealing in the entire world, for Minnesota's climate since the glacial ice melted has not been conducive to the weathering of solid rock. Even where a mossy carpet must be peeled back, the rock is generally well preserved. But with the good comes the bad: the same glaciers that helped to fashion such good outcrops generally covered them with glacial sediment, as if to try to keep them in their pristine state. This blanket of glacial till, meltwater deposits, and lake clays, as much as 150 m thick in northwestern Minnesota, has covered an estimated 99% of Minnesota's rocks (Figure 2-5). In northwestern Minnesota, not a single rock outcrop is to be found in 45,000 km^2! The most abundant exposures are in Minnesota's Arrowhead region. In the southwestern part of the state, the Minnesota River has eroded down through the glacial sediment and has exposed some of the world's oldest rocks along its narrow valley.

Where rocks cannot be seen, geologists must resort to sparse well data. If a law existed requiring the submission of rock samples from every water well or exploration drillhole that is drilled into bedrock, geologists would have much more information than they have now. However, indirect geophysical methods, mainly aeromagnetic and ground gravity measurements, do allow earth scientists to make educated guesses about the type of rock present beneath the glacial cover.

By using all of these approaches, geologists have recently been able to prepare fairly sophisticated geologic maps for the entire state. The first geologic map of the state, obviously based only on a limited knowledge of outcrops, was compiled in 1872 by N. H. Winchell, of the Minnesota Geological and Natural History Survey. It was colored by hand—in watercolors—so the reproduction in Figure 2-6 does not do it justice. Geological interpretations had changed by 1900, when the next state map was compiled by Winchell. In 1932, a more detailed state map was compiled by F. F. Grout and others, but its large size (it comprises two over-size sheets) does not allow it to be reproduced here. In the 1960s, enormous strides in our geological knowledge were made with reactivation of the Minnesota Geological Survey under the directorship of P. K. Sims, and in 1970 the fourth state map was published. Continued work and reinterpretation resulted in a revision in 1976 by G. B. Morey, of the Minnesota Geological Survey (Plate 2). Compare Figure 2-6 and Plate 2 to see how a century of geology has changed the geological map.

In the pages that follow, many of the geologists who made important contributions to the understanding of the geology of Minnesota will be mentioned. After all, it is geologists and geophysicists who have brought and who will continue to bring Minnesota's fascinating past to light. However, it will be impossible to refer to them all, for this book is meant to be a geological history and not a history of geologists. A few old references to the geology of Minnesota are of interest. Early explorers, including Father Louis Hennepin (in 1680), Jonathan Carver (in 1766), William Keating (in 1817), Henry Schoolcraft (in 1823), and G. W. Featherstonhaugh (in 1835), wrote notes about the geological features. Their work is discussed further in Winchell and Upham (1884). An early (1852) geological report by D. D. Owen, of the United States Department of the Treasury, entitled "Report of a Geological Survey of Wisconsin, Iowa and Minnesota and Incidentally a Portion of Nebraska Territory," was the real start of organized studies. This early work was probably stimulated by the discovery of iron ore near Negaunee, Michigan, in 1844.

The major impetus for geological studies in Minnesota has come through the Minnesota Geological Survey, which was established in 1872 as the Geological and Natural History Survey, with N. H. Winchell as director from its inception until 1900. During that time, 24 annual reports were published. In 1900, the survey essentially ceased to exist, but the United States Geological Survey moved in and did some monumental works. Finally, in 1911, the Minnesota Geological Survey was reestablished as part of the University of Minnesota. Since that time, numerous reports and maps have been published by the survey; a complete list is available from the Minnesota Geological Survey, 1633 Eustis St., St. Paul 55108. The 100th anniversary of the survey, 1972, was commemorated with a 632-page volume of 57 articles written by 33 geologists and edited by P. K. Sims and G. B. Morey. This *Geology of Minnesota: A Centennial Volume* is a milestone publication.

The survey has seen good years and not so good years, basically dependent upon funding. From 1911 until 1947, the annual budget was less than $8,000 a year, obviously

Figure 2-3. Horizontal sedimentary rocks of Paleozoic age in Goodhue County, in southeastern Minnesota. (Courtesy of David Stone.)

Figure 2-4. Aerial view showing the eroded low-relief surface (peneplain) of Minnesota. This photo, taken in the Boundary Waters Canoe Area, is representative of the entire state.

not enough to fund much geologic study. From 1947 through 1961, the budget was about $20,000 per year. Three directors—W. H. Emmons, F. F. Grout, and G. M. Schwartz—guided the survey through those lean years and got amazing mileage out of few dollars. In 1961, the legislature, evidently for the first time since the late 1800s, realized the importance of a geologic survey in a state that has a major mining industry, that has much additional mineral potential, and that should also be concerned about such

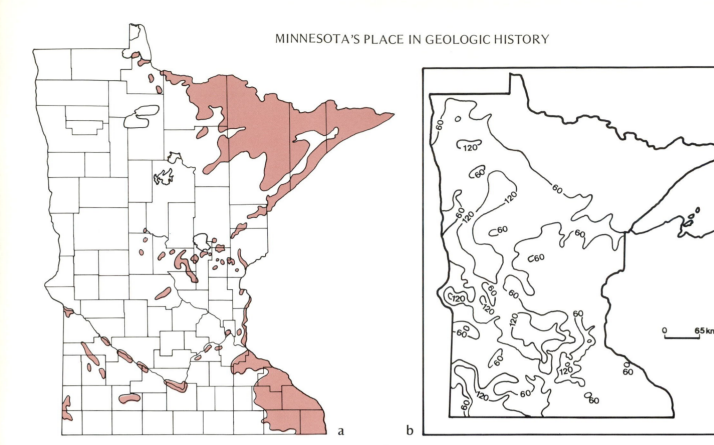

Figure 2-5. (a) Map of Minnesota showing the areas (shaded) in which bedrock is at the surface or beneath less than 15 m of glacial sediments. Actually, very little bedrock is exposed over most of the state. In the northwestern part of the state, the glacial sediments are as much as 150 m thick. (Courtesy of the Minnesota Geological Survey.) (b) Map showing thickness of glacial drift. Thickness lines are in meters. (After a map by S. Tufford in Ackroyd, Walton, and Hills, 1967.) (c) No bedrock here—only swamp!

things as water supplies, sand and gravel, urban planning, and environmental issues. The budget increased quite steadily and between 1965 and 1975, under the leadership of P. K. Sims, was between $150,000 and $300,000 annually. Since 1975, with Matt Walton as director, it has climbed to over a million dollars a year and virtually every facet of Minnesota's geology is under study to some degree.

Thus, from 1852 until today, the geology of Minnesota has been investigated in a systematic way. By foot, horse, canoe, boat, automobile, and airplane, but mostly by foot, hundreds of geologists over more than a century and a quarter have explored this state (Figures 2-7 and 2-8). Their laboriously gathered data have made possible the maps and written scientific reports that other geologists have then used. Obviously, it is the combined works of very many earth scientists over the years that have made the writing of this book possible.

Lest readers think that these geologists were somewhat myopic scientists who were uninterested in, say, the arts, some poetry written by A. C. Lawson, of the Canadian

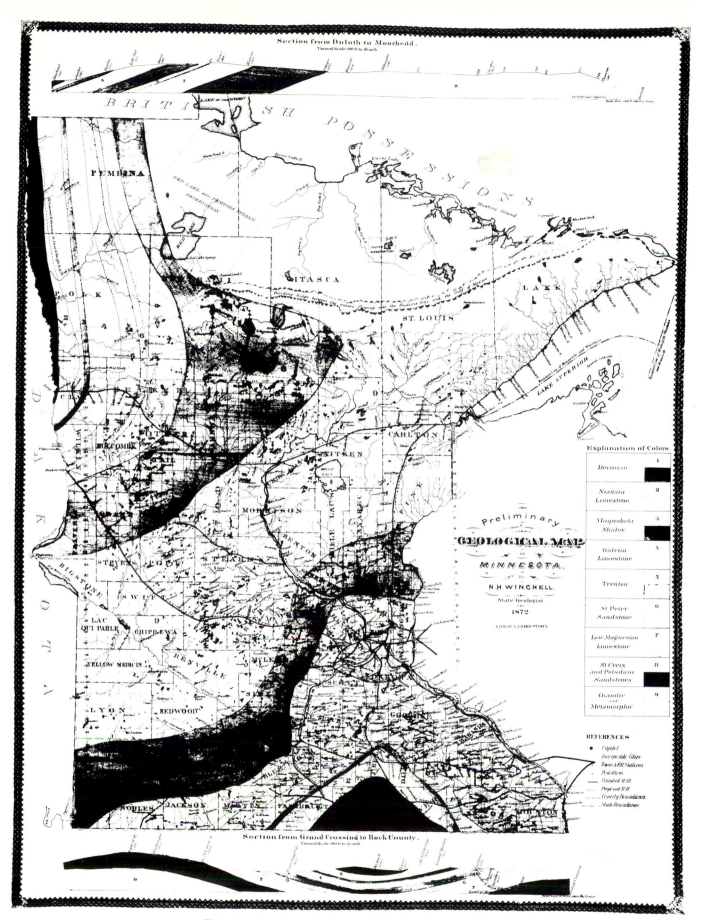

Figure 2-6. The first geological map of Minnesota was compiled by N. H. Winchell and was published in 1872.

Figure 2-7. Early photo of geologists working in the field in northeastern Minnesota. (From Winchell et al., 1899.)

Figure 2-8. State geologist P. K. Sims studying ancient rocks of the western Vermilion district near Tower in 1969. The lines or streaks on the rock face are structural lineations indicating that the folded rocks of this area are plunging steeply beneath the surface.

Geological Survey, is included here. Why include Lawson's poetry and not that of a Minnesota Geological Survey or United States Geological Survey geologist? Because Lawson did much work in the late 1800s and early 1900s on rocks of the Ontario-Minnesota border area, because he later taught at the University of California, and because his poetry was published.*

MENTE ET MALLEO
("By Thought and Dint of Hammering")

Dedicated to the Logan Club on the Occasion of its First Annual Symposium

By thought and dint of hammering
Is the good work done whereof I sing,
And a jollier crowd you'll rarely find,
Than the men who chip at earth's old rind,
And often wear a patched behind,
By thought and dint of hammering.

All summer through we're on the wing,
Kept moving by the skeeter's sting;
From Alaska unto Halifax,
With our compass and our little axe,
We make our way and pay our tax
By thought and dint of hammering.

We crack the rocks and make them ring,
And many a heavy pack we sling;
We run our lines and tie them in,
We measure strata thick and thin,
And Sunday work is never sin,
By thought and dint of hammering.

Across the waters our paddles swing,
O'er wind and rapids triumphing;
Through mountain passes our slow mules trudge
As if they owed us a heavy grudge,
And often can't be got to budge,
By thought and dint of hammering.

To the stars at night our thoughts we bring
But no maiden fair to our arm doth cling;
She, at Ottawa, with smiling lips,
The other fellow's ice cream sips,
You can't prevent these feminine slips
By thought and dint of hammering.

To array the "chiels that waunna ding"
Is our winter's work far into spring;
Some people think us wondrous wise,
Some maintain we're otherwise:
We're simply piercing Nature's guise
By thought and dint of hammering.

*From Alcock (1947) by permission of the Geological Survey of Canada.

PART II

Geologic History

Minnesota's geologic history is developed step by step in the five chapters of this part of the book. Precambrian time, the span of time between the origin of the earth (about 4,500 million years ago) and the development of invertebrate life preserved as abundant fossils (about 600 million years ago), represents seven-eighths of the earth's past. The long Precambrian Period is divided into three portions—early, middle, and late—and the history of each in Minnesota encompasses a separate chapter.

The eras of life—Paleozoic (ancient life), Mesozoic (middle life), and Cenozoic (recent life)—are essentially described in a single long chapter. This is primarily because only a portion of each era is represented in Minnesota's rock record.

The last 2 million years or so saw the waxing and waning of the glaciers of the Great Ice Age. Because the glacial deposits of this ice age were the last major deposits to be formed in Minnesota, they blanket the state and obscure the older rocks. The history that led to the deposition of these well-exposed sediments certainly warrants a chapter of its own, in spite of the fact that many geologists consider them simply as overburden.

Historians study past writings to gather data so that they can interpret the past. Many geologists are, in fact, historians, deciphering more ancient pages of the record—the rock layers—and then explaining the long history of the earth and thereby also of a small but important portion of the earth —Minnesota. Missing "pages," although not as enlightening as existing ones, also yield valuable information about Minnesota's geologic heritage.

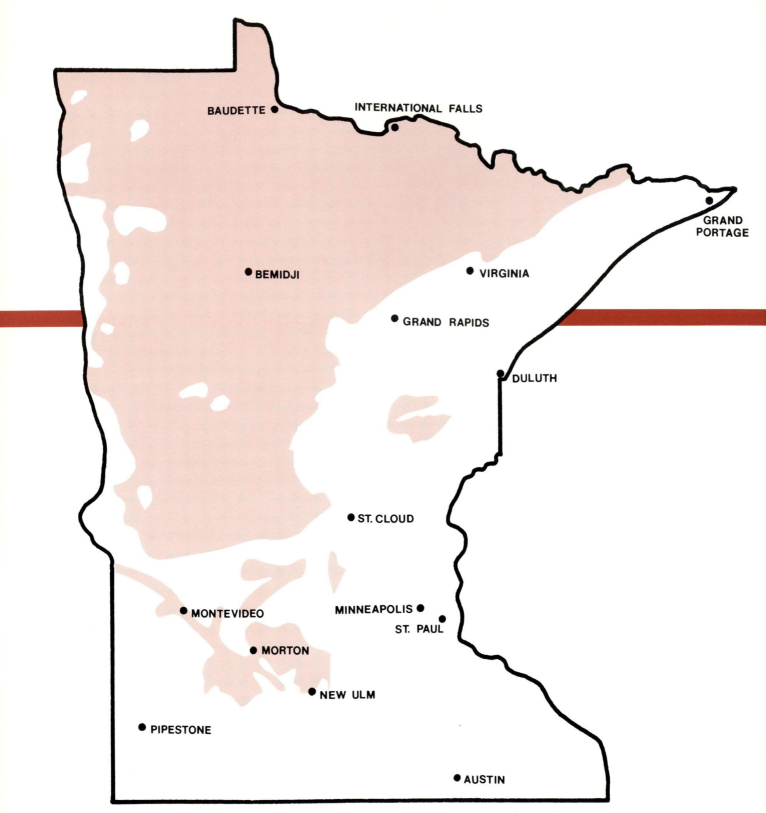

Map showing portions of Minnesota (shaded) where the bedrock is Early Precambrian (Archean) in age.

3

Early Precambrian Time (4,500 to 2,500 Million Years Ago)

The rocks of Early Precambrian age are exciting, if for no other reason than their great antiquity. These rocks contain clues to the conditions and events of a distant part of the Earth's past. Interpreting these old clues is indeed exciting. For example, the study of many rocks from the northern part of the state indicates a history of explosive volcanism. Although the fires are now quiet, the "brimstone" remains for close scrutiny by the earth scientist.

Lower Precambrian rocks underlie much of Minnesota, but they are mostly covered with glacial sediment. Figure 3-1 shows diagrammatically their relations to younger rocks in northeastern Minnesota. These old rocks, quite representative of the oldest rocks on the Canadian Shield, can be divided into three main groups. The oldest rocks are gneisses, the next oldest are volcanic-sedimentary accumulations, and the youngest are granitic complexes that make up several large batholiths. The oldest gneisses are at least 3,600 million years old with some as young as 3,000 million years, and the "younger" granites are 2,500 to 2,700 million years old. It is disconcerting to geologists that the volcanic-sedimentary accumulations, which are definitely older than the granites that cut them and metamorphose them, generally yield about the same age dates as the granites—2,700 million years. This is due to three things: (1) the volcanism and intrusion probably did occur close together in time; (2) the resolution of age-dating techniques is generally less than that necessary to separate the events; and (3) the granitic invasions heated up the older rocks, thereby resetting the latter's geologic clocks to the age of metamorphism. Earth scientists, by consensus, end Early Precambrian time with the 2,500-million-year-old granites.

Unfortunately, the rocks do not tell the entire story of the earth's early days. The earliest part of the earth's long history is still lost in obscurity. Astronomers think that the earth and the other planets in the solar system originated, along with the sun, from an interstellar cloud of dust and gases that collapsed, contracted, and rotated. The dense center of this mass became hot, and thermonuclear reactions were initiated in this new star—the sun. As smaller bodies within the cloud cooled, the planets with their smaller satellites were formed. When did the earth thus form? Radiometric age dates of meteorites, which come from the asteroid belt between Mars and Jupiter, and interpreted age dates on rocks brought back from the moon by the

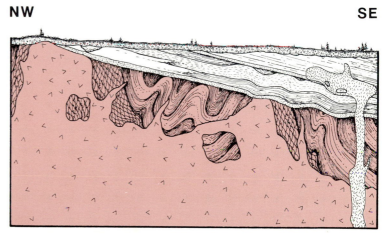

Figure 3-1. Generalized geologic cross section of northeastern Minnesota. Imagine it as a vertical slice of the rock crust or as a high cliff. Lower Precambrian rocks are shaded, whereas younger rocks are not. The rocks that look like fish scales are pillowed greenstones (mostly Ely Greenstone), the folded and bedded rocks are the Knife Lake Group and the Lake Vermilion Formation, and the remaining checked pattern represents younger granitic rocks. Not drawn to scale.

astronauts indicate an age for the solar system, and hence for the earth, too, of about 4,500 million years.

Any very early atmosphere of helium and hydrogen and other lightweight gases consisting of elements that are common elsewhere in space would have been lost before earth materials had coalesced into a dense body with a higher gravitational attraction. But, as the earth continued to evolve, *outgassing* of its interior via volcanism was going on and a different kind of atmosphere was built up. The exact composition of this retained atmosphere is unknown, but the best guesses suggest that the gases included nitrogen, hydrogen, methane (CH_4), ammonia (NH_4), carbon dioxide (CO_2), and water vapor. Oxygen was probably not there in any quantity. It was to be added to the atmosphere later by biologic processes or by the photochemical disassociation of water vapor (caused by ultraviolet radiation from the sun) or by both. The oxygen-rich atmosphere we know today will be discussed again in the next chapter.

The outgassing of the earth's interior via volcanism probably also accounted for the earth's water. Whereas modern volcanoes emit much water vapor, this water appears to be largely rainwater that has worked its way into the crust and is flushed out by hot magma. However, even today, a small part of that water is thought to be from the interior. The early water naturally accumulated upon lower parts of the earth's surface. Ocean basins are floored by basalt and lie lower than dominantly granitic continents partly because basalt is denser (heavier) than granite. Thus, the oceans came to be, and even today they hold more than 97% of the earth's water.

THE ELUSIVE ORIGINAL CRUST

Let us get down to earth with age dates. The oldest rocks yet found on the earth have been dated as "only" 3,800 million years old. This brings the earth's history right home, for it was gneisses from the Minnesota River Valley that were first determined to be 3,600 million years old (Goldich and others, 1970). Since then, the older dates have been obtained on rocks from Greenland and Labrador.

Geologists in shield areas have been searching for the "original granitic crust" of the earth for the last hundred years. As the early earth cooled, lightweight granitic material should have risen to the top of heavier materials, thus forming a "scum" or crust on the earth. But, wherever geologists found what they thought might be "original granitic crust," more detailed studies showed that it was intruding and metamorphosing adjacent volcanic and sedimentary rock. Thus, those granites were not part of the early crust but instead were the products of younger magmatic activity. Minnesota's granites were no exception—they, too, proved to be younger.

When geologists began doing careful work in the Minnesota River Valley in the 1960s, they realized that the granitic gneisses at Montevideo and Morton were intruded by granites. Therefore, they asked whether these rocks could be considerably older than 2,500 to 2,700 million years—the dates so commonly obtained on the granites of the Canadian Shield. Several geochronologists (geologists who specialize in age-dating) thought that they might indeed be older. Other gneisses, including the McGrath Gneiss of central Minnesota, might be of comparable age, although ages of only 2,700 million years have as yet been obtained (see Figure 2-9).

If some Minnesota River Valley gneisses are 3,600 million years old, were *these* rocks part of the original crust? Not likely, for the gneisses appear to be highly metamorphosed sedimentary and volcanic rock (Figure 3-2). Such layered rocks need some type of crust to be laid down upon. Thus, the "original granitic crust" still eludes earth scientists. Perhaps further detailed work will find rocks older than 3,800 million years. Or perhaps the original crust has been completely obliterated and recycled by earth processes. Or perhaps there never was an original granitic crust. If not, the earliest crust would probably have been made of basaltic rocks, like those that floor the ocean basins today.

COSMIC BOMBARDMENT?

What were the first 900 million years of Minnesota's as yet unrecorded history, from 4,500 to 3,600 million years ago, like? The clues, at present, all lie beyond the earth. Mercury, the moon, Mars, and the many other smaller moons are pockmarked with craters. Many craters appear to be the result of meteorite impact rather than volcanic explosion. Because of the earth's position near these many bombarded spheres, it seems logical that the earth, too, received countless hits at the same time, an estimated 4,000 million years ago, even though the earth's early atmosphere may have provided some protection. If that were true, why does not the earth's surface, like that of the moon, look like a battlefield? Because the earth, with the abundant water that has earned it the nickname of "the water planet," is unique in the solar system. Rain, running water, standing bodies of water, vegetation, chemical weathering caused by the water and the atmosphere, and sedimentation have worn away or covered most signs of craters.

The earth does have dozens of meteorite craters, but they are apparently relatively young. While many suspected "craters" in Minnesota have been investigated, none has been demonstrated to be an impact structure. For example, the large and circular—but shallow—Lake Roseau, near Roseau, in northwestern Minnesota, is probably a glacial feature formed by a melting ice block beneath glacial sediment (see Chapter 11). The closest good candidate for a meteorite crater is a deep circular lake near Kenora, on the Ontario side of Lake of the Woods. However, the crater is in rocks about 2,700 million years old, so this depression, too, must have been formed later than the early craters on the moon. Although it cannot be proven that Minnesota was once extensively cratered, it seems likely that it was. And this cratering may have been a factor in obliteration of the early crust.

ANCIENT VOLCANISM

The Canadian Shield, like the other exposed shields of the world, contains dozens of *belts* or strips of volcanic and sedimentary rocks (Figure 3-3). Whether the rocks were actually deposited in elongate patterns is in some doubt. It is possible that volcanic centers were randomly scattered over the entire shield. In this interpretation, the present pattern is simply the result of fortuitous preservation of downwarped areas, with erosion having removed the volcanic rocks from the areas in between or metamorphism having changed them. Regardless of the original distribution, the preservation of the volcanic-sedimentary belts is certainly due to downwarping along elongate zones, for each belt is essentially a large downfolded and/or downfaulted block. The areas between individual belts consist of granite or granitic gneiss. In fact, large portions of the Canadian Shield consist of these rock types. A glance at the geologic map of Minnesota (see Plate 2) shows that this is also true for the northern half of the state.

One Canadian school of thought has developed the concept of superbelts. According to this idea, broad superbelts of dominant gneisses alternate with broad superbelts of dominant volcanics. The smaller volcanic-sedimentary belts lie within the volcanic superbelts. Both types of superbelts contain numerous intrusions of younger granites. Although this concept evokes debate among geologists, it is mentioned here because the superbelts can be extended into Minnesota (Figure 3-4). One might interpret the gneisses of the Minnesota River Valley to be the southernmost of the gneissic belts, but the gneisses here are considerably older than the oldest ages (3,000 to 3,200 million years) so far obtained in the Canadian gneiss superbelts. The gneissic superbelts are of importance because they could be uplifted remnants of the *basement*, the rock upon which the volcanic-sedimentary edifices were constructed. This does not, however, necessarily suggest that they are part of the "original crust."

The Minnesota geologic map shows several volcanic-sedimentary belts, although many of them are belts only in the crudest sense of the word. Unfortunately, most of these belts are covered by a hundred meters or more of glacial deposits and are known only from a few well samples and from geophysical responses of the buried rocks. Parts of only three belts are exposed, and our knowledge of Minnesota's ancient volcanism and sedimentation is largely based on these areas. The two northern belts are poorly exposed on Rainy Lake and 80 km to the west in the Birchdale-Indus area, but are well exposed in adjacent Ontario, and observations there add much to the interpretations. The other belt,

Figure 3-2. The Montevideo Granite Gneiss of the Minnesota River Valley. The black bands probably represent original basaltic material, and the lighter bands may represent original sediments. This gneiss has been dated as at least 3,600 million years old.

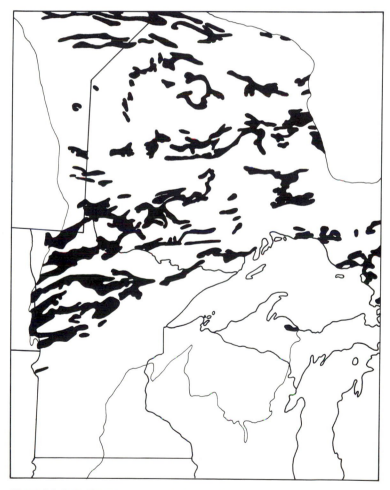

Figure 3-3. Map of the Canadian Shield showing greenstone belts of volcanic and sedimentary rocks. Note that several are located in Minnesota. Unfortunately, most are deeply buried beneath a thick cover of glacial sediments. The areas between the belts consist of rocks of granitic composition.

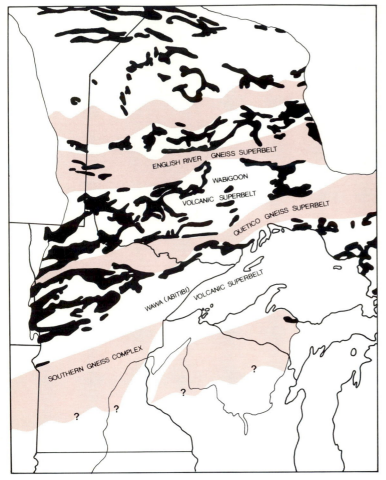

Figure 3-4. Map of Minnesota, Ontario, and adjacent areas showing alternating volcanic superbelts and gneissic superbelts. The gneissic superbelts contain some rocks that are older than the volcanic ones. The Southern Gneiss Complex contains some of the oldest rocks dated anywhere in the world so far.

better exposed, is in the Vermilion district and extends northeastward from Tower and Soudan through Ely to Saganaga Lake, on the Canadian border. These three belts will be described in more detail in the regional geology portion of this book, the northernmost belt in Chapter 11, the Rainy Lake belt in Chapter 10 in the section on Voyageurs National Park, and the Vermilion belt in Chapter 10 in the sections on the Boundary Waters Canoe Area and the western Vermilion district. In this chapter, however, we will discuss their origins.

A generalized model of the predeformational stages in the development of Minnesota's volcanic-sedimentary belts is shown in Figure 3-5. About 2,700 million years ago, during the initial stage, basaltic lava poured out of long rifts or cracks on the sea floor. The lack of fossils in the very old rocks of the belts prevents a positive determination of a marine environment, but the great lateral extent of the volcanic-sedimentary belts is suggestive of very large bodies of

water. And surely most of the basalts cooled under water, for *pillow* structures (Figure 3-6; Plate 8) are common. Modern pillows have been studied off Hawaii by divers and have been photographed along the Mid-Atlantic Ridge by submersibles. Some of Minnesota's pillowed basaltic lavas probably cooled at depths as great as 1,000 m, for they lack the gas cavities, or vesicles, that are formed by gases escaping from the lavas wherever the water pressure is not so great as to keep them within the lava. The pillowed lavas were extruded during stage 1 of Figure 3-5.

What were the pillowed basalts of Minnesota, which locally total as much as 5,000 m in thickness, deposited upon? We do not know, for the bases of the pillowed sequences are not exposed. Younger granites have either engulfed the lower portions of the volcanic sequences or have been moved into fault contact with them. A likely choice for a basement beneath the sequence is granitic gneiss, as found in the gneiss superbelts, but that, too, is speculation.

Upon this widespread basaltic platform, more restricted volcanic edifices of andesitic to rhyolitic rocks were constructed, as in stages 2a and 2b of Figure 3-5. Andesites and rhyolites contain less iron, magnesium, and calcium and more potassium, sodium, and silica than do the basalts. The volcanoes that produced these were more like what people commonly picture when volcanoes are mentioned. Rather than pouring out of cracks as did the basalts, these volcanic materials were thrown explosively out of central vents. Such *pyroclastic* ("fire particle") debris accumulated as beds composed of angular particles (Figure 3-7). While the initial volcanism of this stage was likely submarine, some of the volcanic material would have been thrown out on land wherever the edifice was built high enough to protrude above sea level as an island. The coarser particles, as much as 1 m in diameter, can be used to locate approximately the explosive vent areas.

All during the buildup of these volcanic piles, but especially after the piles grew above sea level, erosion was wearing them down. This is illustrated by stage 3 of Figure 3-5. On the volcanic islands, sediment was generally carried to the seashores by streams. The sediment was then moved away from the shallow water near the volcanic centers out onto the deeper ocean floor by submarine landslides, most of which incorporated much water and became turbidity currents. These landslides, probably triggered largely by earthquakes, were sporadic intruders into the deep-water environment in which fine clay was continually but slowly raining down from slightly muddy near-surface waters and from volcanic ash dropped into the sea. The evidence for a turbidity current mechanism is preserved in the textures of sandstone beds. These beds are graded, with the coarsest grains on the bottom and finer grains higher in the bed. Such beds, theoretically, are deposited by swirling turbid currents from which the bigger, heavier grains settle out first

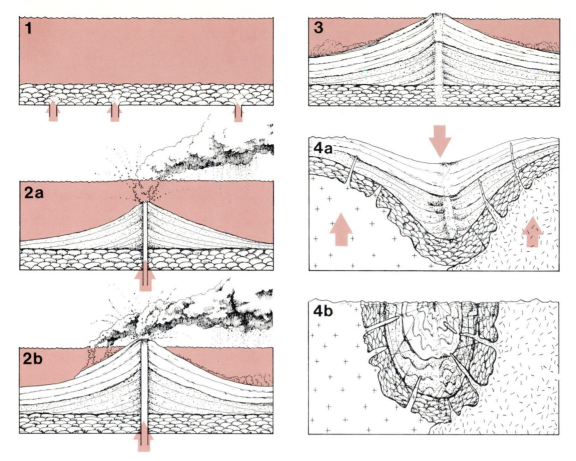

Figure 3-5. Stages in the development of a part of a volcanic-sedimentary (greenstone) belt. Stage 1 illustrates the extrusion of basaltic lava from long fissures beneath the sea; the lava is fluid and moves laterally as pillows form. Stage 2 illustrates the development of an explosive volcanic center, first beneath the sea (2a) and then above the sea (2b); the explosiveness is largely due to the higher viscosity of the more felsic magmas. Stage 3 shows erosion of the pile and deposition of sediments. Stage 4a shows the upwelling of adjacent granitic batholiths and the concurrent subsidence of the volcanic-sedimentary pile. Stage 4b illustrates the present folded, faulted, and eroded volcanic-sedimentary belt. The diagram is highly generalized.

Figure 3-6. Pillow structures in the Ely Greenstone at Ely. Whereas these pillows are exposed in three dimensions, most outcrops are flat, glaciated exposures illustrating only two dimensions of the pillows. Pillows form as hot lava meets cold water, causing the exposed lava to cool rapidly to a glass while the interior remains hot and fluid, continually breaking out to form new globules or pillows until the unit has cooled to a solid. See Plate 8.

Figure 3-7. Explosive fragmental (volcaniclastic) rock near Tower, in the western Vermilion district. Note that the large fragment has a lighter-colored edge. This may be the result of metamorphism of the more rapidly cooled and hence finer-grained edges of the fragments.

a

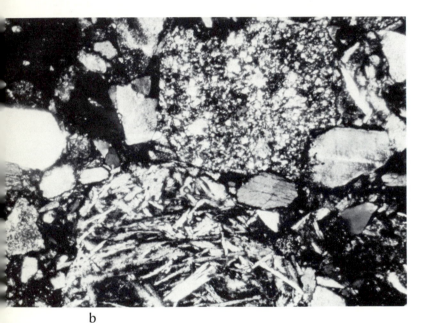

b

Figure 3-8. (a) Graded sandstone bed of volcanic material now tilted to a vertical position. The excellent grading, from coarse grains at the base to finer grains at the top, is probably the result of deposition from a turbidity current that moved into deeper water off an explosive volcanic center. The darker beds are now slate but were muds deposited slowly in quiet water. The location is the Pike River, a few kilometers west of Tower, in the Vermilion district. (b) Photomicrograph of the coarse part of a graded sandstone bed showing volcanically derived fragments. Mafic volcanic (basalt) grain is at lower center, felsic (dacite) volcanic grain is at upper right, and white quartz grain is at upper left, with numerous smaller grains of plagioclase and hornblende. The sample is from Knife Lake in the Boundary Waters Canoe Area in the eastern Vermilion district. The field of view is about 2 millimeters (mm) across.

and the smaller, lighter grains last, as can be seen in Figure 3-8. The total thickness of flows, pyroclastics, and reworked pyroclastics in the Vermilion district may be as much as 10,000 m, and similar volcanic piles in Canada are estimated to have been even thicker.

A three-dimensional model of the volcanic-sedimentary environment is shown in Figure 3-9. Note that some intrusive rocks are also shown, for most volcanic-sedimentary piles have some intrusive rock associated with the surficial rocks. Whereas most of the sediment was derived from the erosion of the explosive volcanic rocks, minor coarse-grained granitic material may have been provided by the exposed intrusions where the overlying volcanic cover was eroded away. Studies to date indicate that nearly all the sediment in Minnesota's volcanic-sedimentary belts is volcanogenic, largely derived from the volcanic rocks. However, some sediment in the eastern Vermilion district was derived from the Saganaga Batholith, which intruded the volcanic-sedimentary pile. No evidence of older gneissic source rocks has been found in the Lower Precambrian sediments of Minnesota.

The determination of the location of Early Precambrian explosive volcanic centers is of great interest to geologists, for many of Canada's and the rest of the world's large base metal deposits (e.g., of copper, zinc, and lead) are volcanic in origin and located near vents. Potential for such deposits in Minnesota's volcanic piles is discussed in Chapter 8.

MODERN VOLCANIC ANALOGUES

Are there modern analogues of these volcanic-sedimentary piles? Yes, the modern island arcs of the world, such as the Aleutians, Japan, the Philippines, and the West Indies, have similar rock types, chemistry, and size. Did the Early Precambrian volcanic arcs form in the same way? Perhaps, but it is hard to prove. Modern island arcs are located near boundaries of large lithospheric plates, composed of crust and upper mantle, where a heavier oceanic plate is moving downward (i.e., being subducted) beneath a lighter continental part of an adjacent plate. Though some geologists have suggested such a plate tectonic origin for the Lower Precambrian volcanic piles, most geologists are somewhat hesitant to apply the theory to such old rocks without better evidence. Perhaps there were more and smaller plates with less movement in Early Precambrian time.

GRANITIC BATHOLITHS
AND MOUNTAINS

After the cessation of volcanism, the volcanic-sedimentary terranes were subjected to intense folding. Lavas, pyroclastic beds, and sedimentary beds were severely folded, many assuming a nearly vertical position, as in Figure 3-10. Small-scale folds presumably mimic the larger fold systems of

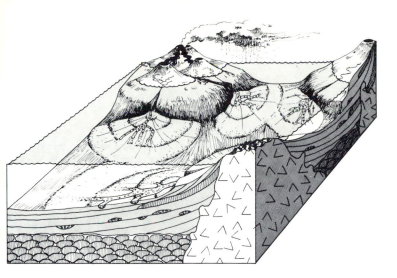

b

Figure 3-9. Generalized model of environment of deposition of rocks from the Vermilion district and other volcanic-sedimentary greenstone belts of Minnesota. Erosion of multiple volcanic centers produced sediments at the bases of volcanic islands. Sediments probably accumulated as submarine *fans* with coarse conglomerates in fan channels and graded sandstones further downslope on the fans or outside of the channels. Note that explosive volcanic centers are built on a pillowed greenstone platform, that the erosion has exposed a granitic body (indicated by checks) that had intruded into the volcanic pile, and that volcanism and erosion were going on side by side. Some volcanic centers may have been submarine.

which they are a part (Figure 3-11). This is fortunate, for it allows the geologist to deduce rock structures even in areas of little outcrop. Accompanying the folding, and certainly a partial cause of it, was the intrusion of large bodies of granitic magma that solidified as the batholiths shown on the geologic map (see Plate 2). These bodies rose from the depths, tens of kilometers below the earth's surface, into the volcanic-sedimentary piles. Pieces of volcanic rock preserved in the granites verify that they rose through preexisting rock, and complex relationships of granite and preexisting country rock confirm the hot origin (Figure 3-12). They did not reach the surface but cooled beneath at least a few kilometers of volcanic and sedimentary rocks, as evidenced by their coarse textures (Figure 3-13).

a   c

Figure 3-10. (a) Metamorphosed beds of mafic volcaniclastic rocks (*tuffs*), now greenschist, standing in a vertical position. The location is Rainy Lake. (b) Metamorphosed beds of conglomerate and graded sandstone standing in a vertical position. The location is Ensign Lake in the Boundary Waters Canoe Area. (c) Metamorphosed beds of felsic volcaniclastic rocks (tuffs) in a steeply tilted position. The location is near the village of Indus, on the Rainy River in northernmost Minnesota.

29

a

b

c

d

Figure 3-11. (a) Highly folded iron-formation (alternating beds of dark gray hematite, red jasper, and white chert) near the Soudan Mine at Soudan in the western Vermilion district. Detailed studies of this outcrop indicate that the rocks have been folded two times. This has been a much photographed outcrop for nearly a century. (b) Highly folded volcanogenic sandstone beds a few miles west of Tower on Highway U.S. 169 in the western Vermilion district. Detailed studies have shown that the rocks have been folded two times. (c) conglomerate and sandstone, with the pebbles of the conglomerate highly elongated by deformation (squeezing) of the rocks. Pebbles were probably relatively equidimensional when deposited. These are on Ensign Lake in the Boundary Waters Canoe Area. (d) *Boudins* (sausagelike structures) of granite in vertical beds of biotite schist. The boudins were once continuous thin sills of granite. The stretching and rupture of the sills is the result of intense compression of the rocks in one direction with consequent elongation of the rocks in a perpendicular direction. These are on the highway between Little Fork and Ray, in the western part of the Vermilion Granitic Complex, in northern Minnesota.

Figure 3-12. Black inclusions of metamorphosed basaltic country rocks (now amphibolite) in granite at the Laurentian Divide on Highway 169 just north of Virginia, on the southern margin of the Giants Range Batholith. Geology students refer to this site as "Confusion Hill."

As the granites rose, the nearby volcanic and sedimentary rocks sank. Dikes of granitic rock cut across the older rocks, and later phases of granitic magma cut across granites of earlier phases (Figure 3-14). The granites metamorphosed the volcanic and sedimentary rocks with which they came into contact, and the entire folding-intrusive event raised the temperature of the volcanic-sedimentary rocks as well, subjecting them to regional directional pressures. The result was the metamorphism of all the volcanic and sedimentary rocks under relatively low-grade metamorphic conditions. All the rocks shown and discussed in this chapter have been metamorphosed to varying degrees. Because minerals, and hence rocks, are stable only in the environments in which they formed, the new conditions caused new minerals to form. Figure 3-15 shows some more prominent metamorphic minerals. For example, the plagioclase, olivine, and pyroxene that made up the pillowed dark gray to black basalts were converted to chlorite, epidote, and actinolite, all green minerals. Thus, the basalts became metabasalts, commonly known as greenstones. Because of this, and the fact that many other metavolcanic rocks also became green, the volcanic-sedimentary belts are often referred to as greenstone belts.

Precisely what caused the formation of numerous large batholiths is conjectural. Did some material melt at depth to form magma? This may have occurred in the upper mantle, the thick zone beneath the thin crust (which is only 5 to 60 km thick), with partial melting of the most easily melted fraction of the mantle rock. By studying the ratio of two isotopes of strontium, strontium-87 and strontium-86, geochemists can determine whether the rock that melted to form the granite may already have been included in

a

b

Figure 3-14. (a) Granitic dikes crossing dark biotite schists near Baudette in northern Minnesota. (b) A pegmatite dike of coarse feldspar and quartz cutting finer-grained granite on an island in Lake of the Woods.

Figure 3-13. Coarse granite with large phenocrysts of potassium feldspar. The pencil and boot (size 9½) provide scale. This is on Minnesota Highway 72 between Red Lake and Baudette in northern Minnesota and is one of only a few exposures of a large granitic intrusion.

earlier geologic events. Strontium-87 forms from rubidium-87 by radioactive decay, and a higher than normal ratio of strontium-87 to strontium-86 would indicate that an earlier period of decay had already enriched the parent rock of the magma with strontium-87. In general, low ratios have been obtained for both granites and volcanics, indicating derivation directly or nearly directly from the mantle.

Probably not long after solidification of the granitic batholiths—again let us say about 2,700 million years ago because geochronologists cannot generally obtain a fine resolution of dates in rock so old—the greenstone-granite terrane was subjected to new stresses that caused movement along numerous faults. Great pieces or blocks of the earth's crust were moved up or down or horizontally relative to

a

b

Figure 3-15. (a) Metamorphic rock (schist) showing original beds and coarse crystals of garnet and sillimanite developed along certain beds. The location is Voyageurs National Park. (b) Staurolite crystals that grew during metamorphism of a muddy rock to schist.

adjacent blocks. Some of these faults can be seen in Figures 3-16 and 3-17. Such faults are very abundant.

This combination of folding, intrusion, and faulting produced mountain ranges throughout what is now northern Minnesota and over the rest of the Canadian Shield. By projecting the now eroded and tilted beds upward, we can guess that these mountains may have been at least a few kilometers high. Similar projections of tilted beds downward, coupled with geophysical measurements on greenstone belts in Canada, suggest that the metavolcanic-metasedimentary rocks of the belts project downward at least a few kilometers. This is an appreciable depth, but it is not very deep when compared to the 60-km thickness of the earth's crust beneath Minnesota. This mountain-building episode of 2,700 million years ago, apparently worldwide in scope, is called the Algoman Orogeny in the United States and the Kenoran in Canada.

All of these events—the volcanism, intrusion, metamorphism, folding, and faulting—probably occurred within a time span of 50 to 100 million years, or about 2,700 million years ago ±50 million years. Except for the volcanism, which of course took place at the surface, most of these events occurred at various depths. It was the later erosion, during Middle Precambrian time, of the Algoman mountains that exposed these deeper rocks for study.

Minnesota today is rather quiet. A few earthquakes still occur, as described in Chapter 12, but the days of geologic violence in Minnesota are long gone. Had you been here 2,700 million years ago, you would have witnessed spectacular explosive volcanism with darkened skies and incandescent ashfalls, like those produced by the eruption of Mount St. Helens in Washington in 1980. Accompanying this volcanism, and the later faulting, were numerous earthquakes. You would have felt the earth tremble, repeatedly. You were just born too late.

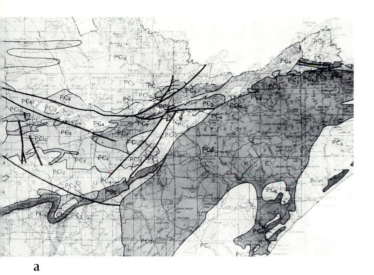

a

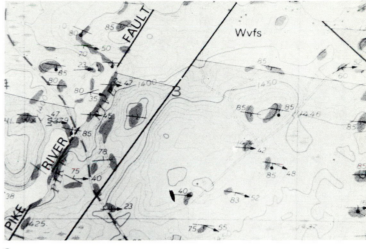

a

b

b

Figure 3-16. (a) Geologic map of the Vermilion district showing large faults (dark lines) cutting the bedrock. The longest and northernmost is the Vermilion Fault, which can be traced westward to North Dakota by geophysical means. This fault has had 19 km of horizontal movement along it, with the northern block moved eastward and upward relative to the southern block. The rocks north of the fault are part of the Quetico Gneiss Superbelt and those to the south are part of the Wawa (Abitibi) Volcanic Superbelt. The next fault to the south is the Haley Fault. The southernmost fault, the Dark River Fault, even cuts the younger rocks of the elongate Mesabi Range; the evidence suggests several periods of movement along this fault, which originated in Early Precambrian time. Note the two main sets of faults, one trending north-northeast and the other trending west-northwest. The Lower Precambrian rocks are as follows: $P\mathcal{C}i_2$ (lightest gray) is granite; $P\mathcal{C}_3$ (dark gray) is metavolcanics: $P\mathcal{C}_4$ (medium gray) is metasediment; $P\mathcal{C}_5$ is granitic rock. The other symbols represent younger rocks. The squares are townships, 9.7 km (6 mi) on a side. (From Sims, 1970.) (b) Pegmatitic granite dike in schist, cut by two small faults. Note that the displacements are obvious where a fault cuts the dike, but not where a dike is not present. The location is Voyageurs National Park, on a small island in Namakan Lake, just northwest of Kettle Falls.

Figure 3-17. (a) Photograph of part of a detailed geologic map showing two northeast-trending faults. The rocks on the right of the Pike River Fault have moved northward about 280 m relative to the left-hand side. This fault can be traced for several miles as a faint line on aerial photographs. The top one-third of the map area (light gray) is underlain by volcanic sandstone, and the lower two-thirds is underlain by graywacke and slate, as in Figure 3-8a. The dark gray spots are rock outcrops. The strike-dip symbols with a long line and a perpendicular shorter line give the trend of the bedding and the direction of dip (short line), with the angle of dip from the horizontal given by the number. A black dot on the dip line shows the top direction of a graded bed. The arrows with numbers give the direction and amount of plunge of small folds and structural lineations, as in *Figure 2-8.* The field of view is about 2 km across. (From the geologic map of the Tower 7.5′ Quadrangle by Ojakangas, Sims, and Hooper, 1978.) (b) Field photograph of the fault in Figure 3-17a. The low cliff on the right side of the photo is the fault surface. It forms the east bank of the Pike River. Whereas large fault zones commonly contain broken rock and calcite or quartz veins and pods, this fault is a simple vertical plane. The location is near the dam and bridge on County Highway 77 about 1 km north of Highway 169 west of Tower.

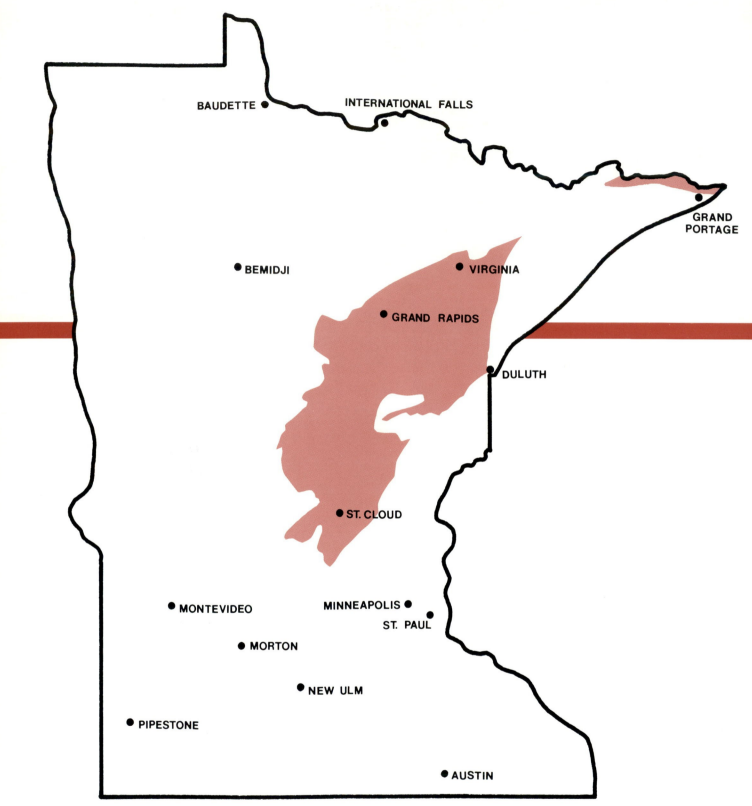

Map showing portions of Minnesota (shaded) where the bedrock is Middle Precambrian in age.

4

Middle Precambrian Time (2,500 to 1,600 Million Years Ago)

History is forever in the making. Events of one era may well influence events of a distant, later era. This fact can be illustrated no better than here in Minnesota. Iron-bearing sediment deposited in a Middle Precambrian sea about 2,000 million years ago, probably in part by the action of seemingly insignificant microscopic bacteria, has had a most profound effect on the socioeconomic framework of today's Minnesota. For that matter, Minnesota's iron played a very important part in shaping the United States into an industrial giant.

Middle Precambrian time could well be called the age of iron, in a geological sense rather than in the archaeological Iron Age sense. Yet the deposition of iron was but one of many events of Middle Precambrian time, which extends from the Algoman mountain-building episode near the end of Early Precambrian, 2,500 million years ago, to shortly after the end of another mountain-building event, the Penokean (the Hudsonian of Canada), about 1,800 million years ago. Thus, the Middle Precambrian lasted from 2,500 to 1,600 million years ago. That is quite a while.

Rocks of Middle Precambrian age are found in east-central and northeastern Minnesota in a basin now partially outlined by the iron-formations of the Mesabi, Gunflint, and Cuyuna ranges (Figure 4-1). All three of these iron-formations were probably once connected, or nearly so, but the Upper Precambrian gabbro complex swallowed up the iron-formation between the Mesabi and Gunflint ranges.

The derivations of the names of the three ranges are interesting. *Mesabi* is a Chippewa word for "giant." *Gunflint* came from Gunflint Lake, so named by early explorers because of the abundance on the lakeshore of flint (probably mainly red chert or jasper rather than black flint), which was used in the old flintlock guns of the 1700s to ignite gunpowder and to start fires. The Cuyuna certainly has the most unique name, suggested by the wife of Cuyler Adams, a surveyor who searched and found the iron ore of the area. The "Cuy-," of course, was from Cuyler, and "Una" was the name of Cuyler's dog, his faithful companion during his explorations. Adams reportedly preferred a plain "Una," but C. K. Leith, of the University of Wisconsin, used *Cuyuna* in a geological publication in 1903 and the name stuck.

The Middle Precambrian rocks of Minnesota can be divided into five main groups. The oldest group is very poorly known. A few drillholes have indicated the presence of these sedimentary rocks (dolomite, quartz sandstone, and other types) in east-central Minnesota on the Cuyuna Range beneath younger Middle Precambrian rocks, and a few exposures of probably correlative rocks occur near the village of Denham, 80 km southwest of Duluth. The next three groups, all sedimentary rocks, are found on the three iron ranges. In order of age, they are quartz sandstones (now quartzites), iron-formations, and dark gray sandstones and mudstones (now largely black slates). The fifth and youngest Middle Precambrian rock group consists of various granitic rocks that were intruded into the other rocks in east-central Minnesota.

LONG EROSION BY RUNNING WATER AND ICE

The uplifted volcanic, sedimentary, and granitic rocks of the 2,700-million-year-old Early Precambrian mountains were immediately attacked in earliest Middle Precambrian time by erosive processes, and especially by the never-ending action of running water. However, a special type of erosion

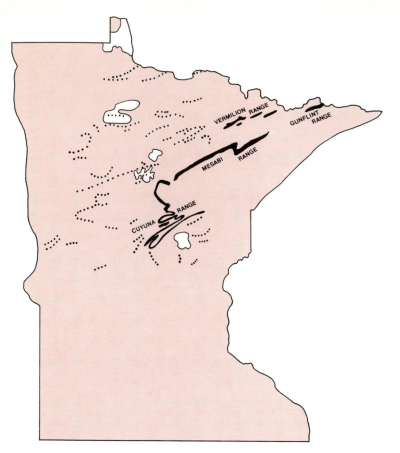

Figure 4-1. Map of Minnesota's iron ranges. The Gunflint Range continues into Canada. The lines of dots represent magnetic anomalies caused by buried iron-formations, none of which appears to be of economic value.

and deposition also occurred, perhaps for the first time in the earth's history, at several locations—in what is now Ontario north of Lake Huron and both east and west of Hudson Bay, in what is now the Upper Peninsula of Michigan, and in what is now southeastern Wyoming. At these places, deposits of the earth's oldest recorded glaciation are preserved. These rocks in Ontario—*tillites*, or lithified glacial till—have been bracketed by age dates on related igneous rocks and appear to have been formed by glaciers some time between 2,400 and 2,100 million years ago. Many geologists think that the glacial deposits at all these locations were the result of a single widespread continental glaciation (Figure 4-2). If so, what is today Minnesota too would have been glaciated and may have been partially covered by these ancient deposits. Thus, glaciers of this most ancient of ice ages would have played a part in the erosion of Minnesota's Lower Precambrian rocks. This is conjectural, however, for ancient glacial deposits have not been found in Minnesota and are not very likely to be found, either, based on the known ages of Minnesota's exposed rocks. If such ancient glacial deposits were once present, they were probably eroded away in later times. Yet there is a possibility that such deposits exist deep in the subsurface of east-central Minnesota.

Whether sedimentary rocks of this age, 2,100 to 2,400

million years, are present in the subsurface is of prime economic interest. Why? Because north of Lake Huron in the Blind River-Elliot Lake area, beneath the tillites and other sediments, are basal quartz pebble conglomerates that contain uranium. These congomerates are the products of the long weathering and erosion of the Lower Precambrian rocks and are the key rocks in one of the world's richest uranium provinces. Rounded but unweathered grains of pyrite (FeS_2) and uranium minerals, especially uraninite (UO_2), suggest deposition in an environment that was deficient in oxygen; the pyrite grains would have been rusted if exposed to oxygen, and the uranium minerals would have been changed to other uranium minerals. Therefore, it must be asked whether or not the conglomerates could have been deposited *before* an oxygen-rich atmosphere was present on earth. Many earth scientists think so. Could such conglomerates also have been deposited and preserved in east-central Minnesota? Many geologists hope so. Whereas the first premise may never be proven, the second can be tested with a few drillholes a thousand or so meters deep.

ANCIENT QUARTZ SANDS

Resting upon the Lower Precambrian granites and volcanic-sedimentary rocks on the Mesabi Range is the Pokegama

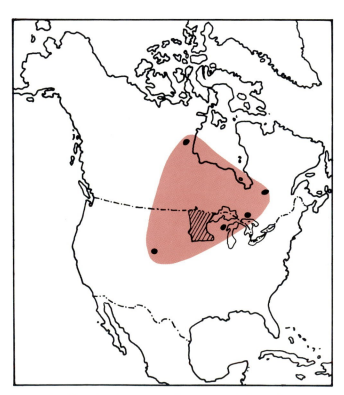

Figure 4-2. Map showing the possible extent of Middle Precambrian glaciation in North America. The five dots show locations where rocks interpreted to be ancient glacial deposits have been found. (After Young, 1970.)

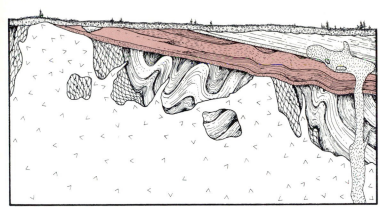

Figure 4-3. Generalized geologic cross section of northeastern Minnesota. Middle Precambrian rocks are shaded. Note that three main sedimentary units are present. The lowest is the Pokegama Quartzite. The lensoid one where the layers approach the surface is the Biwabik Iron Formation; as discussed later in this chapter, it could extend under Lake Superior. The thick upper one represents the Virginia and Rove formations. Note that the Middle Precambrian sequence rests upon eroded Lower Precambrian rocks; this erosional surface was originally sub-horizontal. The tilting and folding of the Middle Precambrian rocks obviously must have occurred after their deposition between an estimated 2,000 to 1,800 million years ago. Not drawn to scale.

Quartzite, a quartz sand and mud unit as much as 100 m thick. It is shown as the basal Middle Precambrian rock unit in Figure 4-3, a highly generalized composite geologic cross section of the rocks of northeastern Minnesota. Note that it rests on a rather featureless erosional surface, for like all basal quartz sand units the Pokegama was the result of long weathering and erosion. In this case, the erosional interval may have lasted longer than it did north of Lake Huron, from 2,700 to about 2,000 million years ago. A thin quartz-pebble conglomerate is at the base of the Pokegama. The Pokegama is exposed in only a few places, such as on the Mississippi River just west of Grand Rapids, on the Prairie River just north of Grand Rapids, and at a point about mid-way between Eveleth and Virginia (Figure 4-4). Interestingly, the Pokegama resembles the better-exposed Palms Formation, a correlative unit on the Gogebic Range in Wisconsin and Michigan. The Palms appears to have been deposited in a tidal flat environment in the same Middle Precambrian sea as was the Pokegama.

On the Gunflint Range in Minnesota and adjacent Ontario, the very thin Kakabeka Quartzite is the basal quartzose unit. On the Cuyuna Range, drillholes have penetrated the Mahnomen Quartzite, a similar unit.

IRON-FORMATIONS

Each of these quartzite units is overlain by iron-formation (see Figure 4-3), a rock made up largely of chert and hematite or magnetite, with other iron minerals usually present as well (Figure 4-5). On the Mesabi, the iron-formation is called the Biwabik Iron Formation; on the Gunflint, it is the Gunflint Iron Formation; and, on the Cuyuna, it is the Trommald Iron Formation. These iron-formations contain abundant rounded pellets and *oolites*, concentrically ringed grains indicative of reworking in shallow waters (Figure 4-6). Especially in the Gunflint, many of the chert-rich beds contain fine grains of hematite that color the chert red, making the variety of chert called jasper.

Minnesota's three iron ranges are not unique in the Lake

a

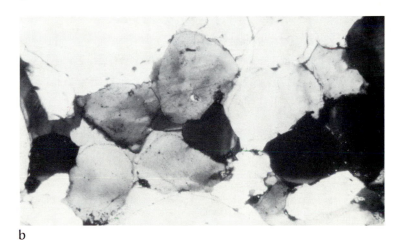

b

Figure 4-4. (a) Outcrop of Pokegama Quartzite on the Prairie River near Grand Rapids. Note the crossbedded nature of the original sands. (b) Photomicrograph of the Pokegama Quartzite. All of the grains are quartz, the mineral that best survives extensive weathering and abrasion. Note the rounded grain boundaries beneath the quartz cement. The field of view is about 2 mm across.

a

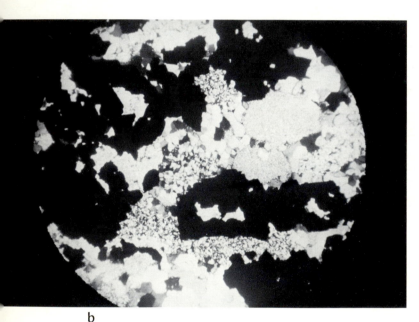

b

Figure 4-5. (a) Biwabik Iron Formation in roadcuts on U.S. Highway 53 at Eveleth. Note the prominent bedding. (b) Photomicrograph of a typical iron-formation. The black grains are hematite and magnetite, the large gray grains are iron silicates, and the fine-grained gray and white portion of the view consists of chert (fine-grained quartz). The field of view is about 2 mm across.

Superior region. The Gogebic Range in Wisconsin and Michigan and the Marquette and Menominee ranges and the Crystal Falls-Iron River district in Michigan have also been important (Figure 4-7). All of these iron-formations are about the same age, all overlie quartzites, and all were succeeded by the same types of black muds and sands, which will be described later. The iron-formations were probably deposited in the same marine basin, with the iron minerals deposited relatively near shore in the shallow sea.

Whereas the iron-bearing beds of the Michigan and Wisconsin ranges and the Cuyuna are now steeply tilted or folded (Figure 4-8), those of the Gunflint and Mesabi ranges are gently tilted, dipping at about 9° or 10° toward Lake Superior (Figure 4-9). As discussed in Chapter 8, on metallic mineral resources, these configurations had an influence on mining. High-grade iron ores were formed from the taconite by natural processes where the taconite beds were exposed at the surface.

The world's other major iron-formations—in Labrador, western Australia, Russia, Venezuela, Brazil, and Africa—are nearly all about the same age as those in the Lake Superior region, about 2,000 million years or so. These formations account for most of the world's iron ore. Never before and never since has there been such a period of deposition of iron materials. Why did this happen on a worldwide scale at essentially a single time? Logically, the answer must involve a worldwide mechanism. That mechanism may have been a change in the earth's atmosphere about 2,000 million

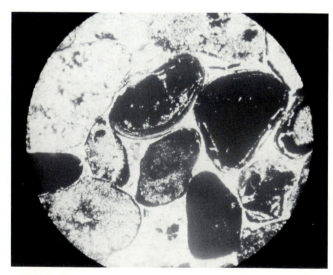

Figure 4-6. Photomicrograph of Biwabik Iron Formation showing rounded dark grains of hematite. Note the layering in some of the grains. Such shapes and structures indicate a high-energy shallow-water environment. However, it is likely that the round grains were originally iron carbonates ($FeCO_3$) that were later altered by oxidation to hematite (Fe_2O_3). Iron minerals can change quite readily to other iron minerals, depending on the acidity of the groundwater and the abundance of oxygen. The field of view is about 2 mm across.

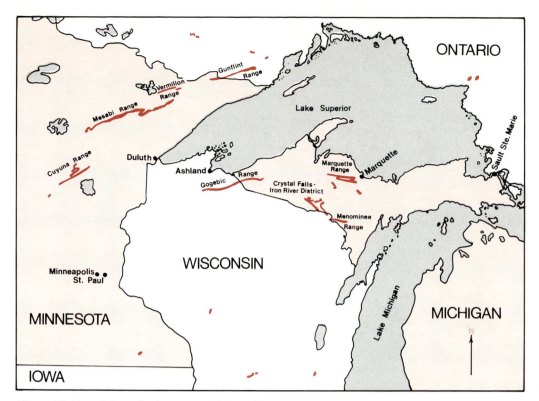

Figure 4-7. Map of the major iron ranges of the Lake Superior region. Only the larger ranges are labeled. Of the labeled ranges, only one does not consist of Middle Precambrian iron-formation. The iron-formation of the Vermilion Range is Lower Precambrian.

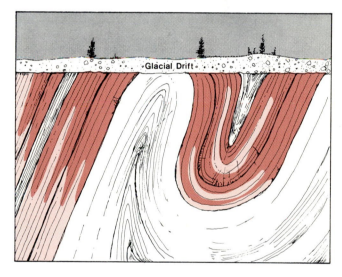

Figure 4-8. Diagrammatic cross section of folded Trommald Iron Formation (light shading) and natural high-grade iron ore (dark shading) of the Cuyuna Range. (After a diagram by the United States Steel Corporation.)

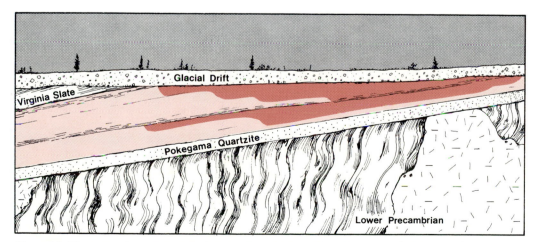

Figure 4-9. Diagrammatic cross section of gently dipping, Biwabik Iron Formation (light shading) and natural high-grade iron ore (dark shading) of the Mesabi Range. Note that the high-grade ore was obtainable simply by removing the glacial drift. This led to the big open-pit operations. (After a diagram by the United States Steel Corporation.)

years ago, from a carbon-dioxide-rich one to an oxygen-rich one.

What would have caused such a change in atmosphere? Plants did. Scientists think that green marine plants, which use carbon dioxide and sunlight to make their food and give off oxygen as a waste product, had developed in sufficient quantities by 2,000 million years ago to have enriched the

a

b

Figure 4-10. (a) Large algal colonies preserved as mounds on a bed surface in the Erie Mining Company pit at Hoyt Lakes. Similar mounds are found in modern oceanic lagoons in western Australia, and hence these algae may have lived in shallow, intertidal waters. (b) Drill cores showing algal structures. The cores are about 3.8 cm in diameter. See Plate 5.

atmosphere with oxygen. Actually, part of the evidence is preserved in the iron-formations themselves. Beds of fossil algal structures, probably originally composed of calcium carbonate ($CaCO_3$), but since replaced by silica (SiO_2), are present at the base and near the middle of the iron-formation (Figure 4-10; Plate 5). D. G. Darby (1972) has described both megascopic and microscopic Precambrian life forms in Minnesota. In Michigan, such fossils are abundant in some carbonate beds that are slightly older than the iron-formations. Land plants were not to evolve for another 1,500 million years or so, but there is abundant evidence of marine plant life on microscopic as well as on megascopic scales (Figure 4-11). Thus, these early photosynthesizers probably changed the gases in the atmosphere as well as those dissolved in seawater.

How does this addition of oxygen relate to the origin of iron ore? Iron and silica freed by weathering and erosion of the ancient Lower Precambrian rocks should have been building up in the seawater over millions and millions of years. Then, when oxygen became abundant enough, the iron rather suddenly (geologically speaking) combined with the oxygen and was chemically precipitated in shallow waters as the common iron oxides—hematite (Fe_2O_3), limonite ($Fe_2O_3 \cdot 2H_2O$), and magnetite (Fe_3O_4)—as well as iron carbonate ($FeCO_3$) and various iron silicates. At the same time, abundant silica in the water was precipitated as the mineral quartz (SiO_2), and beds of these mixed materials were formed (see Figure 4-5a). But perhaps volcanism, rather than weathering, was the source of the iron and the silica. Beds of volcanic ash are found in the iron-formation on the Gogebic and Cuyuna ranges, and various volcanic rocks are associated with iron-formations in Michigan. Although the Mesabi Range does not contain definite volcanic rocks, it does contain many muddy beds that may have been, at least in part, volcanic ash. Recall that all the Lake Superior iron-formations seem to have been deposited within the same basin; thus, volcanism anywhere in the region could have put iron and silica into the seawater.

Actually, the detailed origin of iron-formation is a subject of much debate among the experts, and the hypotheses presented above may well be too simple. There are several other kinds of iron minerals in addition to the common iron oxides, including iron sulfide (pyrite), iron silicates (several types), and iron carbonate (siderite). As acidity and oxygen content change, a specific iron mineral may become unstable and may quite easily be changed to another iron mineral. Thus, changes could have occurred soon after deposition or during the original conditions of formation. Yet, the textural evidence of rounded pellets and concentrically banded grains of hematite and chert (see Figures 4-5 and 4-6) suggests a definite environment for the deposition of the chemical precipitates—shallow water where waves and currents periodically or constantly agitated and reworked the minerals.

INSTABILITY AS A PRECURSER
TO MOUNTAIN BUILDING

Although the iron-formations of the Lake Superior district (see Figure 4-7) all dip down toward the center of what was apparently a Middle Precambrian basin located approximately where Lake Superior is now positioned, most geologists do not think that iron-formation underlies the entire basin. On the other hand, iron-formation was deposited over an entire now well-exposed basin of the same age in Australia, and that could have been the case here as well. Drillholes several kilometers south of the Mesabi Range indicate that the iron-formation is still present as a thick sedimentary unit, but no deep drillholes have been put down to test the more central parts of the basin.

If iron-formation was not being deposited nearer the center of the basin, what was being deposited when iron minerals were forming near shore? Probably not much, other than the same fine mud that makes up part of the iron-formation itself. In fact, the episode of iron precipitation may have been, geologically speaking, quite rapid. The precipitates could have diluted or masked the muds, which were probably of both volcanic (ash) and weathering origin. As the iron mineral precipitation slowed and finally ceased, deposition of mud, albeit slow, became the dominant process. This transition is evident on each range and is illustrated diagrammatically in Figure 4-3. On the Gunflint Range, the Gunflint Iron Formation is overlain by the Rove Formation; on the Mesabi Range, the Biwabik Iron Formation is overlain by the Virginia Formation; and, on the Cuyuna Range, the Trommald Iron Formation is overlain by the Rabbit Lake Formation. Similar relationships also exist on the south side of this Middle Precambrian basin in Wisconsin and Michigan.

The Virginia Formation was exposed in several iron mines but does not crop out in natural exposures. The Rabbit Lake Formation was observed only in drill cores and mines of the Cuyuna Range. The Rove is quite inaccessible except on U.S. Highway 61 a few miles north of Grand Portage, at Grand Portage itself, and in Ontario. Unfortunately, however, the similar and correlative Thomson Formation, which was probably deposited nearer the center of the basin, is well exposed in the vicinity of Cloquet; Carlton; the village of Thomson, for which it was named; and Jay Cooke State Park (Figure 4-12).

These originally muddy formations, where observed, are very revealing of the environment of deposition. Whereas the muddy rocks are not too useful for this, interbedded graywacke sandstone and siltstone beds are. Graywackes are dark-colored sandstones containing appreciable amounts of fine-grained mud between the sand-sized grains, which

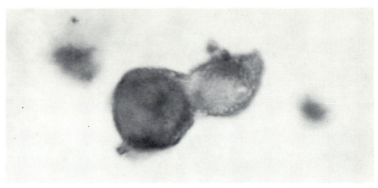

a

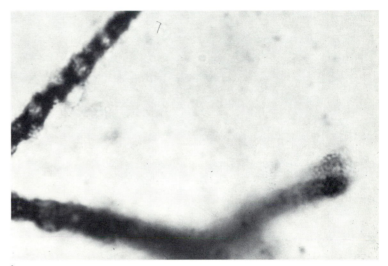

b

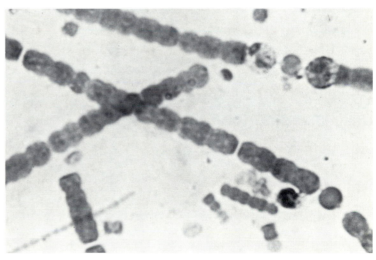

c

Figure 4-11. (a) Microscopic view of primitive unicellular algae from the Gunflint Iron Formation in Ontario. Each cell is about 0.007 mm across. (Courtesy of D. G. Darby). (b) Microscopic view of filamentous blue-green algae from the Gunflint Iron Formation in Ontario. Segments are 0.002 to 0.003 mm across. (Courtesy of D. G. Darby. (c) Microscopic view of modern filamentous blue-green algae. Compare this with Figure 4-11b. This specimen is slightly larger than the previous one. (Courtesy of D. G. Darby.)

41

a

b

consist of many minerals and rock fragments. Unlike quartz sandstones, which indicate long weathering and abrasion, graywackes indicate that rapid erosion, deposition, and burial occurred, and thus the sediments were not reworked. The presence of graywackes thus strongly suggests a rapidly subsiding and relatively deep basin in which the sediment was dumped.

The dumping of these graywackes was accomplished by turbidity currents, the same mechanism that deposited the Lower Precambrian sandstones described earlier (see Figure 3-8a). Many of the graywacke beds display a grading in grain size, from coarse on the bottom to fine on the top. Various internal bedding characteristics are also diagnostic. A few of the beds display markings on their bottom surfaces, or soles, that are also characteristic of turbidite sequences (Figure 4-13).

These sole marks, along with crossbeds in the upper parts of some graded beds and in associated siltstone beds, are used to determine the directions in which the depositing currents were moving. Such *paleocurrent* ("old current") studies on the Rove and Thomson formations in Minnesota and on the correlative Tyler Formation on the Gogebic Range in Wisconsin and Michigan show that the paleocurrents were generally moving toward the south in the Minnesota part of the basin and generally toward the north in Wisconsin and Michigan. Such data are basic in developing paleogeographic models such as the one shown in Figure 4-14. Actually, the sediment making up the graywackes was

Figure 4-12. (a) Tilted graywacke sandstone and siltstone beds of the Thomson Formation in the St. Louis River Valley near the villages of Thomson and Carlton. Rows of holes are the result of the weathering out of carbonate concretions (better-cemented parts of the rock) that formed after deposition of the sediment on the sea floor. (b) Photomicrograph of graywacke sandstone from the Thomson Formation. Only about half of the grains are quartz; the others are mainly feldspars and rock fragments. Note the poor sorting, with grains of many sizes present. The sandstone has undergone little abrasion, as indicated by both the angularity and the composition of the grains. Compare this with Figure 4-4b. The field of view is about 2 mm across.

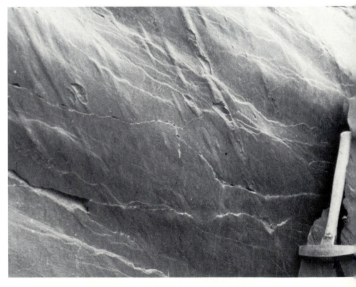

Figure 4-13. Elongate markings on the bottom of a tilted sandstone bed in the Thomson Formation. These were probably formed by objects such as pebbles being dragged along by the current, making grooves in the mud that were then filled with sand and hardened and now stand out in relief. By measuring such marks and geometrically rotating the bed back to horizontal, geologists can determine the direction of the paleocurrent that deposited the bed.

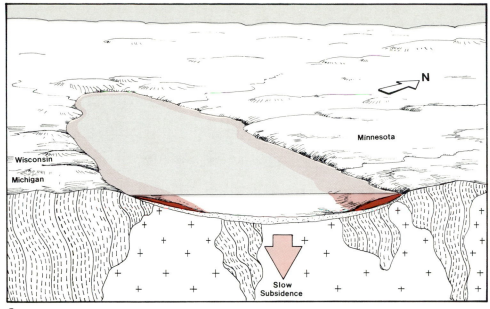

a

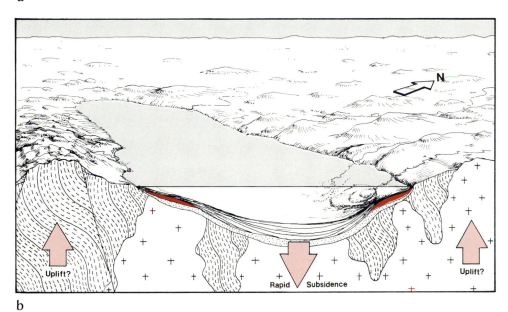

b

Figure 4-14. (a) North-south cross-sectional model showing a probable early stage in the development of the Middle Precambrian basin in which the Middle Precambrian sediments, including the iron-formations, were deposited. The quartz sands that overlie the eroded Lower Precambrian surfaces are represented by the thin, white, dotted layer. The iron-formations, which may have formed near shore, are shown by the dark layers. Note that the basin was probably subsiding slowly. Note the vertical exaggeration; the basin was probably on the order of 150 km wide, whereas the quartz sands and iron-formation are only a few hundred meters thick. (b) Later stage of the same model showing rapid subsidence and the deposition of a thick sequence of dark muds and sands. Note that adjacent landmasses may have been rising and that sediment deposition may have been on submarine fans such as are found in the modern oceans.

probably resedimented. That is, the sediment was carried to the sea by rivers and accumulated near shore, later to be moved down the steeper slopes by periodic turbidity currents.

Figure 4-14 is intended to depict only the western end of the Middle Precambrian basin. If the basin were extended another 160 km to the east, to the vicinity of Marquette,

Michigan, and to the south, it would become more complicated. The southern shoreline of the basin would be highly irregular, with numerous large islands of Lower Precambrian rocks that may have resulted from blockfaulting, which was going on concurrently with sedimentation. Also, volcanic rock units are much more abundant in the sedimentary sequence, and such volcanic rocks extend far to the south

43

a   b

Figure 4-15. (a) Fold in the Thomson Formation. The location is the west bank of the St. Louis River below the dam just east of Carlton. (b) Small anticline in the Thomson Formation. Note the nearly vertical cleavage manifested as fine cracks. The location is the east bank of the St. Louis River below the dam just east of Carlton. This photo was taken from the other side of the highway bridge shown in Figure 4-15a.

in Wisconsin and may contain the recently discovered massive copper and zinc sulfide deposits at Crandon, Ladysmith, and Pelican. Thus, whereas the part of the basin illustrated in Figure 4-14 was unstable in that it was rapidly subsiding, the southern part of the basin was highly unstable, with volcanism much more dominant.

In terms of plate tectonic theory, the volcanics may have formed in volcanic arcs caused by the subduction of a denser oceanic plate, moving northward and being subducted beneath a less dense continental plate. This, however, is difficult to prove. In terms of older geosynclinal theory, which hypothesizes deposition in rapidly subsiding troughs usually adjacent to a continental margin, the Minnesota sediments were deposited nearer the continent than the volcanic-bearing sequences in Michigan and Wisconsin. Some geologists have suggested that this same geosynclinal basin extended all the way into Labrador, where similar deposits are present. If so, these intervening Middle Precambrian rocks would have been eroded away.

THE PENOKEAN
MOUNTAIN-BUILDING EVENT

Out of this basin in what is now east-central Minnesota, but especially to the east in what are now Wisconsin and Michigan, new mountains rose near the end of Middle Precambrian time. This mountain-building event, the Penokean, is now being precisely dated. The dates range from 1,900 to 1,800 million years ago. Complicating the picture are some younger granites in Wisconsin, apparently unrelated to the major event, which yield age dates of 1,450 to 1,500 million years. Perhaps there are two events here, closely spaced. A number of geochronologists and geologists are working on the Penokean, and more should be known in the next few years.

The Penokean mountain-building event caused the folding and metamorphism of Precambrian sedimentary rocks. Those units deposited in the northern part of the basin on the Mesabi and Gunflint ranges were hardly affected; the metamorphism is of very low grade, meaning that the rocks

were affected by only slight temperature and pressure increases. The beds now have a slight tilt or dip of about 10° to the south.

The Thomson Formation, nearer the center of the basin, however, was folded into broad open folds a few kilometers across, elongated in an east-west direction so that the beds dip to the north or south. Smaller outcrop-size folds are readily visible at a few locations (Figure 4-15). Virtually all outcrops of the Thomson Formation consist of tilted beds that are on the flanks of large folds (see Figure 4-12). The Thomson Formation is also more metamorphosed than the Virginia and the Rove formations. The clay minerals were changed to micas, either muscovite, biotite, or chlorite, depending upon the original iron content of the clays. These micas, microscopic in size through most of the formation, all grew perpendicular to the direction of dominant pressure, as is true of the micas in nearly all metamorphic rocks. In the Thomson Formation they are generally oriented in east-west vertical planes, indicating that the pressures were strongest in a north-south direction. Because the greatest folding and mountain building occurred still farther south, it can be assumed that the pressures were from the south. Since the microscopic micas are all parallel to each other, they create planes of weakness parallel to the micas. These are manifested as slaty cleavage, and the rocks are now slates, as illustrated in Figure 4-16.

Farther south in the Thomson Formation, from Moose Lake south to Denham, the micas are no longer microscopic. The rock has become a mica schist. Still farther south in central Minnesota, near Little Falls, some rock that resembles the Thomson is now a staurolite-muscovite schist, as shown in Figure 4-17. Though it is apparent that there is a distinct increase in metamorphic grade from north to south in the formation, the picture is complicated by metamorphic *hot spots*. In the Upper Peninsula of Michigan, where exposures are more abundant than in east-central Minnesota, some classic concentric metamorphic *aureoles* or zones of specific minerals have been distinguished, with successively higher-temperature metamorphic minerals appearing nearer

Figure 4-16. Bedding and cleavage in slate of the Thomson Formation. Beds dip to the left, and cleavage is nearly vertical. The bedding is clearly visible on the cleavage planes. Roofing slates and the old-fashioned slate boards came from such rocks. Rock cleavage forms when original clay minerals recrystallize to micas, which are oriented perpendicular to the pressures that caused the folding and recrystallization. The flat surfaces reflect the microscopic mica orientation. The location is the east end of dam over the St. Louis River just east of Carlton.

Figure 4-17. Photo of staurolite schist (possibly correlative with the Thomson Formation) on the Mississippi River about 11 km south of Little Falls. Note the large staurolite ($HFe_2Al_9O_8Si_4O_{16}$) crystals. This rock has undergone a much higher degree of metamorphism than the Thomson Formation farther north.

Figure 4-18. Quarry in Middle Precambrian granite 8 km south of Isle, in central Minnesota.

the centers of the hot spots. The staurolite-muscovite schist at Little Falls may be part of such an aureole. The iron-formation of the Cuyuna Range became highly folded at this time (see Figure 4-8).

Middle Precambrian granitic bodies are present in the Penokean belt and are a part of the picture of increased temperatures and pressures; most were also deformed somewhat during the deformation of the Thomson. In central Minnesota, several different types of 1,700- to 1,800-million-year-old granitic rocks are exposed (see Plate 2) in the vicinity of St. Cloud and Mille Lacs Lake. These rocks are beautifully exposed in quarries (Figure 4-18).

G. B. Morey of the Minnesota Geological Survey and P. K. Sims of the United States Geological Survey (1976) have suggested that the more highly deformed and more metamorphosed Middle Precambrian rocks, including the metamorphic hot spots, were all deposited upon an ancient gneissic basement (the southern gneissic complex of Figure 3-4) whereas the more gently folded and less metamorphosed rocks in northern Minnesota, including most of the Thomson Formation, were deposited upon a granite-greenstone basement. During the Penokean mountain-building event, the gneissic basement behaved much more plastically and perhaps melted more easily than the terrane to the north, thus accounting for the higher deformation and metamorphism of the overlying sedimentary rocks. The boundary between these two basement terranes is suggested to be a major fault zone. This fault zone shows up on the geologic map in western Minnesota (Plate 2) and has been named the Great Lakes Tectonic Zone because it may extend north of Lake Huron as well (Sims et al., 1980). That this fault zone, which may have originated in Early Precambrian time, is still active is documented by the occurrence of five of Minnesota's last seven earthquakes along this boundary, as discussed in Chapter 12.

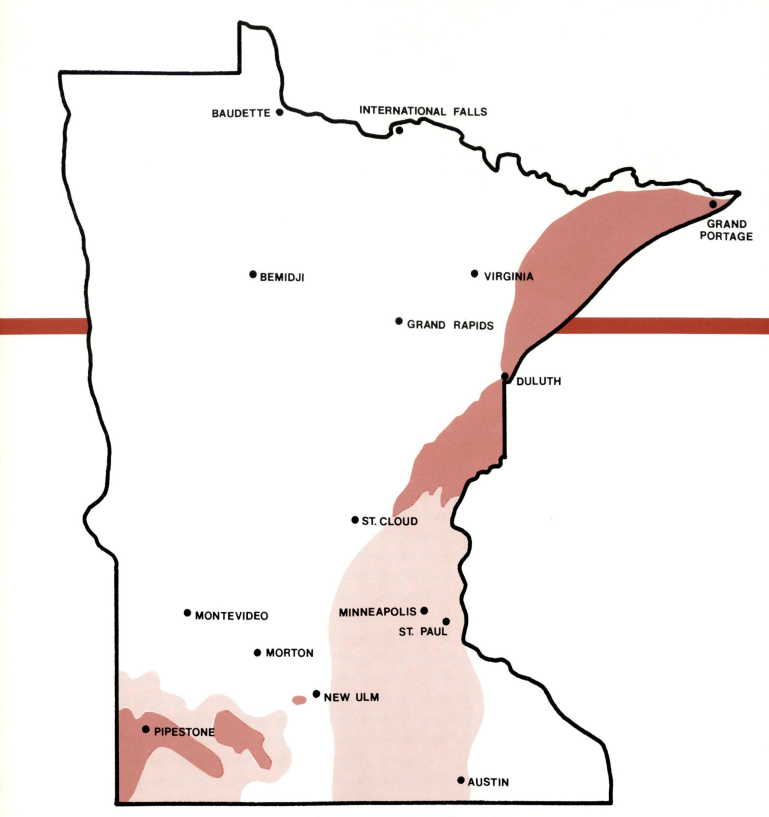

Map showing portions of Minnesota (shaded) where the bedrock is Late Precambrian in age.

5

Late Precambrian Time (1,600 to 600 Million Years Ago)

Late Precambrian time extends from about 1,600 million years ago to the beginning of the Paleozoic ("old life") Era, an age of abundant fossils about 600 million years ago. Thus, like the Middle Precambrian, the Late Precambrian spans a billion years. A lot can happen in a billion years, and a lot did!

Upper Precambrian rocks in Minnesota can be divided into three main groups. The oldest consists of quartz sandstones, the next oldest consists of 1,100-million-year-old lava flows and gabbroic intrusions, and the youngest consists of postvolcanic sandstones and other sediments. These are shown diagrammatically on the composite generalized cross section of northern Minnesota (Figure 5-1) and are arranged schematically in Figure 5-2.

EROSION AND DEPOSITION OF QUARTZ SANDS

By the start of the Late Precambrian, erosion had already begun to wear down the Penokean mountains of the Lake Superior region, including those of the east-central portion of what is today Minnesota. This erosion continued for a few hundred million years. Meanwhile, the older greenstone-granite terrane to the north and the gneiss-granite terrane to the south were still being slowly weathered, eroded, and lowered, as they had been for at least the previous 1,000 million years.

The net result of this long weathering and erosion during the first part of Late Precambrian time was quartz sand, similar to that which made up the Pokegama Quartzite and correlative formations of Middle Precambrian age. These Upper Precambrian quartz sandstones survive in Minnesota as the 1,500-m-thick Sioux Quartzite of southwestern Minnesota and southeastern South Dakota (Figure 5-3), the 8-m-thick "Nopeming Quartzite" of the Duluth area (Figure 5-4), and the 60-m-thick Puckwunge Formation of northeasternmost Minnesota (Figure 5-5). Each of these three quartzose units has a thin basal conglomerate composed largely of resistant pebbles of quartz, chert, and iron-formation.

These formations pose many problems for geologists. They are too old to contain fossils, and, like most sediments, they cannot be readily dated by radiometric techniques. Partly because of a lack of fossils, their environments of deposition are difficult to determine. Most of the sedimentary structures, such as crossbedding and ripple marks, could have been formed in any of several different environments, such as rivers, deltas, lakes, or oceans. There is some evidence in the top one-third of the Sioux Quartzite that tides may have been important; its juxtaposed crossbeds pointing in opposite directions are a feature readily formed by the ebb and flow of tides.

Similar quartz sandstone units rest on Lower or Middle Precambrian rocks in Wisconsin, Michigan, and Ontario (see Figure 5-2). Wherever the basal contact of the formation can be seen in a single exposure (but this is rare) or where the relationships of the sandstones and the older units can be ascertained from a number of outcrops, there is abundant evidence of the preceding erosional event. The widespread regional distribution of these quartz sandstones suggests that they are erosional remnants of a blanket of quartz sand deposited in a broad sea that slowly encroached from south to north over a flat, eroded, vegetationless, and sand-covered surface. In support of the hypothesis of deposition in a sea are limestones above the quartz sandstones in Ontario just north of Lake Superior: most limestones are formed in marine environments. Alternatively, the sands

47

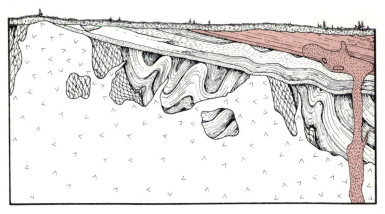

Figure 5-1. Generalized geologic cross section of northeastern Minnesota. Upper Precambrian rocks are shaded. Note that three main Upper Precambrian rock units are present. The lowest is the Puckwunge Formation (a quartz sandstone), the thick one is the North Shore Volcanic Group, and the intrusive unit is the large sill that makes up the Duluth Complex. Note also that a buried surface of erosion separates the Upper Precambrian sequence from the Middle Precambrian sequence. This surface was probably subhorizontal when formed.

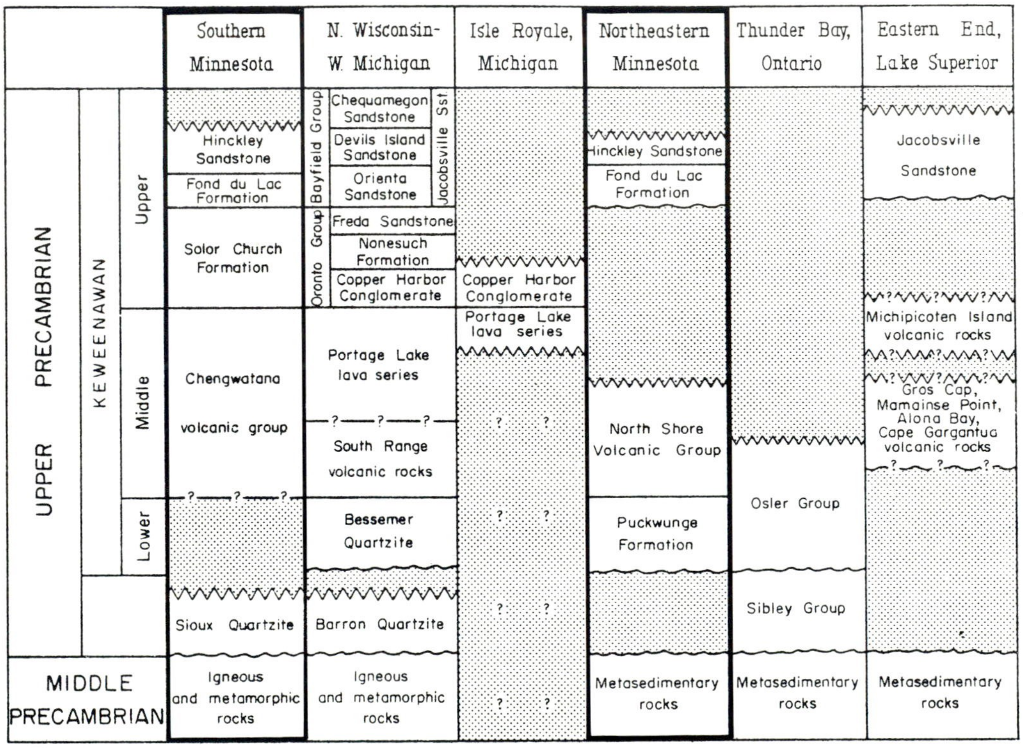

		Southern Minnesota	N. Wisconsin–W. Michigan	Isle Royale, Michigan	Northeastern Minnesota	Thunder Bay, Ontario	Eastern End, Lake Superior
UPPER PRECAMBRIAN	KEWEENAWAN — Upper	Hinckley Sandstone / Fond du Lac Formation / Solor Church Formation	Bayfield Group: Chequamegon Sandstone, Devils Island Sandstone, Orienta Sandstone / Oronto Group: Freda Sandstone, Nonesuch Formation, Copper Harbor Conglomerate	Jacobsville Sst / Copper Harbor Conglomerate	Hinckley Sandstone / Fond du Lac Formation		Jacobsville Sandstone
	KEWEENAWAN — Middle	Chengwatana volcanic group	Portage Lake lava series / ?—?—? / South Range volcanic rocks	Portage Lake lava series	North Shore Volcanic Group		Michipicoten Island volcanic rocks / Gros Cap, Mamainse Point, Alona Bay, Cape Gargantua volcanic rocks / ?—?—?
	KEWEENAWAN — Lower	?—?—?	Bessemer Quartzite	? ?	Puckwunge Formation	Osler Group	
		Sioux Quartzite	Barron Quartzite	? ?		Sibley Group	
MIDDLE PRECAMBRIAN		Igneous and metamorphic rocks	Igneous and metamorphic rocks	? ?	Metasedimentary rocks	Metasedimentary rocks	Metasedimentary rocks

Figure 5-2. Correlation chart of rock units of Late Precambrian age in the Lake Superior region. The oldest units are on the bottom and the younger are on the top. Units of the same age lie on the same horizontal level; for example, the Sioux Quartzite of southern Minnesota and the Barron Quartzite of Wisconsin are thought to be the same age. Jagged lines mean that the top of the unit has been eroded off. The dotted areas indicate that rocks of that age are missing, usually because they were never deposited, but possibly because they were eroded. Straight horizontal lines between units indicate no erosion between those units, whereas wavy lines indicate erosional surfaces. No connotations of relative thicknesses are implied. (From Craddock, 1972.)

could have been deposited on a broad *alluvial plain* (a plain resulting from stream deposits) formed by a number of major rivers flowing southward off the large landmass that lay to the north. Or perhaps both environments were involved, as shown in Figure 5-6. However, with the lack of age dates and fossils, age correlations of the widely separated units are difficult to prove. Rhyolite in a drillhole in the Sioux Quartzite in northwesternmost Iowa has been dated at approximately 1,470 million years, and rhyolites just

beneath the Baraboo Quartzite in south-central Wisconsin have been dated at 1,760 million years. Thus, the quartz sandstones all have ages somewhere between 1,760 and 1,100 million years, the age of the overlying lava flows. The Sioux may be closer to 1,700 million years old and, if so, would be late Middle Precambrian in age.

A paleomagnetic technique has also been applied to the study of Upper Precambrian rocks. This method determines the location of the magnetic north pole and has provided one of the main lines of evidence for continental drift and plate tectonic theory in rocks less than 200 million years old. The magnetic north pole apparently reverses itself from time to time; that is, a compass needle that now points north toward the magnetic north pole would, after a decrease in the strength of the earth's magnetic field and a reversal of the magnetic pole, point toward the south pole. Three such changes have been noted in Upper Precambrian rocks of the Lake Superior region. Not all the quartz sandstone units show the same polarity, and on this basis it has been suggested that they cannot all be the same age and hence are not part of the same blanket deposit. Now, does that clear or muddy the water of this ancient sea, if it was a sea at all?

Some of the quartz sand grains in the quartz sandstone units (for example, the Sioux) display some very special clues to their history. The original rounded grains can be seen beneath two distinct layers or rims of quartz "cement," as in Figure 5-7. The inner layer has been rounded by abrasion during transport, whereas the outer layer is not rounded. This suggests that the quartz grain once resided, after first being transported and rounded, in an older sandstone, where it was cemented to other grains, and that it was later eroded, transported, rounded again, deposited, and cemented once again as part of the Sioux Quartzite. Perhaps the basal Middle Precambrian sands, such as the Pokegama, were the source of such multicycle grains.

The quartzose composition and the well-rounded grains in these sandstones suggest that much weathering and abrasion took place on a low-lying, peneplained, structurally stable landmass like the one depicted in Figure 5-6. Rivers round sand only with great difficulty, and thus it seems that the sand either was rounded in a shallow sea or was already somewhat rounded in an earlier cycle of erosion and sedimentation. However, the stability and probable tranquility of this first portion of Late Precambrian time was soon to be shattered by crustal rupture, earthquakes, and volcanism.

RIFTING OF NORTH AMERICA

In recent years, the old but intriguing idea of drifting continents has been transformed into modern plate tectonic theory. Much evidence, especially from the floors of the oceans, supports the hypothesis that the individual continents were all parts of a single supercontinent, Pangaea, about 200 million years ago. Then, presumably due to little understood convection currents in the upper part of the earth's mantle beneath the thin crust, Pangaea was split asunder at long faults or rifts into six large lithospheric plates and many smaller ones.

The forces that set the plates in motion relative to one another are still at work. For example, North America and Europe are about 6 cm farther apart each year. The Mid-Atlantic Ridge, with its active volcanism and abundant

a

b

c

Figure 5-3. (a) Crossbedded Sioux Quartzite of southeastern Minnesota. Bedding is horizontal. (Photo by R. Weber.) (b) Basal conglomerate of the Sioux Quartzite near New Ulm. (Photo by R. Weber.) (c) Ripple-marked Sioux Quartzite. (Photo by R. Weber.)

Figure 5-4. "Nopeming Quartzite" (light colored) overlain by a pillowed basalt flow. A fine-grained siltstone beneath the pillows apparently was deformed while still soft, suggesting that this first flow of the North Shore Volcanic Group was extruded into the same body of water in which the sand and silt were accumulating. This photo was taken behind Grandview Golf Course near Nopeming, just west of Duluth.

Figure 5-5. The Puckwunge Formation of northeasternmost Minnesota. Note the crossbedding. A study by A. F. Mattis (1972) suggests that the main currents that deposited the sand were moving from northwest to southeast.

earthquakes, is one line along which the spreading is occurring. As the plates on both sides of the ridge move apart, basalt probably generated by partial melting of upper mantle rock pours out along this line to cool and form new basaltic crust. As the spreading continues, the strip of most recent basalts is split into two, and more basalt comes between them. Such is the origin of the paleomagnetic strips, according to plate tectonic theory.

There is even fairly good evidence for the breakup of an earlier supercontinent about 450 million years ago, with the fragments eventually reuniting to form Pangaea. Is there some evidence of still older crustal rifting and spreading? Yes, and some of the best evidence is in the Lake Superior region, including Minnesota.

About 1,200 or 1,100 million years ago, North America began to spread apart along a rift that extended from what is now eastern Lake Superior all the way to what is now Kansas. Fluid basaltic lava moved into this zone, much as it is doing today along the Mid-Atlantic Ridge. One big difference, however, is that this lava poured out onto land rather than under water, for only the lowermost lava layers are pillowed (see Figure 5-4). This lava is like the lava that poured out onto Iceland, which is on the Mid-Atlantic Ridge, and also like that of Hawaii.

But the most important difference is that the spreading stopped after only a few tens of millions of years and what could have developed into a broad ocean never did. Duluth, Holyoke, Hinckley, Anoka, and Wells thus missed the chance to be seaside cities on the east coast of "Continent West." Similarly, Prescott, Austin, West Concord, Hampton, Hayfield, and Wangs, would have been part of the Minnesota Riviera on the west coast of "Continent East." The total spread, based in part on the width of the basaltic zone, has been estimated at 50 to 80 km. This zone of basalts has created a long magnetic and gravity anomaly—both are high due to the magnetite in the basalts and the greater density of the basalts compared to the adjacent rocks. This feature, known as the midcontinent gravity high, is the largest such anomaly on the North American continent (Figure 5-8).

VOLCANIC ROCKS OF LAKE SUPERIOR

The volcanic rocks formed in this spread center are exposed only in the vicinity of Lake Superior and sporadically as far south as Pine City, 115 km from Duluth. Elsewhere along the high, their presence has been verified by deep drilling. The wave-eroded northern shore of Lake Superior and the many North Shore stream valleys that have been cut into the lava flows contain an abundance of excellent outcrops. In fact, the North Shore is practically one long outcrop from Duluth to the Pigeon River on the Canadian border and beyond (Figure 5-9).

There are hundreds of individual flows in the North Shore

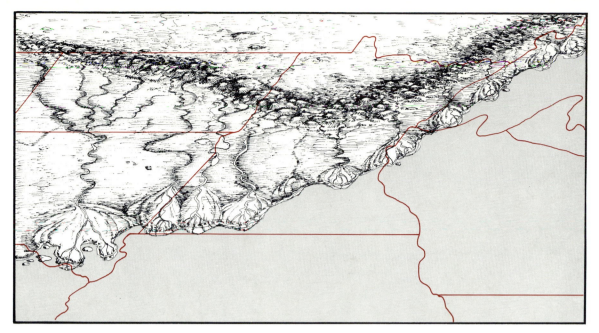

Figure 5-6. Reconstruction of how Minnesota and the surrounding region may have appeared at some time between about 1,600 and 1,100 million years ago, with rivers carrying sediment into a sea. The quartz sands that now make up the Sioux Quartzite and the Puckwunge Formation, as well as other units in Wisconsin and Ontario, may have been deposited in either a river or a marine environment, or both.

Volcanic Group; these vary in thickness up to many tens of meters (Figure 5-10). The total thickness of the volcanic pile is more than 8,000 m, but J. C. Green (1977), who has done the most detailed work on them, has delineated six discrete lava plateaus in the Lake Superior region. That is, not all flows were stacked up on top of another. Instead, they are more like a few tilted stacks of pancakes adjacent to each other, with some stacks older than others. (Green has made comparisons with other younger volcanic terranes, such as Iceland, and has studied Hawaii and the Columbia Plateau, in the northwestern United States, as well.) When the hot lava issued forth from cracks, it flowed to topographic lows and commonly formed what must have been lava lakes and ponds, now individual flows. Some can be traced for 65 km.

The oldest flows in Minnesota are found in two places, just west of Duluth and near Grand Portage, resting upon the quartz sandstones of the Nopeming and Puckwunge formations described earlier in this chapter. The lowest flows near Duluth (see Figure 5-4), and in Wisconsin and Michigan as well, are pillowed. The youngest flows on the North Shore are near Tofte. From Duluth northeast to Tofte, the flows dip southeastward at 10° to 20°, and, from Grand Portage southwest to Tofte, the flows dip southwestward in similar fashion. Thus, the flows become progressively younger as one nears Tofte from either direction. The entire volcanic pile was extruded about 1,100 million years ago, and some detailed age dating (based on the decay of uranium-238 to lead-206) by L. T. Silver and J. C. Green (1972) suggests that the volcanism lasted 20 million years, from about 1,140 to 1,120 million years ago.

Erosion of the sequence has exposed many flow tops (Figure 5-11). Between some flows are stream-deposited sedimentary units, some with crossbedding, made up of pebbles, sand, and silt eroded from the already extruded flows (Figure 5-12). A study of crossbeds in dozens of interflow sandstones on the North Shore by M. A. Jirsa (1980) indicates that the streams that carried and deposited the sand were in general flowing southward toward the present axis of Lake Superior. This would suggest that the

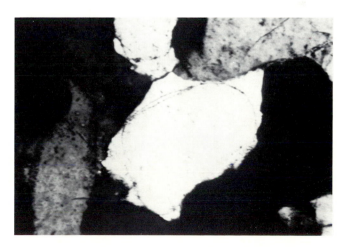

Figure 5-7. Photomicrograph of a quartz grain in the Sioux Quartzite. Note the grain and then a quartz layer with a rounded edge (indicating abrasion during transport) and, on the outer edge of the grain, a younger, unabraded overgrowth. The abraded inner overgrowth indicates erosion from an older sandstone. Such multicycle grains have been found in other Upper Precambrian quartzite units of the Lake Superior region.

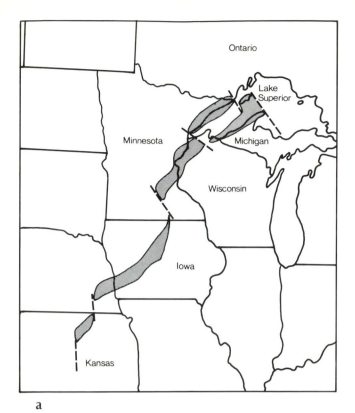

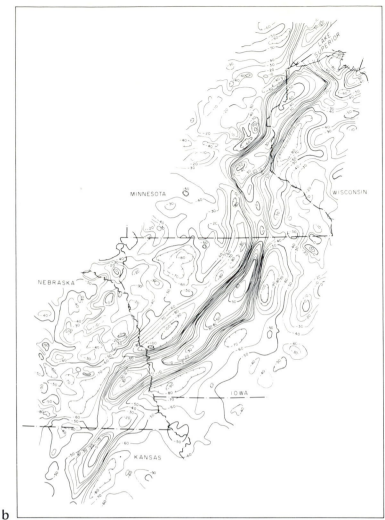

Figure 5-8. (a) Map showing the midcontinent gravity high. This zone contains basalts that extruded during a rifting or splitting apart of North America 1,200 to 1,100 million years ago. (After Chase and Gilmer, 1973.) (b) Simple Bouguer anomaly map of the midcontinent gravity high. Contour lines are in milligals, a unit of measure for gravity. Positive values indicate dense rocks beneath the surface and negative values indicate less dense rocks. (From Craddock, Mooney, and Kolehmainen, 1969.)

a

b

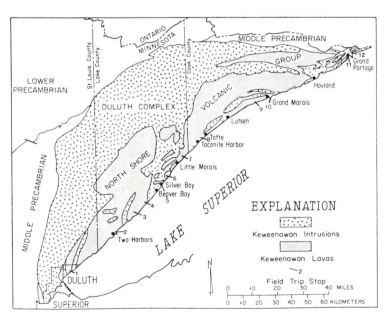

Figure 5-9. Geologic map of the North Shore of Lake Superior showing the North Shore Volcanic Group and Keweenawan (Upper Precambrian) intrusions. Numbers refer to field trip stops described in Chapter 10. (After Green, 1979.)

Figure 5-10. Lava flows on the northern shore of Lake Superior northeast of Silver Bay. Palisade Head in the distance and Shovel Point, from which the photo was taken, are both part of a 60-m-thick rhyolitic lava flow, one of the thicker flows in Minnesota.

52

a

actual fissures or faults from which the lavas issued were located somewhere north of the North Shore. W. S. White (1957), in a study of interflow sandstones on the South Shore in Michigan, found that the depositing currents had flowed northward, again toward the axis of Lake Superior. However, rather than concluding that the rifts from which those lavas were extruded lay to the south of the South Shore, he concluded that the fissures that fed that plateau, or subbasin, were in the center of what is now Lake Superior. As lava was extruded, according to his hypothesis, the underlying area sagged because of the removal of underlying magma. Thus, streams flowed down the slopes toward the center of the sag. With the next eruption pending, the center rose because of the rising magma, and the lava moved both northward and southward. The actual locations of the fissures are not yet known, but it seems obvious that removal of magma from beneath what is now Lake Superior caused the overlying lava flows to be tilted toward the center of the regional Lake Superior basin, that is, generally southward on the North Shore and northward on the South Shore.

Whereas most of the flows are basaltic, the North Shore Volcanic Group includes some rhyolites and lavas of intermediate composition as well. All the flows contain *amygdules* or blebs of minerals that filled original *vesicles* or gas cavities formed by escaping gases as the lava was crystallizing (Figure 5-13). In an individual flow, the vesicles are most

b

Figure 5-11. (a) Irregular top of a lava flow a few kilometers northeast of Duluth. Its billowy, irregular nature suggests that the lava was fairly viscous. The top was covered by the next flow, but erosion of the overlying flow has again exposed it. (b) Ropy top of a lava flow at Sugarloaf Landing. This type of flow top suggests a more fluid lava than that which formed the flow in Figure 5-11a. Note the subvertical pipe amygdules in the base of the overlying flow. A pencil provides the scale.

53

Figure 5-12. Sedimentary rock unit deposited upon one lava flow (not in the photo) and buried by another lava flow. The red siltstones and sandstones are probably river deposits. This is a roadcut on U.S. Highway 61, a few kilometers southwest of Grand Marais.

abundant near the top because the light gases were rising through the lava. Some vesicles near the bottoms of flows were pipe vesicles, which were elongate owing to the flow of much gas (largely water vapor) from along the bases of flows as they moved over wet terrane; they are now pipe amygdules (Figure 5-14).

Individual amygdules contain different minerals, including white calcite, apple-green epidote, pale green prehnite, dark green chlorite, and various pinkish radiating zeolite minerals. The most popular amygdules, however, are agates made of fine concentric bands of quartz and the zeolite thomsonite. Thus, world-famous Lake Superior agates (Plate 10) found on the beaches and in the gravel pits are nothing more than old amygdules, weathered out of the lava flows. The agates and the other amygdules were probably formed after the flows were buried by younger flows, as minerals precipitated out of the warm waters associated with volcanism and burial. Most outcrops of lavas have abundant cracks or joints, due either to cooling or to later tilting, and many of these are also coated with apple-green epidote.

In Minnesota, but especially on the Keweenaw Peninsula of Michigan and on Isle Royale, an additional very intriguing mineral has been deposited in vesicles, between pebbles of conglomerate, and in miscellaneous veins and cracks in the rocks. This mineral is easily recognized as copper-colored native copper—pure metallic copper—or by the green copper oxides that usually cover weathered surfaces. Some contains a bit of silver. Copper is discussed again in Chapter 8.

At various places in the North Shore Volcanic Group, and in nearby older rocks, there are nearly vertical basalt dikes that were presumably feeder dikes to flows higher in the sequence (Figure 5-15). West of Duluth, similar dikes cut the Middle Precambrian Thomson Formation, indicating that that area, too, was once covered by flows (Figure 5-16). Evidently, the magma rose along cracks and minor faults related to the major rifts or fissures.

Many large sills, intrusive rock bodies parallel to the lava flows, are present on the North Shore. These may be related to the gabbro complex and are mentioned in the next section.

Figure 5-13. Gas cavities or vesicles (now amygdules) in lava flows. These have been filled with calcite by waters moving through the flow.

There are also lava flows farther south in Minnesota, in the vicinity of Taylors Falls, Pine City, and Hinckley. These southernmost volcanics are called the Chengwatana Volcanic Group, after a Chippewa village near Pine City, which is the English translation of the name.

INTRUSIVE GABBROIC ROCKS:
THE DULUTH COMPLEX

A large arcuate area north of Lake Superior extending from Duluth to Ely to Pigeon Point is underlain by the dominantly gabbroic rocks of the Duluth Complex (see Figure 5-8; Plate 2). This large body of rock was long called the Duluth Gabbro, for it was thought to have been formed by a single large mass of mafic magma that forced its way up along the contact of the flows and the older rocks to the north. F. F. Grout of Minnesota, basing his interpretation on the knowledge available back in 1918, thought the gabbroic mass to be dish-shaped, underlying Lake Superior and cropping out both north and south of the lake. He named this form a *lopolith*. Curiously, the original lopolith is no longer considered a lopolith, for geophysical measurements indicate that the gabbroic rocks extend only a few kilometers to the south of the North Shore. Thus, the gabbro complex is a sill perhaps 16 km thick (Figure 5-17).

However, the sill is a complex one. Instead of a single intrusion, there are several major phases to this body of gabbroic rocks. Many of the rocks are not true gabbros, which by definition contain plagioclase, pyroxene, and olivine. Rather, the rocks range from ultramafic peridotites (rocks made of pyroxene and olivine) to various types of gabbro to anorthosites of pure plagioclase to granodiorites to red granitic rocks. Thus, the name Duluth Complex is

Figure 5-14. Pipe amygdules filled with a zeolite mineral. Amygdules were originally long cavities caused by the upward movement of gases at the base of a lava flow.

Figure 5-15. Vertical basalt dikes cutting a lava flow southwest of Leif Ericson Park in Duluth.

a

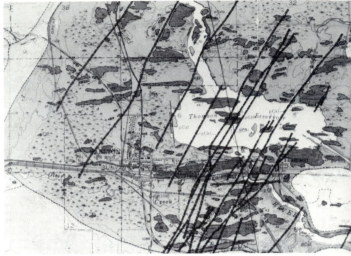

b

Figure 5-16. (a) Photo of a basaltic dike cutting Middle Precambrian Thomson slates at Thomson Dam, near Carlton. (b) Map of the Thomson Dam area showing a swarm of basaltic dikes. Their parallelism suggests that the magma rose along a set of parallel joints or minor faults. According to J. C. Green (1977), these dikes were probably feeders to overlying lava flows of the North Shore Volcanic sequence that have since been eroded away. (From Wright, Mattson, and Thomas, 1970.)

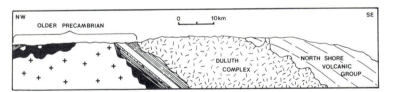

Figure 5-17. Highly generalized cross section of the Duluth Complex, showing rocks above and beneath. Note that the complex was intruded between the North Shore Volcanic Group and the older Middle Precambrian rock units. Here it appears as a simple sill, but it is in reality a series of intrusions that probably were emplaced in rapid succession about 1,100 million years ago and later were cut by a series of faults. (After Weiblen and Morey, 1980.)

appropriate, even though the early studies of the gabbros by Grout (1918, 1923) made the Duluth Gabbro famous as one of the largest such intrusions in the world.

The rocks in this complex are coarse grained, indicating that they cooled at depth beneath a cover of insulating rocks (Figure 5-18). Certainly, the greatest part of this cover consisted of lava flows several kilometers thick. Some rocks have been "baked" or metamorphosed by the gabbroic magmas. Near what is now the northern margin of the Duluth Complex, the hot mafic magma engulfed large blocks of Middle Precambrian iron-formation and black mudstone and graywacke. Thus, the Biwabik Iron Formation and the Virginia Formation on the west are not now continuous with the Gunflint Iron Formation and the Rove Formation to the east, although they once were. (This can be seen on Figures 5-9 and 4-1). Obviously, the hot magma had an effect on nearby preexisting rocks, metamorphosing them so that mineral assemblages stable under these conditions of higher temperature and pressure were formed. Thus, Reserve Mining Company's "asbestos" problems did not begin in the 1950s but instead started back in Late Precambrian time

1,100 million years ago, when the iron minerals of the eastern end of the Biwabik Iron Formation were transformed into the amphibole mineral cummingtonite (see Chapter 8). The gabbros also contain large quantities of low-grade copper and nickel ores. These are described in Chapter 8.

There are also several sills and smaller gabbroic intrusions in the North Shore Volcanic Group, all probably related to the Duluth Complex. These include the Endion, Northland, and Lester sills in Duluth, the 60-m-thick sill at Silver Creek Cliff just northeast of Two Harbors (Figure 5-19), and the even thicker Lafayette Bluff Sill a few kilometers farther northeast.

A NEW SEDIMENTARY BASIN

As the large volume of gabbroic magma moved up into the flow sequence from beneath the spread center, sagging, tilting, and faulting of the lava flows probably became even more pronounced than during the earlier volcanism. This created a topographic low upon the tilted lava flows in the vicinity of Lake Superior and, in fact, intermittently all the way to Kansas (see Figure 5-8). As happens wherever a topographic low exists, running water began to carry sediment into the low area, a new basin of sedimentation, from the surrounding highlands (Figure 5-20). Sedimentary rocks accumulated in the new basin for an unknown period of time, but probably for millions of years. Whereas the sedimentary rocks formed at this time are only a few hundred meters thick in Minnesota near Lake Superior, they are thicker to the south beneath the Twin Cities and perhaps 5,000 m thick in Wisconsin near Ashland (see Figure 5-2).

The lowest sedimentary rock unit deposited in the Minnesota portion of the basin is the Solor Church Formation, which is not exposed in outcrop. A search for natural reservoirs (porous sandstone with suitable impermeable cover rocks and anticlinal folds) for the storage of large quantities of natural gas for the Twin Cities region resulted in the drilling of several deep holes in southeastern Minnesota. Thus, samples of the buried Solor Church rocks became available for detailed study. G. B. Morey (1972, 1974), of the Minnesota Geological Survey, has studied and summarized what is known about this unit. It is as much as 1,000 m thick and consists largely of reddish to greenish immature

a

b

c

Figure 5-18. (a) Gabbro outcrop near Ely. Note the bedding, which is due to the settling out of the magma of olivine-rich layers and plagioclase-rich layers. (b) Close-up of coarse-grained gabbro. The rock is dark gray to black when fresh; this rock surface has been partially altered to light-colored clay minerals by weathering. (c) Big inclusions of light-colored anorthosite in gabbroic rocks. Anorthosite consists almost exclusively of plagioclase. The place at which the anorthosite cooled prior to being carried up by gabbroic magmas is a bit of a mystery, discussed again in Chapter 10. The location is Silver Bay on Highway 61.

a

b

Figure 5-19. (a) Silver Cliff, a few kilometers northeast of Two Harbors. This feature exists because of the intrusion of a massive sill into the lava sequence. Note the huge columnar joints in the sill and the subhorizontal flows beneath the sill. A close inspection of this site shows that the flows were baked by the hot sill. (b) Huge columns formed as the Silver Cliff sill cooled. In such cooling, joints form perpendicular to the cooling surface; thus, they are near vertical in a subhorizontal body such as a sill or a flow and are subhorizontal in a vertical dike. Here they are steep. The baked lava flows into which the magma intruded are visible at the base of the sill.

sandstones with an abundance of feldspar and volcanic rock fragments; conglomerates, siltstones, mudstones, and even thin limestones are also present. The nature of the Solor Church rocks suggests that most deposition was on an alluvial plain of meandering streams, floodplains, and lakes.

Morey noted a 30-m-thick regolith, or weathered zone, on top of the formation. From this he determined that the formation had been exposed and weathered for a long period of time before the overlying Hinckley Sandstone was deposited. Note from Figure 5-2 that the unit that should overlie the Solor Church is the Fond du Lac Formation rather than the Hinckley. However, movement along the faults during the time of sedimentation resulted in some blocks being elevated during sedimentation and therefore not receiving sediment and other blocks being dropped down and thus receiving sediment. This is illustrated in Figure 5-21. In fact, movement along some of these faults apparently continued through Cambrian time, at least locally southeast of where the Twin Cities are today.

The lowest exposed rock in the sedimentary sequence is a thin quartz-pebble conglomerate, the basal part of the Fond du Lac Formation (Figure 5-22). Where the conglomerate is exposed in Jay Cooke State Park west of Duluth, it rests upon the steeply dipping and eroded Middle Precambrian Thomson Formation just northwest of the main part of the basin. The quartz pebbles were derived from quartz veins in the underlying Thomson Formation. For the most part, the Fond du Lac consists of brown to red feldspathic

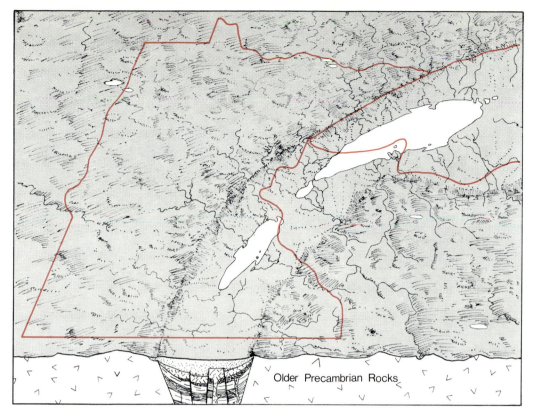

Figure 5-20. Reconstruction of how the Lake Superior region may have appeared after cessation of volcanic activity. A new basin formed largely on the depressed lava flow sequences. The faults separating the flows from adjacent rocks are hypothetical. Once the topographic low appeared, streams began filling it with sediments, locally to a thickness of 5,000 m. Note the lakes. Their presence is suggested by the more quartzose nature of rock units such as the Hinckley Sandstone, which may have resulted from the cleaning up of feldspathic sands in a shallow lake with considerable wave and current action.

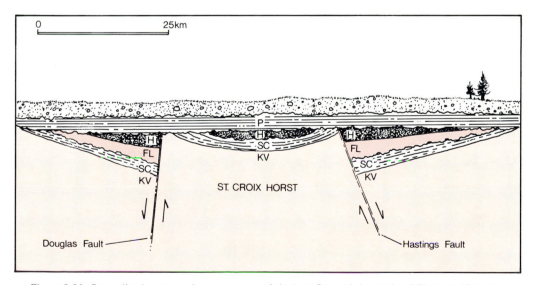

Figure 5-21. Generalized cross section across part of the Late Precambrian basin of Figure 5-20 in the vicinity of the Twin Cities. Note the faults and the uplifted block or horst. Minor faults and blocks are not shown. Vertical dimensions are not to scale. KV indicates Keweenawan volcanics, SC indicates the Solor Church Formation, FL indicates the Fond du Lac Formation, H indicates the Hinckley Sandstone, and P indicates Paleozoic sedimentary rocks. Glacial drift is on top. Wavy lines depict buried surfaces of erosion, or unconformities. (After Morey, 1972.)

59

sandstone, siltstone, and shale, as exposed in the village of Fond du Lac and along the St. Louis River just north of Fond du Lac. The sandstones are crossbedded and contain numerous layers of mud-chip conglomerate, with the mud chips probably eroded from the floodplains next to the ancient river channels. Mudcracks further suggest a periodic, perhaps seasonal, drying in a shallow-water environment. These features are characteristic of sediments deposited by rivers. The distribution of the Fond du Lac Formation all the way from Duluth to southern Minnesota, mostly in the subsurface, suggests a broad plain of merging river deposits. A 600-m-long drill core from a hole drilled near Moose Lake in the search for uranium in the late 1970s shows 175 sequences, fining upward from conglomerate to silt or mud. This is also suggestive of river deposits. Crossbedding in the formation where it is exposed near Duluth shows that the currents were moving from west to east into the sedimentary basin. Paleomagnetic pole position data (Watts, 1981) suggest an age of between 950 and 1,040 million years for the Fond du Lac, which, being a sedimentary unit, cannot be easily dated by radiometric methods.

Overlying the brown to reddish feldspathic Fond du Lac Formation is the buff to tan Hinckley Sandstone, which contains more than 95% quartz (Figure 5-23). The Hinckley may have formed in a lake that existed in the basin as streams continued to bring in water and sediment. In such a shallow-water environment, wave and current action could have abraded and eliminated the softer feldspar and rock fragments, leaving the more resistant quartz as rounded grains in a well-sorted sand (Figure 5-24). Crossbedding and ripple marks are abundant. The Devil's Island Sandstone in the Apostle Islands area of Wisconsin is also rich in quartz and lies above feldspathic sandstones of the Orienta Formation (see Figure 5-2). These formations were probably once continuous with the Hinckley and the Fond du Lac.

The Fond du Lac has a total exposed thickness of about 90 m, but wells have penetrated 600 m. The Hinckley is much thinner, about 30 m thick in outcrop and as much as 150 m thick in wells. The correlative formations in Wisconsin (Figure 5-2), which may be as much as 5,500 m thick, were deposited in the center of the basin rather than near the edge as were the Fond du Lac and the Hinckley.

a

b

Figure 5-22. (a) Basal quartz-pebble conglomerate of the Fond du Lac Formation just west of Duluth in Jay Cooke State Park along the Little River. The dark material between pebbles is rusted pyrite. A small distance from this exposure, nearly vertical beds of Middle Precambrian Thomson Formation are exposed. Thus, the Fond du Lac at this locality rests upon a buried surface of erosion. (b) The Fond du Lac Formation exposed on the St. Louis River northwest of the village of Fond du Lac. Note the resistant sandstone beds and the less resistant, more gently sloping beds of shale. These are ancient river deposits, formed near the edge of the basin depicted in Figure 5-20.

Figure 5-23. Hinckley Sandstone exposed in a roadcut on Interstate Highway 35 at the Kettle River north of Sandstone. Note the bedding and crossbedding.

a

b

Figure 5-24. (a) Photomicrograph of Fond du Lac Sandstone. Note the abundance of rock fragments (fine grained and speckled), feldspar (striped or lined), and quartz (white, gray, or black). The field of view is about 2 mm across. (b) Photomicrograph of the Hinckley Sandstone. Note the pure quartz composition and rounded grains. Could it have been formed by the reworking of Fond du Lac sands? Some geologists say it could have. The field of view is about 2 mm across.

Upper Cambrian marine sandstones in a roadcut south of Taylors Falls.

6

Post-Precambrian Time (600 to 2 Million Years Ago)

GEOSYNCLINES AND NORTH AMERICA

While rifting, volcanism, and sedimentation were occurring in the mid-continental region, western, eastern, and northern North America were experiencing the early stages in the development of major downwarped zones called *geosynclines*. These zones, in the oceans and peripheral to the Precambrian Canadian Shield and the rest of the central stable continent, or *craton*, measured thousands of kilometers long and a few hundred kilometers wide. The Cordilleran Geosyncline was developing in the west; the Appalachian Geosyncline, in the east; and the poorly studied Franklin Geosyncline, in the far north (Figure 6-1). The first two are of great importance, for they eventually became subjected to uplift of such magnitude that the Rocky Mountains and Appalachian Mountains were the result.

The Appalachian and Cordilleran geosynclines, while they subsided, received sediment. The more they subsided, the more sediment was carried in from the adjacent interior of the North American continent. Back in the 1800s, geologists argued about whether the crust in these zones subsided because of the weight of the sediment or whether the sediment accumulated there because the crust was subsiding. It was an argument like that about the chicken and the egg: which came first? New evidence suggests that subsidence along such zones or geosynclines is related to the movement of major segments (the lithospheric plates) of the earth's crust. Sedimentation accompanied, and usually kept pace with, the subsidence.

Sedimentation in the geosynclines may have begun about 1,200 million years ago and continued through the remainder of the Precambrian, which ended 600 million years ago. With a 600-million-year span of time thus available, the sediments should have accumulated to great thicknesses. In fact, they are 15,000 m (15 km!) thick in the Idaho-Montana

region. Near the top of the Precambrian rock sequences in both the Rockies and the Appalachians, ancient glacial sediments, or tillites are present as a record of what is apparently the earth's second major continental glaciation, about 800 to 600 million years ago. The glaciers moved from east to west in Utah, outward from the continental interior. However, there is no tillite of this age preserved in Minnesota or anywhere in the continental interior; it may once have been there but has been eroded away. Sedimentation did not cease just because the Precambrian ended. In the geosynclines, sediments continued to be deposited without interruption for several hundred million years more, until the geosynclines finally were elevated to form the Appalachians 225 million years ago and the Rockies 65 million years ago.

SHALLOW SEAS, LIFE, AND SEDIMENT

One of the profound facts of the earth's history is the inconstancy of the oceans in their distribution over the surface of the planet. And nowhere is that fact better documented than in the sedimentary rocks that were deposited upon the continental landmass of North America during the Paleozoic Era. The Late Precambrian marine geosynclinal sedimentation did not, of course, directly affect Minnesota, which was the site of nonmarine volcanism and sedimentation in the basins associated with the midcontinent rifting. But the seas in the geosynclines were not to remain stationary. During Middle and Late Cambrian (not Precambrian) time, about 550 to 500 million years ago, much of the North American continent underwent a slow subsidence. As this occurred, the oceans spilled out of the geosynclines and advanced across the continental interior. What had been dry land above sea level became part of the sea floor.

Although seas had come and gone before, the marine

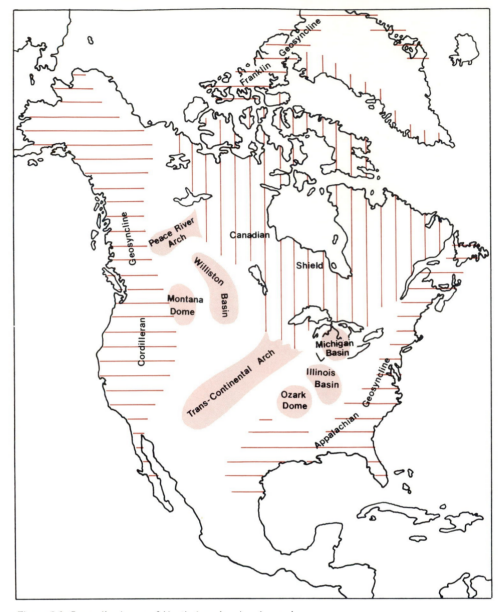

Figure 6-1. Generalized map of North America showing major structural elements of Early Paleozoic time. The Canadian Shield is the stable nucleus of the continent and generally remained above sea level. The Appalachian and Cordilleran geosynclines were zones of rapid subsidence that were later uplifted to form mountain ranges. The basins were areas of lesser subsidence during sedimentation, whereas the domes and arches were areas that were either above sea level or were rising during sedimentation. Between the domes and basins was stable shelf area that subsided very slowly. Therefore, Paleozoic sedimentary rocks are thickest in the geosynclines, thick in the basins, thinner on the shelf, and thin or absent on the arches and domes and the Canadian Shield. Note that Minnesota was thus destined to receive only thin deposits of Paleozoic sediments, and only over part of the state. (After Clark and Stearn, 1960.)

waters that spread across the craton beginning about 550 million years ago were very different in one important aspect from earlier seas—they were populated by a rich assemblage of plants and animals that had evolved from primitive beginnings more than 3,000 million years ago. Paleontologists think that the development in life forms, both plants and animals, had been going on at least since the algal colonies and microscopic life forms had lived in Middle Precambrian seas.

So bountiful and diverse are the forms preserved as fossils in the sediments of Late Cambrian age, in comparison to rocks of Late Precambrian age, that their seemingly abrupt appearance calls for a special explanation. Some would argue for a special creation. A distinctive difference between many of the Cambrian life forms and their predecessors was the acquisition of the ability to secrete hard body parts in the form of internal skeletons and shells. It is these durable parts, some of which have survived burial for hundreds of millions of years, that give us such a rich fossil record in Cambrian and younger rock strata.

Several evolutionary hypotheses have been developed to explain the spectacular explosion in the development of living forms. All are based on the assumption that some radical change in the physical, chemical, or biological environment triggered a burst of evolutionary activity. In one view, the important change was the oxygen content of the atmosphere. The early atmosphere contained no free oxygen. It was not until the development of plants and photosynthesis that oxygen was discharged into the atmosphere in large enough quantities to accumulate faster than it was used up by natural chemical processes, especially the oxidation of elements such as iron and carbon. As the oxygen content built up, via this atmospheric "pollution" by plants, oxygen gas (O_2) became available for respiration by animals. Some of the free oxygen became ozone (O_3), a high-altitude shield that continues to protect us against the deadly ultraviolet radiation from the sun.

Such changes in life forms were gradual, and the first oxygen users probably developed around plant colonies in marine waters deep enough to protect them from the harsh effects of ultraviolet radiation. The oxygenated waters in these "oases" eventually became dispersed throughout the sea, and the animal communities themselves were able to migrate, even into shallower water. Three new conditions in this environment would have favored the development of a protective shell for soft body parts: (1) a higher intensity of ultraviolet radiation; (2) greater physical stresses from wave and current energy; and (3) higher predator stresses in the less spacious, and therefore more crowded, near-shore ecological niches. The spread of shallow seas across vast continental areas during the early part of the Paleozoic, accompanied by the dimming of radiation hazards as the ozone layer developed, not only induced a population explosion but created pressure to diversify as well. Life in the modern oceans is most abundant in water shallow enough for sunlight to penetrate. By the principle of uniformitarianism, we assume that beds with an abundance of fossils indicate shallow seas, probably less than 100 m deep.

Throughout the Paleozoic and Mesozoic eras, the interior of North America was subjected to repeated broad and gentle subsidence and uplift. With each subsidence came a new marine advance and a unique assemblage of life forms. The gradual appearance of thousands of new species of plants and animals and the disappearance or extinction of many of them are dramatically documented in the sedimentary tombs left behind by the restless seas. As uplift occurred, the seas withdrew, only to return with a changed life assemblage when the next subsidence occurred. It was thus that Minnesota's fossil heritage came to be.

After the cessation of sedimentation during Late Precambrian time, what is now Minnesota was part of a high topographic trend that straddled the midcontinent. This transcontinental arch (see Figure 6-1) became an important barrier to transgressing ocean waters, as well as a source of sediments to be carried by streams into the surrounding seas.

A major marine transgression beginning in Middle Cambrian time on the far continental margins eventually brought a mediterranean sea into Minnesota, but the onlap was slow, with the sea encroaching at perhaps the rate of a kilometer or two each 100,000 years. What causes such extensive and periodic floods of ocean water across continental areas? Different mechanisms may be involved. An increase in the absolute volume of ocean water would result in a spillover onto low continental areas. Such changes in volume have been directly linked to the large-scale melting of glacial ice and the return of water to the ocean reservoirs at the termination of ice ages. However, if all of the present glaciers were to melt, sea level would rise by only 30 to 60 m, not enough for a major inundation that would reach Minnesota. Another explanation calls upon the slow subsidence, or sinking, of continental areas relative to sea level with no change in ocean volume. Similarly, uplift or subsidence of large portions of the ocean floor would cause changes in sea level. Although both melting of glaciers and vertical movements in the crust have contributed to the periodic redistribution of marine waters on the planet, a hypothesis involving the latter process seems to better explain the profound and repeated invasions and retreats of the oceans across the interior of North America during the last 600 million years.

Vertical movements in the central portion of North America were, for the most part, neither profound nor uniform. The rise and fall of various parts of the continent produced two fundamental types of structures—arches and basins (see Figure 6-1). The basins became effective sediment traps, and their former existence is deduced from the characteristic thickening of strata in comparison to surrounding areas. Arches, in contrast, either received a lesser volume of sediment cover or were high enough to be emergent and actively eroded. The distribution of these structures at any one time was an important influence on patterns of deposition, depositional environments, and the distribution of marine plants and animals. The basins, especially, became repositories for economically important materials, including petroleum, salt, gypsum, limestone, and coal. Unfortunately, the small Twin Cities Basin, which underlies the Twin Cities, is a minor feature. (It will be discussed in the section on the geology of the Twin Cities metropolitan area, in Chapter 14.) As can be seen from Figure 6-1, the major basins do not extend into Minnesota.

The sedimentary sequence left behind in Minnesota by the invading *epicontinental* seas (seas on the continent) is impressive in its complexity, but by no means does it represent the entire Paleozoic Era. Minnesota was apparently

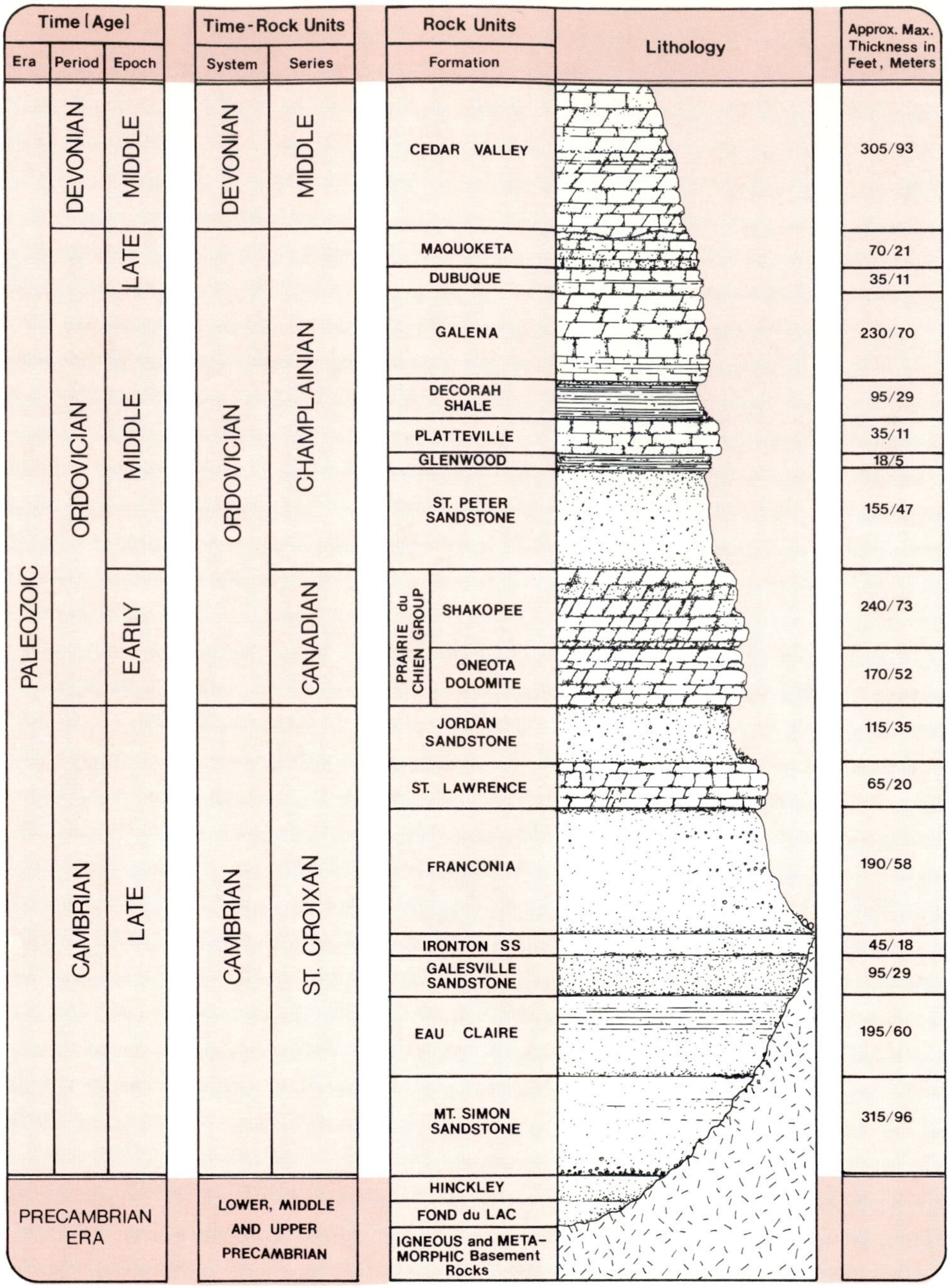

Time (Age)			Time-Rock Units		Rock Units	Lithology	Approx. Max. Thickness in Feet, Meters
Era	Period	Epoch	System	Series	Formation		
PALEOZOIC	DEVONIAN	MIDDLE	DEVONIAN	MIDDLE	CEDAR VALLEY		305/93
PALEOZOIC	ORDOVICIAN	LATE	ORDOVICIAN	CHAMPLAINIAN	MAQUOKETA		70/21
PALEOZOIC	ORDOVICIAN	LATE	ORDOVICIAN	CHAMPLAINIAN	DUBUQUE		35/11
PALEOZOIC	ORDOVICIAN	MIDDLE	ORDOVICIAN	CHAMPLAINIAN	GALENA		230/70
PALEOZOIC	ORDOVICIAN	MIDDLE	ORDOVICIAN	CHAMPLAINIAN	DECORAH SHALE		95/29
PALEOZOIC	ORDOVICIAN	MIDDLE	ORDOVICIAN	CHAMPLAINIAN	PLATTEVILLE		35/11
PALEOZOIC	ORDOVICIAN	MIDDLE	ORDOVICIAN	CHAMPLAINIAN	GLENWOOD		18/5
PALEOZOIC	ORDOVICIAN	MIDDLE	ORDOVICIAN	CHAMPLAINIAN	ST. PETER SANDSTONE		155/47
PALEOZOIC	ORDOVICIAN	EARLY	ORDOVICIAN	CANADIAN	PRAIRIE du CHIEN GROUP — SHAKOPEE		240/73
PALEOZOIC	ORDOVICIAN	EARLY	ORDOVICIAN	CANADIAN	PRAIRIE du CHIEN GROUP — ONEOTA DOLOMITE		170/52
PALEOZOIC	CAMBRIAN	LATE	CAMBRIAN	ST. CROIXAN	JORDAN SANDSTONE		115/35
PALEOZOIC	CAMBRIAN	LATE	CAMBRIAN	ST. CROIXAN	ST. LAWRENCE		65/20
PALEOZOIC	CAMBRIAN	LATE	CAMBRIAN	ST. CROIXAN	FRANCONIA		190/58
PALEOZOIC	CAMBRIAN	LATE	CAMBRIAN	ST. CROIXAN	IRONTON SS		45/18
PALEOZOIC	CAMBRIAN	LATE	CAMBRIAN	ST. CROIXAN	GALESVILLE SANDSTONE		95/29
PALEOZOIC	CAMBRIAN	LATE	CAMBRIAN	ST. CROIXAN	EAU CLAIRE		195/60
PALEOZOIC	CAMBRIAN	LATE	CAMBRIAN	ST. CROIXAN	MT. SIMON SANDSTONE		315/96
PRECAMBRIAN ERA			LOWER, MIDDLE AND UPPER PRECAMBRIAN		HINCKLEY		
PRECAMBRIAN ERA			LOWER, MIDDLE AND UPPER PRECAMBRIAN		FOND du LAC		
PRECAMBRIAN ERA			LOWER, MIDDLE AND UPPER PRECAMBRIAN		IGNEOUS and META-MORPHIC Basement Rocks		

Figure 6-2. Minnesota's Paleozoic sedimentary rock column. Each formal unit name includes the word "sandstone," "shale," "dolomite," or "formation" (if the unit contains several rock types); for brevity the word "formation" is omitted. Not shown are most Precambrian rock units. (After Webers and Austin, 1972.)

never completely covered by the shallow seas. This statement is based on the evidence of the rock record. The presence of sedimentary rocks with marine fossils is proof that a given area was beneath sea level. That is, when a surface is beneath the sea, sediment will accumulate upon that area. If the sediment is lithified to solid rock, there is a chance that the record will be preserved.

Conversely, if an area is above sea level, it undergoes erosion. Of course, some sediment may accumulate in rivers, lakes, and swamps, but because these sediments are above sea level their eventual erosion is likely. Thus, there is usually no sedimentary record of a topographically high area, and its history can only be interpreted as one of erosion. Another possibility also exists: a given area may have been below sea level, received marine sediment, and been uplifted and eroded. If erosion of the marine sediments were complete, it would be difficult if not impossible to distinguish such a history from one of nondeposition.

It is assumed here that the lack of marine sediments of a given age in Minnesota means that such sediments were never present and that Minnesota was above sea level during the given time interval. There is abundant evidence, especially the thinning of rock units toward ancient shorelines, to support this assumption as more likely than the assumption of the one-time presence and subsequent removal of such sedimentary layers. Yet, there are some large dolostone boulders in the glacial till of northern and southwestern Minnesota that seem to be locally derived from erosional remnants of more extensive Paleozoic rocks.

Comparison of the fossil content and the physical tracing of the beds to areas outside of the state indicate that rock records of parts of just three Paleozoic periods are represented: the Late Cambrian, most of the Ordovician, and the Middle Devonian (see Figure 1-11). The rest of the Paleozoic rock record is missing either because of nondeposition or because of erosion. Even so, the maximum thickness of all the formations present totals more than 600 m (Figure 6-2). Rocks of this age are distributed mainly in southeastern Minnesota, where they crop out extensively along the Mississippi River Valley and its tributaries (Figure 6-3). A more detailed geological map of the St. Paul area, the 2° sheet by R. E. Sloan and G. S. Austin (1966), available from the Minnesota Geological Survey, shows the distribution of each formation.

Marine waters from the profound transgression that had begun to flood the continental margins in Late Precambrian time, about 1,200 million years ago, finally invaded what is now Minnesota about 550 million years ago. This Late Cambrian sea followed the Hollandale Embayment, a long, shallow depression that extended into southeastern Minnesota from a larger basin in what is now Iowa. A shoreline of maximum advance was eventually established against the high Transcontinental Arch on the west and north and by uplifted Precambrian rocks on the Wisconsin Dome farther east in Wisconsin (Figure 6-4). Not until Ordovician time did marine waters encroach upon the northwestern part of the state; this a spillover from the deep Williston Basin centered in southern Alberta (see Figures 6-1 and 6-4). In this northwestern sea, 140 m of sediment was deposited; drillholes have penetrated them beneath a thin sheet of Mesozoic rocks and a thick blanket of glacial sediment. Just how much of the highlands of central and northern Minnesota eventually became inundated is not known, because erosion surely has removed some of the evidence. However, the thickness of the present eroded edges of buried sedimentary rocks on both sides of the Transcontinental Arch, almost 120 m, indicates that the shorelines of maximum transgression invaded more deeply into Minnesota

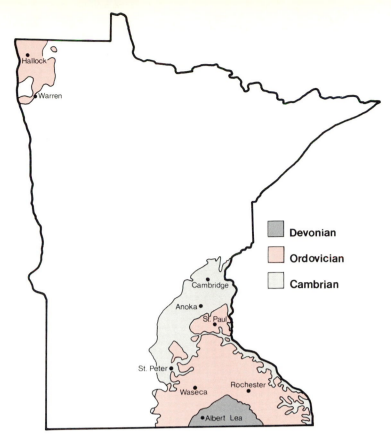

Figure 6-3. Paleozoic sedimentary rocks in Minnesota. The Cambrian rocks are underlain by Precambrian rocks, the Ordovician rocks by Cambrian rocks, and the Devonian rocks by Ordovician.

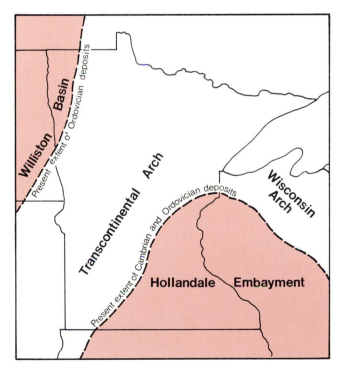

Figure 6-4. Structural features that influenced Paleozoic sedimentation in Minnesota. Note that the Paleozoic rocks originally extended over more of Minnesota but have been eroded away. Most of Minnesota's Paleozoic rocks were deposited in the Hollandale Embayment, a shallow depression and bay in the Early Paleozoic seas. (After Webers, 1972.)

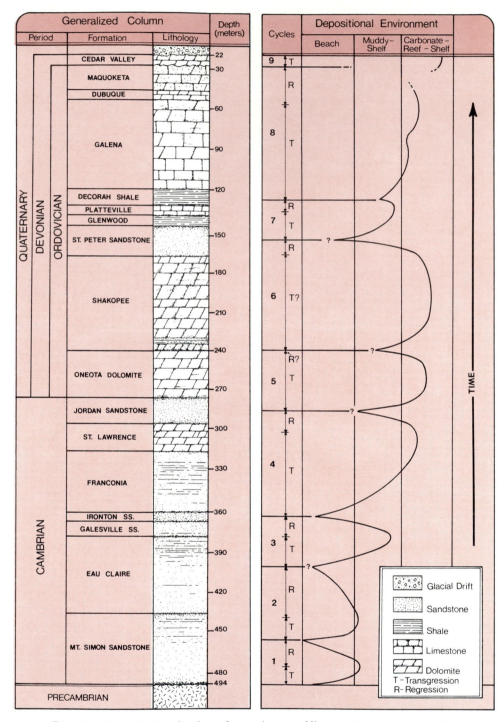

Figure 6-5. Generalized rock column for southeastern Minnesota based on a deep drill-hole at Hollandale. The depositional environment is shown for each formation. The looping line illustrates the transgressions and regressions of the sea. Where the line loops to the right, a transgression (advance) of the sea is indicated, and where the line loops to the left, a regression (retreat) of the sea is indicated. Note that there were repeated cycles of transgression and regression. Each formal unit name includes the word "sandstone," "shale," "dolomite," or "formation" (if the unit contains several rock types); for brevity the word "formation" is omitted. (After Austin, 1972a.)

than the present distribution of rocks of Paleozoic age indicates.

The layers of sediments left behind by the Early Paleozoic seas were first extensively studied in the bountiful exposures along the Mississippi, Minnesota, and St. Croix river valleys. There, and elsewhere throughout their wide distribution in southeastern Minnesota, the bulk of the strata consist of four major rock types: pure quartz sandstone; sandstone with abundant glauconite, a green micaceous mineral that justifies calling this variety greensand; mudstone and shale; and carbonates, mainly dolostone but with some limestone. The stratigraphical relationships, both vertical and lateral, among these various rock units have been used to reconstruct a complex scenario of changing sea levels, with all the attendant shifts of marine environments through time.

Figure 6-5 is a generalized log of the rock units encountered in a deep stratigraphic test well near Hollandale, Minnesota. Note that the sedimentary character of the rock formations changes vertically in the section. Shifts in composition can be directly related to changes in the environment prevailing at the time of deposition. Well-sorted quartz-rich sand is the kind of sediment generally found along beaches and in the shallow waters just offshore. Waves and currents roll and wash, winnow and sort, smooth and round, until all but the largest and toughest sand grains are removed. The finer mineral and rock particles are carried in suspension seaward, where they eventually settle out, along with organic debris, to form muddy sediments in deeper water where wave energy cannot reach to stir them up again. Warm, shallow seas are also conducive to the development of organic communities that precipitate lime, which eventually accumulates as carbonate banks such as the Bahamas.

The vertical changes in the bedrock at Hollandale have been related to cycles of marine advance and retreat during the time represented by the thickness of rock found there. The right-hand column of Figure 6 portrays the sequence of marine transgressions and regressions as interpreted by G. S. Austin (1972b), a geologist who has extensively studied the Paleozoic rocks of Minnesota. He defined nine cycles based on the recurring sedimentary sequences of sandstone, shale, and limestone in southeastern Minnesota during Early and Middle Paleozoic time; these are shown in Figure 6-5 by the looping curved line.

A good example of one such cycle of changing depositional patterns is the sequence of sedimentary strata exposed along the Mississippi River bluffs throughout the Twin Cities metropolitan area (Figure 6-6). At the base near river level is the white St. Peter Sandstone, a pure quartz sand. Higher up, just beneath a vertical ledge near the top of the bluffs, is an indentation composed of soft green-gray shale called the Glenwood Formation. This thin, easily eroded shale reflects a change in depositional environment from

one of high energy to one of less agitated water. The overlying caprock of rusty to gray carbonate is the Platteville Formation. Its fossiliferous nature and its composition both indicate an offshore environment in a warm climate.

Although the changing character of sedimentary rocks through thick vertical sequences is a valuable record of past changes in environment, careful mapping of the areal extent of each rock type is necessary for the reconstruction of the boundaries of the natural systems. All of the paleogeographic maps in this book are approximations based on the distribution of earth materials left behind to mark the environments in which they formed. Of course, the record is incomplete, and yes, the true meaning of some of the record is yet to be deciphered. Still, the reality of profound changes in the geological history of Minnesota cannot be contested.

Figure 6-6. Mississippi River bluff along Warner Road in St. Paul. Most of the rock exposure consists of white St. Peter Sandstone. Just beneath the overhang near the top is the Glenwood Shale. At the top is the ledge-forming Platteville Formation. The figure at lower left provides the scale.

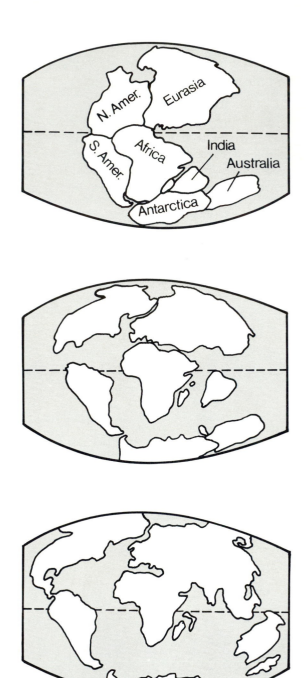

MINNESOTA AND NORTH AMERICA IN THE WORLD

In Chapters 2 and 3, it was stated that each continent has a Precambrian shield. Apparently, each shield acted as the core of that continent, around and upon which younger sediments were deposited, especially during the Paleozoic and Mesozoic eras. In Chapter 1, the rudiments of modern plate tectonic theory were presented, and in the chapters on Early and Late Precambrian time, plate tectonic theory was again briefly mentioned. The reader could legitimately ask why this Minnesota exposé does not present evidence for the positioning of North America, and hence Minnesota, relative to the other continents, throughout geologic time. That would be a fair question, but the answer is not so fair. Geologists do not yet have, and they may never have, evidence that would enable them to answer the question for all portions of geologic time. Fairly definite positions can be given for the continents for the last 200 million years of the earth's history, and tentative answers are available for some of the continents for the preceding few hundred million years. These answers are presented in Figure 6-7. But for the first 4 billion years of the earth's history, the answers are not yet very well known.

MINNESOTA'S CLIMATE THROUGH TIME

Another question that might be asked by Minnesotans, especially in January, is "Was Minnesota always this c-c-cold?" Three lines of evidence answer this question. (1) The fossils in the Paleozoic and Mesozoic sedimentary rock layers give indications of the general climatic conditions. For example, fossil corals in the Devonian beds indicate warm subtropical to tropical waters, as modern corals do today. (2) Limestones are forming today in warm subtropical places such as Florida, the Bahamas, and the northern coast of Australia. Minnesota's Paleozoic limestones probably formed under similar conditions, for as we know, the present is the key to the past. (3) Measurements of the magnetism of certain iron-bearing rocks, such as basalt or red (hematitic) sedimentary rock, allow the position of the magnetic pole to be determined at the time the rock was formed. If it is assumed that the magnetic pole was near the earth's rotational pole, as it is today, then the equator can also be positioned. Because the continents have been moved as passengers on lithospheric plates, each continent has a different pole location for the same instant of geologic time. Determinations must be made for each continent by using rocks from that continent in order to locate the relative position of the magnetic north pole through time. In this way, the location of the equator has been determined for each period since the Precambrian, as shown in Figure 6-8.

Figure 6-7. The supercontinent of Pangaea. The top diagram represents its configuration at the end of Paleozoic time, about 225 million years ago. The breakup of Pangaea was well along by mid-Cretaceous time (100 million years ago), as shown in the middle diagram. The bottom diagram shows the configuration during the Quaternary Period, from about 2 to 1 million years ago.

Figure 6-8. Map series showing the location of the equator relative to Minnesota through time. The map at upper left is for Late Precambrian time, and successive maps represent, in order from left to right row by row, Cambrian, Ordovician, Silurian, Devonian, Mississippian, Pennsylvanian, Permian, Triassic, Jurassic, Cretaceous-Tertiary and present times. The relationships are based on paleomagnetic determinations for the locations of the various continents.

The answer to the climatic question now should be obvious. Minnesota was hot during all of Paleozoic time, was fairly warm during much of the Mesozoic, has been cool throughout the Cenozoic, and has been especially cool during parts of the Great Ice Age of the last 2 million years or so. Note that climatic controls during most of Precambrian time have been ignored here; this is because the data are few and less reliable than data for the past 600 million years.

THE PALEOZOIC ERA

The Cambrian Period
(600 to 500 Million Years Ago)

By late Cambrian time, about 550 million years ago, the epicontinental sea had covered the southeastern portion of what is today Minnesota (Figure 6-9). Streams were flowing

a

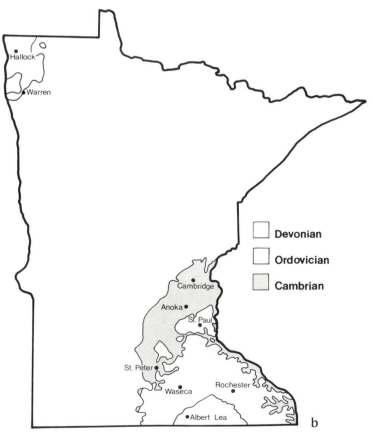

	Devonian
	Ordovician
	Cambrian

b

into the sea off the land in western and northern Minnesota, carrying sand to the seashore. There was no vegetation on land to help minimize erosion, so rain falling on the surface easily dislodged weathered rock.

Minnesota's rock surface had been weathering for eons. Most had been above sea level since the Algoman mountain-building episode 2,700 million years ago, and even the portion that received volcanics and sediments in Late Precambrian time had been weathering for an estimated 500 million years. With the total lack of vegetation on land, it is quite likely that winds moved the sediments back and forth, creating dust storms and broad fields of sand dunes. Surely streams had been carrying the weathering products—quartz sand, clay, iron oxides, and dissolved ions—off the weathered continent into the geosynclines for a long time. Yet the climate from at least 1,000 million years ago into Late Cambrian time was subtropical to tropical, and, because high temperatures and abundant rainfall led to a deep and thorough weathering, the weathered zone, or regolith was probably always quite thick. The long partnership of weathering and erosion had resulted in a rather subdued landmass, upon which the advancing sea readily encroached. That portion of the landmass not inundated was probably relatively low lying, yet high enough to provide streams with the gradients necessary for the transportation of sand.

The rock record of Late Cambrian time is so well developed in southeastern Minnesota and southwestern Wisconsin, with excellent exposures along the St. Croix and Mississippi rivers, that the rocks formed during Late Cambrian time in all of North America are called the St. Croixan Series (see Figure 6-2). The Cambrian sequence consists mainly of quartz sandstones with minor shales. This is to be expected, for quartz sand and clay would have been the major weathering products. The seven Upper Cambrian rock units (formations) of Minnesota that can be mapped are shown in Figures 6-2 and 6-10. Most of the Paleozoic formations of this region are named for towns in southeastern Minnesota or southwestern Wisconsin.

The Mt. Simon Sandstone, the lowest unit, is a white, gray, yellow, or pink, medium-grained quartzose sandstone. Crossbeds are common, and this, coupled with good sorting and rounding of the sands, indicates a high-energy environment of deposition with strong waves and currents. Thin green or red shale interbeds indicate periods of quieter water. Small brachiopods are common, especially in the upper part

Figure 6-9. (a) Paleogeography of North America during Late Cambrian time. Most of the Canadian Shield constituted a large, low-lying, deeply weathered landmass off of which streams carried quartz sand, clay, and ions in solution to the sea. The dashed line is the possible equator. (After Dott and Batten, 1981.) (b) Distribution of Cambrian rocks in Minnesota. Ordovician and Devonian rocks cover much of the Cambrian sedimentary rocks of southeastern Minnesota.

of the formation (Figure 6-11). The lack of fossils in the lower part of the formation leaves open the possibility that river deposits are also present. Unfortunately, sedimentary structures do not as yet provide a unique solution to the environmental question.

The Mt. Simon is quite similar to the Upper Precambrian Hinckley Sandstone and is distinguished in well samples with difficulty, except where a good weathered zone is present on the Hinckley or where fossils are present in the Mt. Simon. The Mt. Simon, as much as 100 m thick, though generally a quartzose sandstone, also contains many beds with noticeable amounts of feldspar.

The Eau Claire Formation, nearly 60 m thick, contains several different rock types, including red shale, gray-green shale, fine-grained quartzose sandstone, and fine-grained glauconitic quartzose sandstone or greensand. The fine sediments, commonly well burrowed by organisms, suggest either relatively deep and quiet water conditions or shallow-water tidal flats.

The Galesville Sandstone, as much as 30 m thick, is a coarser, white to gray quartz sandstone that contains some glauconite. Its coarser grain size and better sorting compared to the Eau Claire indicate a return to higher-energy near-shore or beach conditions.

The Ironton Sandstone, at most 14 m thick, is a quartzose sandstone that contains much fine-grained silty material in additon to quartz. The formation was deposited in a lower-energy environment than the coarser Galesville Sandstone. Some workers have interpreted the lower boundary of the formation as resting, at least locally, upon an erosional surface of the Galesville. This, of course, suggests a withdrawal of the sea and erosion prior to a readvance during Ironton time. In the introduction to this chapter, it was stated that the trends of ancient shorelines, and the approximate locations of the shorelines, can be determined by the thicknesses of the sedimentary rock units in question. Figure 6-12 shows lines of equal thickness (*isopachs*) for the four lowest Cambrian formations considered together. Note that they are thickest in southeastern Minnesota and thinnest to the north and that the contour lines end abruptly where the units have been eroded away.

The Franconia Formation, 30 to 60 m thick, is variable in its nature but is commonly characterized by abundant glauconite, making it largely a greensand unit (Figure 6-13). Glauconite forms on the sea floor in oxygen-poor waters where the rate of sedimentation is very slow. The parent material is usually micalike clay minerals that have much the same platy crystalline structure as the glauconite but that do not contain the potassium and iron of the glauconite. Glauconite is virtually as good an indicator of a marine environment as are marine fossils. The distribution of rock types and their variation in thickness suggest that the Hollandale Embayment (see Figure 6-4) came into existence during Franconia time, whereas deposition of the preceding

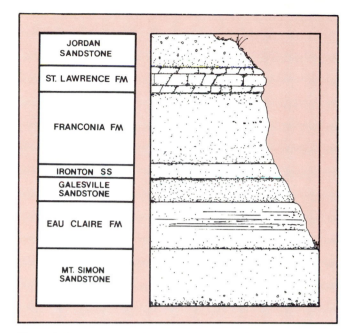

Figure 6-10. Cambrian rock units of Minnesota. Note the dominance of sandstone, shown by the dotted pattern. Carbonates are shown by a brick pattern, and siltstones and shales are shown by a pattern of horizontal dashes.

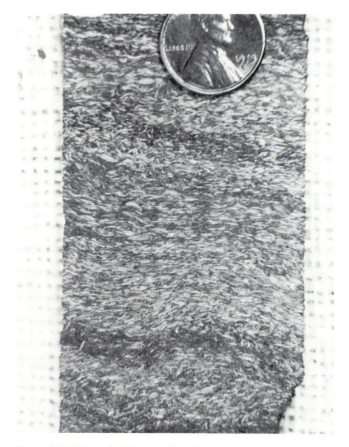

Figure 6-11. Piece of drill core of Mt. Simon Sandstone showing myriad small brachiopod shells in cross section.

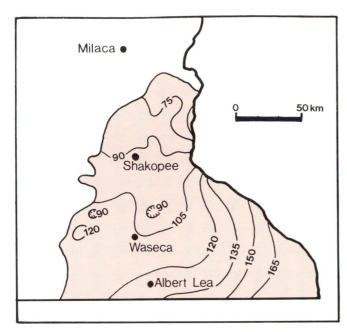

Figure 6-12. Generalized isopach map showing the combined thickness in feet of the Mt. Simon, Eau Claire, Galesville, and Ironton formations. The data base is largely from water wells. The fact that the isopach lines end abruptly in the north and west shows that these are erosional margins. Recent work by the Minnesota Geological Survey has added immense amounts of new detail to this map. (After Austin, 1972a.)

units was not controlled by the embayment; that is, the prominent basin called the Hollandale Embayment had not existed previous to Franconia time.

The St. Lawrence Formation, 20 m thick, is unique in that it is the first major carbonate unit in the Paleozoic column of Minnesota. A pure limestone unit would signify the dominance of chemical-biochemical precipitation of calcium carbonate out of the seawater, a lack of sand and mud reaching that part of the basin of deposition, and probably a depositional site far from shore. The St. Lawrence, however, is not clean but contains clay, silt, sand, and glauconite, indicative of fluctuating conditions. Cleaner carbonates dominate the Ordovician and Devonian portions of the Paleozoic rock column, as will be described later. All of the Paleozoic carbonates, including the St. Lawrence, were originally composed of the mineral calcite ($CaCO_3$), but many were altered by the replacement of about half of the calcium ions (Ca^{2+}) by magnesium ions (Mg^{2+}), making the mineral dolomite $[CaMg(CO_3)_2]$ and the rock dolostone. Magnesium is common in seawater; replacement probably occurred shortly after burial of the original calcite on the sea floor.

The Jordan Sandstone, 25 to 35 m thick and variable in sandstone type, is perhaps best characterized as a white or yellow, medium- to coarse-grained quartz sandstone with well-rounded and well-sorted grains. Also present are pebbles of dolomitic rocks. This indicates a high-energy near-shore environment of deposition, perhaps a beach.

b

Figure 6-13. (a) Franconia Sandstone at Curtain Falls, in Interstate State Park near Taylors Falls. (Courtesy of Carl Steuland.) (b) Crossbedded Franconia Sandstone in a roadcut just south of Taylors Falls.

a

74

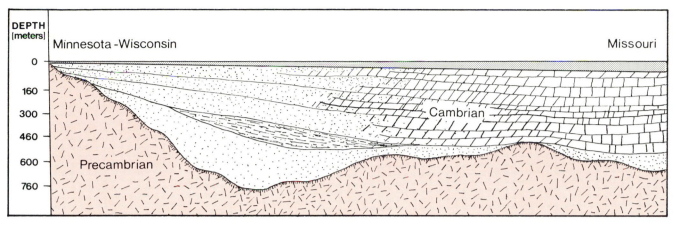

Figure 6-14. Generalized north-south cross section from southern Minnesota and southern Wisconsin to Missouri, illustrating the lateral transition from sand in the north to limestone in the south. Note that the vertical scale is grossly exaggerated relative to the north-south distance of about 1,000 km. If drawn to proper scale, the thickest sedimentary rock sequence on the diagram would be 0.14 mm thick, thinner than a pencil line. (After Kay and Colbert, 1965.)

Cambrian sedimentation in Minnesota can be summed up by saying that sands were the dominant sediments in the warm, shallow seas. Coarser sands suggest deposition nearer the shoreline, and finer sands, silts, and shales were deposited farther from shore. The sands were all derived from adjacent highlands, predominantly from what is now Wisconsin and northern Minnesota but also from the Transcontinental Arch to the west. Perhaps the rounding of quartz sand grains, which are resistant to rounding, had already been partially accomplished on land by wind, the most efficient abrader of sand.

Only one carbonate unit, the St. Lawrence, formed during Cambrian time, for the sands and silts dominated the sedimentary environment. Whereas carbonate precipitation may have been going on throughout Late Cambrian time, the carbonate was completely masked by the more abundant quartz grains. In Missouri, farther from the northern sources of sand, the Cambrian sands pass into carbonates (Figure 6-14).

The Ordovician Period
(500 to 430 Million Years Ago)

The close of the Cambrian Period saw the end of dominant sand deposition. The countless quartz grains that were the survivors of the long and thorough weathering of Precambrian rocks and the continual buffeting by winds on the unprotected, vegetationless surface were finally at rest beneath water. Streams had carried most of them to the sea, and winds probably blew some in. The previously low-lying landmasses that had stayed above sea level during the Cambrian (see Figure 6-9) had been eroded still further and were largely inundated during the Ordovician (Figure 6-15). Thus, quartz sand was no longer available in great quantities, and carbonate deposition, which was already important in

St. Lawrence time prior to the final deposition of Cambrian sands as the Jordan Sandstone, became the dominant rock-forming process. It was to remain so during deposition of most of the remainder of Minnesota's Paleozoic rock column. Moreover, the changeover from Cambrian sand to Ordovician carbonate was, at least locally, an abrupt one (Figures 6-16 and 6-17). The Ordovician formations shown in Figures 6-2 and 6-17 indicate that they are indeed mostly carbonate units. The Oneota Dolomite, 50 m thick, is a tan to gray, fine to medium-grained dolomite (see Figure 6-16), signifying an original precipitation of calcite that was later dolomitized by magnesium-rich brines.

The Shakopee Formation consists of a lower, 20-m-thick, well-sorted and well-rounded quartz sandstone unit called the New Richmond Member and an upper, 70-m-thick, dolomitic member (Figure 6-18). Algal stuctures (stromatolites) and oolites (sand-sized carbonate grains with nuclei and concentric bands of carbonate) are clearly visible in spite of the coarsening of the original carbonate grain size and the obliteration of original textures and fossils that commonly result from dolomitization.

The St. Peter Sandstone, a white to light yellow, medium-grained quartz sandstone as much as 50 m thick in Minnesota, is famous the world over as *the* example of a well-rounded, well-sorted, pure quartz sand (Figure 6-19). It is the "Ivory Soap" of sediments, because it is close to 99.44% pure. Even the grains of the other minerals that constitute the remaining 0.56% of the rock show excellent roundness. The grains, furthermore, are frosted and pitted, the result of either a roughing up during wind transport and abrasion or a chemical etching by carbonate cement. Such cement, if ever present, was dissolved away by acid ground-waters. In fact, most of the St. Peter is virtually uncemented. Along the Mississippi River bluffs in the Twin Cities, St. Peter sand is sometimes erroneously interpreted as sand bars

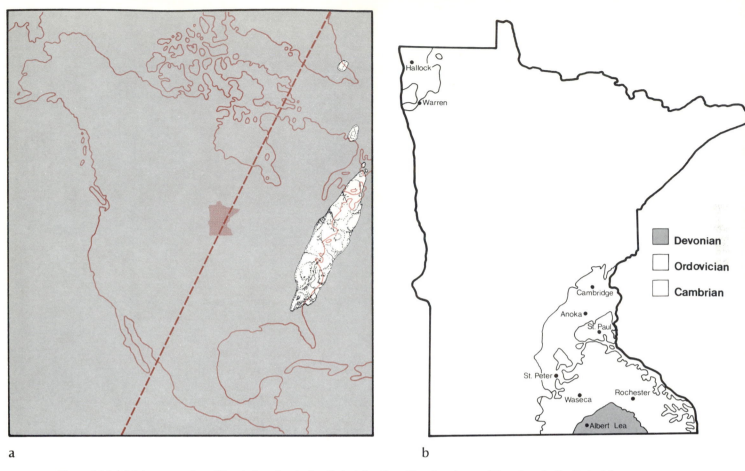

a

b

Figure 6-15. (a) Paleogeography of North America during Ordovician time. The abundance of limestone in the Ordovician sea indicates that the landmass has been eroded to an even lower level since Late Cambrian time. The dashed line is the possible equator. (After Dott and Batten, 1981.) (b) Distribution of Ordovician rocks in Minnesota. Devonian rocks cover some Ordovician rocks along the southern boundary of the state.

a

b

Figure 6-16. (a) Contact of white Jordan Sandstone (of Cambrian age) and the overlying buff Oneota Dolomite (of Ordovician age) on a bluff just above Minnesota Highway 95 at the north edge of Stillwater. The contact shows a transition zone 30 to 40 cm thick. Note the pockmarked nature of the dolomite, a feature characteristic of weathered dolomite surfaces. (b) Algal stromatolite "heads" (colonies) exposed on a quarry floor on County Highway 1 (North Ridge) above the village of La Crescent in lowest Oneota Dolomite. They are indicators of shallow and perhaps tidal-depth marine waters.

76

deposited by the nearby river. Not so! Every sand grain in the bluffs came to rest more than 450 million years ago along the shoreline of the Ordovician sea. The St. Peter was not named for the town of St. Peter, as one might guess, but for the St. Peter River, which is now called the Minnesota River.

The St. Peter represents the last great period of quartz sand deposition in Minnesota during the Paleozoic. Actually, Minnesota's St. Peter is but the northern edge of a great expanse of sand formation that covers 585,000 km^2 in the middle United States and totals 20,000 km^3 in volume. The excellent sorting, rounding, and purity suggest derivation from older sandstones. The most likely older sandstone source would be the Upper Cambrian sandstones exposed when the Ordovician seas temporarily withdrew and the Shakopee and Oneota Dolomites were removed by erosion. At the base of the St. Peter is an erosional surface. Lines of equal thickness drawn for the Cambrian sand formations do show extensive erosion, but for 20,000 km^3 to have been eroded from these units seems unlikely. The only other possible source was the Canadian Shield to the north, but why would the sand supply have again been available, after having been minimal during deposition of the Oneota and the Shakopee? The answers are not yet in. As scientists are so prone to say, further study is necessary.

The Glenwood Formation is a 5.5-m-thick, gray-green shaly unit with a sandy base. It is transitional with the underlying St. Peter Sandstone and was probably deposited

a

b

Figure 6-18. (a) A large roadcut on County Road 8 at Lanesboro. This is an excellent exposure of the Ordovician Shakopee Formation overlying the Ordovician Oneota Dolomite, which is visible in the lower left-hand corner of the photo. The middle part of the cut consists of interbedded dolomite and sandstone and then a pure sandstone member (the New Richmond Sandstone), which shows on the photo as a darker gray. The top part of the cut is a dolomite member of the Shakopee. Note the geologist for scale. (Courtesy of P. J. Squillace.) (b) the Shakopee Formation along the Vermillion River at Hastings. (Courtesy of R. L. Heller.)

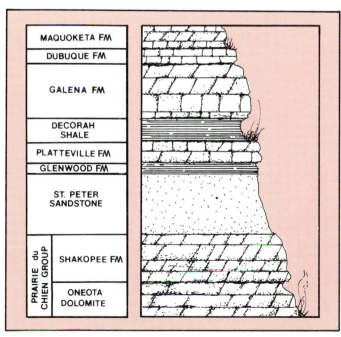

Figure 6-17. Ordovician rock units of southeastern Minnesota. Note the dominance of carbonates; the diagonal pattern represents dolomite, and the vertical pattern represents limestone.

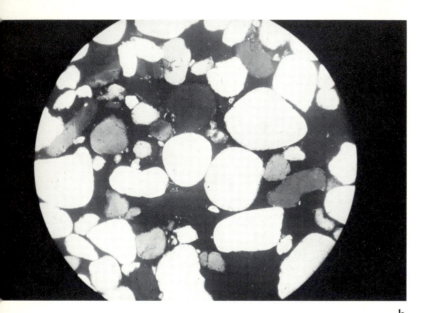

in a deeper-water, lower-energy environment, offshore from the beaches that now make up the St. Peter. Still farther offshore, where sand and mud were not transported by wave and current action, calcite was precipitated to form limestone, in this case the Platteville Formation. Where such lateral changes in sediment type occur within the same horizon, each of the different lithologies is a different sedimentary facies (Figure 6-20a). As a sea transgresses onto a landmass, the more seaward sediment types are then deposited upon the previously deposited, more landward sediments. Therefore, at a given outcrop a sandstone may be overlain by a shale that is overlain by a limestone. Now look at Figure 6-20b. Does not this real-life example make sedimentary facies easier to grasp?

The Platteville Formation, just mentioned as a more seaward sedimentary facies of the Glenwood Formation and the St. Peter Sandstone, is a 9-m-thick limestone, one of the least dolomitized and most fossiliferous limestones in Minnesota. Consequently, it is one of the best-studied units. The lower part of the formation is dominantly limestone (as in the Twin Cities), whereas the upper part contains more interbedded shaly units (Figure 6-21).

The Platteville was deposited on a shallow-marine bank of widespread carbonate probably much like the modern Bahaman platform. Conditions evidently changed slowly, for R. E. Sloan (1972) has traced individual beds less than 1 cm thick for distances of tens of kilometers. At least one special event occurred in the stable Platteville sea: a bed of volcanic ash a few centimeters thick, now altered to bentonite clay, was deposited in the sea and caused widespread death of the biota and perhaps even extinctions, according to Sloan. Life was indeed abundant in the Platteville sea, and several species of brachiopods, cephalopods, gastropods, bryozoa, crinoids, and trilobites are preserved as fossils. Fossils of large cephalopods, as much as 4.5 m long and 25 cm in diameter are not uncommon (Figure 6-22).

The Decorah Shale, a gray-green shale with thin limestone beds, is as much as 24 m thick (Figure 6-23). It has been suggested that the muds deposited in the Decorah sea were eroded off of the Transcontinental Arch. Fossils are exceedingly abundant and diverse in this unit, even more so than

Figure 6-19. (a) St. Peter Sandstone along U.S. Highway 14 a few kilometers west of Rochester. (b) Microscopic view of a thin section of St. Peter Sandstone. Note the well-rounded nature of the grains and the almost total lack of cement. All grains are quartz in various stages of extinction relative to polarized light passing through the grains. The field of view is about 2 mm across. (c) Heavy mineral grains separated from the St. Peter Sandstone. The lighter grain is tourmaline and the other four grains are zircons. Note the well-rounded nature of all the grains, the result of a long history of abrasion. Other minerals have been eliminated by abrasion, weathering, and solutions in the rock. The grains are 0.10 to 0.15 mm in length.

a

b

c

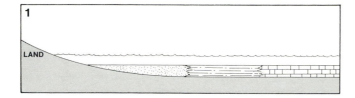

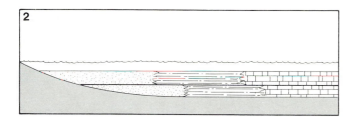

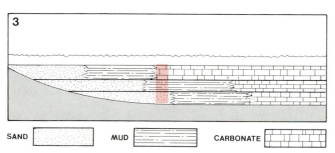

SAND [ ]   MUD [ ]   CARBONATE [ ]

a

b

in the underlying Platteville. The shale indicates a quiet-water environment, but thin beds of "fossil hash" suggest either large storm waves or sorting by tidal currents.

The Galena Formation is a 70-m-thick carbonate unit consisting of dolomite, limestone, and shale. Fossils are common and varied where not obliterated by dolomitization (Figure 6-24). A most unusual species is *Receptaculites* (popularly known as "sunflower coral"), a fossil whose classification is a problem. It is definitely not a coral, nor is it a sponge. It remains without a home in the present classification of living things. The upper member of the formation has a very different faunal assemblage than the lower part; this has been interpreted to mean a change from normal marine waters to hypersaline waters. As in the Platteville, several thin beds of altered volcanic ash (bentonite) are present.

The Dubuque Formation is a 10-m-thick gray-green to yellow-gray to gray unit of crinoidal limestone and shale. The faunas are those of a normal, shallow-marine environment. Two bentonite beds are also present.

The Maquoketa Formation is less shaly than the Dubuque, containing limestone and dolomite. The upper part of the

Figure 6-20. (a) Sedimentary facies. At stage 1, sand, mud, and carbonate are deposited as laterally differing parts of the same bed. At stage 2, representing a later time, the sea has transgressed landward (to the left), again with the three types of sediment deposited. At stage 3, representing a still later time, the sea has transgressed further. The shaded zone of stage 3 shows the sand, mud, and carbonate stacked vertically. Interpretations of such a vertical sequence indicates that the lateral equivalents also exist in the region. (b) Field exposure showing vertical stacking of sand, mud, and carbonates as in Figure 6-20a. The St. Peter Sandstone at the base is overlain by the Glenwood Shale, which in turn is overlain by limestone of the Platteville Formation. (See also Figure 6-19a.) The location is Highway 14 west of Rochester.

Figure 6-21. Limestone of the Platteville Formation. The hammer rests on a thin, muddy layer that may represent weathered volcanic ash.

a

b

Figure 6-22. (a) Fossil cephalopod (*Endoceras*) nearly 2 m long in the Platteville Formation. (b) Closer view of part of the same cephalopod. It appears that it sank into soft carbonate mud on the sea floor when it died. Note the segmented nature of the shell. (c) Artist's sketch of *Endoceras* as it may have appeared when living on the Ordovician sea floor.

c

formation contains more sand than seen since St. Peter deposition, suggesting another minor pulse of uplift of the Transcontinental Arch. Certain parts of the formation are fossiliferous, but nowhere in Minnesota do we find the unique "dwarfed" faunas of Iowa and Illinois in phosphate-rich beds. Presumably, the chemical conditions there restricted the growth of all species; an alternative explanation, not so good, is size sorting by currents.

So far only the marine deposits of southeastern Minnesota have been described. Note on Figure 6-15a that the seas had also entered northwestern Minnesota from the Williston Basin. Drillholes have penetrated two Ordovician formations totaling 90 m in thickness—a quartz sandstone and shale unit named the Winnipeg Formation and the overlying Red River Formation, which is a carbonate unit. Microscopic fossils have been used to date the rocks brought up from the wells.

The Ordovician in Minnesota was, in summary, a peaceful period. The low-lying vegetationless landmass was certainly dull and barren, but there was indeed activity in the ocean, where cephalopods 5 m long, the terrors of the Ordovician seas, prowled the shallow carbonate banks in their quest for food (Figures 6-22 and 6-25). Perhaps early forms of fish were present, too, for small plates from primitive armored fishes have been found in Colorado, Michigan, and southern Ontario.

Now recall the thin beds of volcanic ash that were mentioned as being present in the Platteville and Galena formations. The fine ash probably blew in from volcanoes in the Appalachian Geosyncline, which was experiencing local mountain building in what is now New England. Those ashfalls probably provided the only extraordinary excitement during Ordovician days in Minnesota.

The Silurian Period
(430 to 400 Million Years Ago)

As well as geologists can determine, there are no Silurian rocks in what is now Minnesota. Yet, data from elsewhere in North America suggest that the Silurian seas were widespread (Figure 6-26), and that the Silurian rocks have been eroded away. To the south and north of Minnesota, Silurian rocks are similar to the Ordovician rocks in that they are dominantly composed of carbonates. Numerous coral "patch" reefs dotted the Silurian sea floor and grew upward as much as 80 m to within 1 m of the water surface. Debris downwind from each reef shows that winds blew from what is now the southeast.

a

a

b

b

Figure 6-23. (a) The Decorah Shale (darker) overlain by the basal part of the Galena Formation in a Minnesota Highway 52 roadcut at "Golden Hill" in southern Rochester. (b) Fossil brachiopods and bryozoans in limestones of the upper Decorah Shale. The location is the same as in Figure 6-23a.

Figure 6-24. (a) Dolomites of the Galena Formation in a quarry near Rochester. (Courtesy of David Stone.) (b) *Receptaculites*, or "sunflower coral," from a quarry in the Galena Formation at Rock Dell, in Olmsted County. It has some resemblance to both sponges and corals but fits into neither classification. The chisel is 15 cm long. (Courtesy of David Stone.)

Although plants had existed in the aquatic environments of the earth from Early Precambrian time, the land probably remained barren of any vascular plant life until the Late Silurian, although primitive plants, such as mosses and liverworts, were probably present. Invasion of the land surface had to await development of a vascular system that could distribute water and nutrients to every cell of an organism. More than that, root systems had to evolve to allow plants to derive nutrients directly from soils.

So Minnesota in Silurian time was a low-lying peneplain. Sluggish streams carried runoff waters to the sea, but those waters probably contained only a little mud amid an abundance of ions in solution. Surely sand was a residual weathering product on the land surface, but where it eventually was deposited is unknown. Whether the early plants were

stabilizing the weathered surface is, likewise, unknown. Minnesota, then, featureless and barren, was probably hot—recall that the equator passed through this region.

The Devonian Period (400 to 350 Million Years Ago)

During Early Devonian time, the same conditions that were present during the Silurian probably prevailed. However, Middle Devonian time saw another northward advance of the sea into what is now southern Minnesota, as shown by Figure 6-27 (see also Plate 2). It was to be the last advance for about 225 million years. Thus, the Cedar Valley Formation, another carbonate unit, was deposited (Figure 6-28). This formation is 150 m thick in southwestern Iowa

a

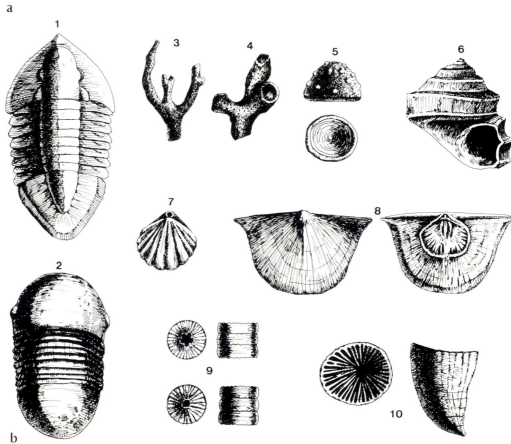

b

Figure 6-25. (a) Panoramic view of the Ordovician sea floor. A cephalopod, two trilobites, and several corals and crinoids are visible. (b) Common fossils in Ordovician rocks of Minnesota: 1 is a trilobite (*Isotelus*); 2 is a trilobite (*Bumastus*); 3 is a bryozoan (*Rhinidictya*); 4 is a bryozoan (*Batostoma*); 5 is a bryozoan (*Prasopora*); 6 is a gastropod (*Trochonema*); 7 is a brachiopod (*Rhynchotrema*); 8 is a brachiopod (*Strophomena*); 9 is a crinoid (columns); and 10 is a horn coral (*Lambeophyllum*). (After Don Wallace, Minnesota Ordovician Fossils of the Twin Cities Area, a one-sheet plate distributed by the Minnesota Geological Survey.)

Figure 6-26. Paleogeography of North America during Silurian time. There are no Silurian rocks in Minnesota, but the Silurian seas probably covered most of North America. The dashed line is the possible equator. (After Dott and Batten, 1981.)

but only half that thick in Minnesota. However, the top of the formation has been eroded and it certainly was thicker at one time.

All of the Cedar Valley in Minnesota was deposited in warm, shallow water, either near or far from shore, or very near shore in a tidal flat and/or lagoonal environment. Thin shale layers are present in the tidal flat deposits. According to J. H. Mossler (1978), who has done an extensive study of the Cedar Valley, the lower half of the formation was deposited during the advance of the sea, and the upper half, including most of the tidal flat deposits, was deposited during the withdrawal of marine waters. Thus, a cycle of marine transgression and regression has been documented by a careful study of the rocks.

Extensive dolomitization has obliterated most of the abundant fossil hash that is an indicator of extensive reworking of shells by wave action. Abundant corals in one zone also indicate shallow water. The Devonian is known as the age of fishes. Numerous bizarre species, including jaw-

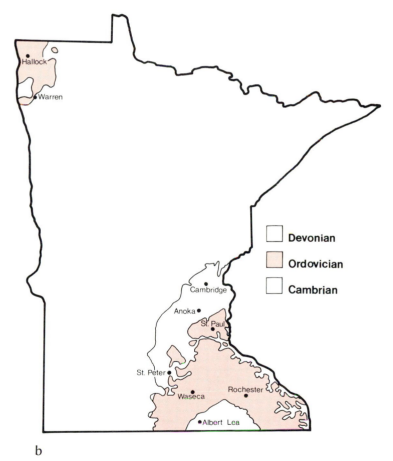

b

Figure 6-27. (a) Paleogeographic map of North America during Devonian time. Note that the seas once again transgressed into Minnesota, for the last time for about 225 million years. The dashed line is the possible equator. (After Dott and Batten, 1981.) (b) Devonian rocks of southeastern Minnesota. Only one Devonian formation, the Cedar Valley, is found in Minnesota.

83

less and armored types, were abundant, but none has ever been caught in Minnesota's Devonian rocks.

What was happening on the land surface that encompassed most of Minnesota? Obviously, we have no record, but we can refer to some remarkable findings elsewhere that can be applied to this area. The landmass was well eroded and low lying. That much we know because the Devonian rocks are composed of carbonate rather than sand and shale. A low-lying landmass, especially near the seashore, would quite likely have been swampy. What would have grown in that swamp? Probably the same things that grew in Late Devo-

nian time in eastern New York. There, Devonian rocks contain fossils of trees, 10 m tall and 30 cm in diameter, in what was evidently a swampy forest. The oldest known land vertebrates, primitive amphibians, are from Devonian rocks of the Arctic. Perhaps they walked over Minnesota too.

Although Minnesota was low lying, not all of North America was quiet. The Appalachian Geosyncline was again experiencing mountain building, as during the Ordovician, but its effects were not felt as far west as Minnesota.

The Mississippian Period
(350 to 325 Million Years Ago)

Once the Middle Devonian sea withdrew, the Canadian Shield, and therefore what is now Minnesota, may have been above sea level for the next 225 million or so years, a time span including the Mississippian, Pennsylvanian, Permian, and Triassic Periods and part of the Jurassic. Some geologists think the shallow Mississippian seas also encroached into Minnesota (Figure 6-29). We prefer the former interpretation. Although no record of specific events during this time has been preserved in Minnesota, study of the rock record elsewhere in North America gives some indication of what Minnesota was like during those 225 million years.

The Mississippian Period in North America was a time of widespread shallow and warm seas (Figure 6-29). This is well documented by the great abundance of fossils in Mississippian limestones to the south and west of Minnesota. Crinoids, or sea lilies, were especially abundant. The equator may have extended from the eastern end of Lake Superior through southern Minnesota and on to what is now southern California. This position of the equator, based on paleomagnetic evidence, once again supports the fossil evidence for a tropical Minnesota.

In the interior of North America, the Mississippian rocks are predominantly limestones. However, from what is now Texas to what is now New York, some sands were being carried northwestward and westward off of locally uplifted portions of the Appalachian Geosyncline. In the far west, a landmass was beginning to rise out of the Cordilleran Geosyncline that was to become, by Cretaceous time, a dominant physiographic feature of western North America.

The abundance of limestone to the south of Minnesota indicates that Minnesota and adjacent areas were so low lying that streams had low gradients (or slopes) and did not achieve the velocities necessary to transport sand or pebbles. Instead, the sluggish streams draining the low areas were probably carrying an abundance of clay and dissolved ions. The presence of calcium bicarbonate ions $[Ca(HCO_3)_2]$, which became more and more abundant in the warm, shallow seas, led to the chemical and biochemical precipitation of calcium carbonate $(CaCO_3)$ as limestone.

Was there life on land in Minnesota during Mississippian time? Probably, but there is no proof of it. By Late Devonian time, amphibians had already made their appearance

a

b

Figure 6-28. (a) The Cedar Valley Formation exposed in a quarry on County Highway 2 a few kilometers southwest of the village of Racine, 25 km southwest of Rochester. (b) Fossils in the Cedar Valley Formation. Most fossils are here internal molds of the original brachiopod shells. The location is the same as in Figure 6-28a.

and were moving about on land. Since amphibians require water in which to lay their eggs and to keep their bodies moist, it is reasonable to assume that any Minnesota amphibians would have stayed close to rivers, swamps, or lakes.

Some land plants had already developed by Silurian time, and forests existed in the Devonian. Thus, Minnesota, at least in its wetter areas, was probably vegetated. But, had you been there, you would have noticed the complete absence of flowering plants. They were not to develop for another 100 million years and were not to become abundant for still another 100 million years.

The Pennsylvanian Period
(325 to 280 Million Years Ago)

The Pennsylvanian Period is known as the age of coal. The interior of North America was so low lying that even a slight rise in sea level caused the sea to advance tens or hundreds of kilometers onto the land (Figure 6-30). Adjacent to the sea was a network of swamps and rivers. Muds and sands were deposited in the rivers and in deltas, where the rivers met the sea.

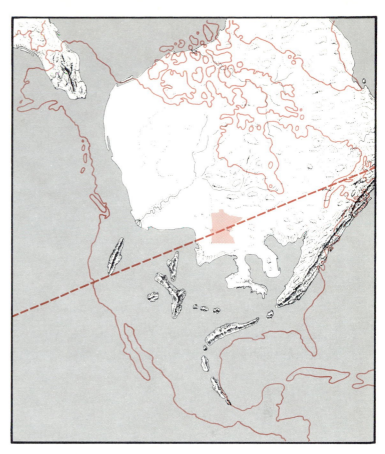

Figure 6-30. Paleogeography of North America during Pennsylvanian time. Whereas the seas did not quite reach Minnesota, it is possible that some river and swamp deposits could have been deposited in the southernmost part of the state, for such deposits exist in northern Iowa. The dashed line is the possible equator. (After Dott and Batten, 1981.)

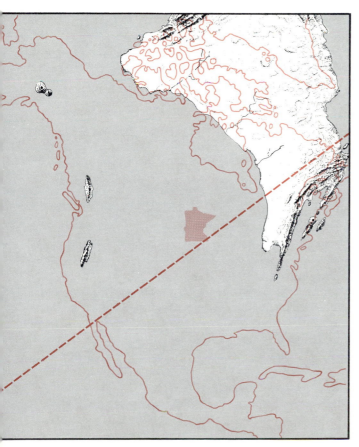

Figure 6-29. Paleogeography of North America during Mississippian time. The seas may or may not have reached Minnesota. The dashed line is the possible equator. (After Dott and Batten, 1981.)

As in Mississippian time, the equator probably cut through what is today Minnesota. Thus, a tropical climate would have prevailed in central North America. Broad swamps were sites of lush growths of plants such as tree ferns and horsetails, some of which grew to heights of more than 30 m. As the plants died and fell beneath the swamp water, incomplete decay resulted in the formation of peat beds. Further burial and compaction transformed the peat into bituminous coal. Large Pennsylvanian coal basins were present in what are now Illinois, Iowa, Missouri, and Michigan and even farther east (see the basins in Figure 6-1). Although Pennsylvanian coal is found in north-central Iowa, quite close to Minnesota, none is present in Minnesota. It is possible that some was once present in the southernmost part of the state, but if so, all traces of it have been removed by erosion.

Certainly southward-flowing streams were draining Minnesota, for crossbedding in Pennsylvanian sandstones of the coal measures indicates dominant southward transport of sand. It is quite likely that amphibians, small, recently evolved reptiles, and an abundance of insects (some with wingspans of 75 cm), were present in Minnesota, especially near the rivers and swamps. But, alas, they left no record:

no Pennsylvanian rocks, if ever present, have survived erosion. It seems that Minnesota and the Canadian Shield were slightly uplifted during Pennsylvanian time, for the Pennsylvanian sediment to the south was no longer dominantly the result of the precipitation of calcium carbonate, as during the Mississippian Period. Instead, muddy and sandy rocks predominated, along with the coal beds. In western Canada, sands were being carried into the sea.

Meanwhile, other changes had occurred on the face of North America. The low-lying Transcontinental Arch, which had been so prominent in Early Paleozoic time, had disappeared, either due to regional subsidence or to erosion, or both. Mountains dubbed the "Ancestral Rockies" had risen in the Colorado-New Mexico region. The landmass in the far west was growing larger. The Appalachian Geosyncline was becoming increasingly active and was beginning to rise out of the sea. In fact, it seems that North America was tilting westward and the sea was withdrawing from much of the eastern third or so of North America.

The Permian Period
(280 to 225 Million Years Ago)

During most of Permian time, what is now Minnesota was part of a large landmass that encompassed the eastern two-thirds of North America (Figure 6-31). The westward tilting of North America evidently continued from the Pennsylvanian through the Permian. The Appalachian

Mountains had risen out of the geosyncline in Late Pennsylvanian and Early Permian time, probably owing to the collision of two continental landmasses. These landmasses, an early North America and an early Europe, would have been passengers on an earlier generation of moving lithospheric plates similar to those referred to in Chapter 1 and shown in Figure 6-7.

Permian seas of the west extended as far east as what is now Kansas. In the west, many Permian rocks are evaporites, that is, salts and gypsum deposited when arms of the Permian sea were drying up. The equator probably extended through Minnesota, as during the Mississippian and the Pennsylvanian. All in all, it appears that much of the interior of North America was at least seasonally hot and dry. Even with such a climate, periodic rains would have flowed to the seas via rivers. The muddy nature of Permian deposits in the area to the west of Minnesota suggests a continuation of the generally low relief so typical of all of Paleozoic time.

What would life forms in Minnesota have been like? Reptiles had already become dominant over amphibians in both size and number. A great variety of reptiles, including some mammallike reptiles and fin-backed types, were probably present. Many such reptiles have been found in Permian rocks of Texas, New Mexico, and Oklahoma. But Minnesota, with no Permian rocks, obviously has no Permian fossils.

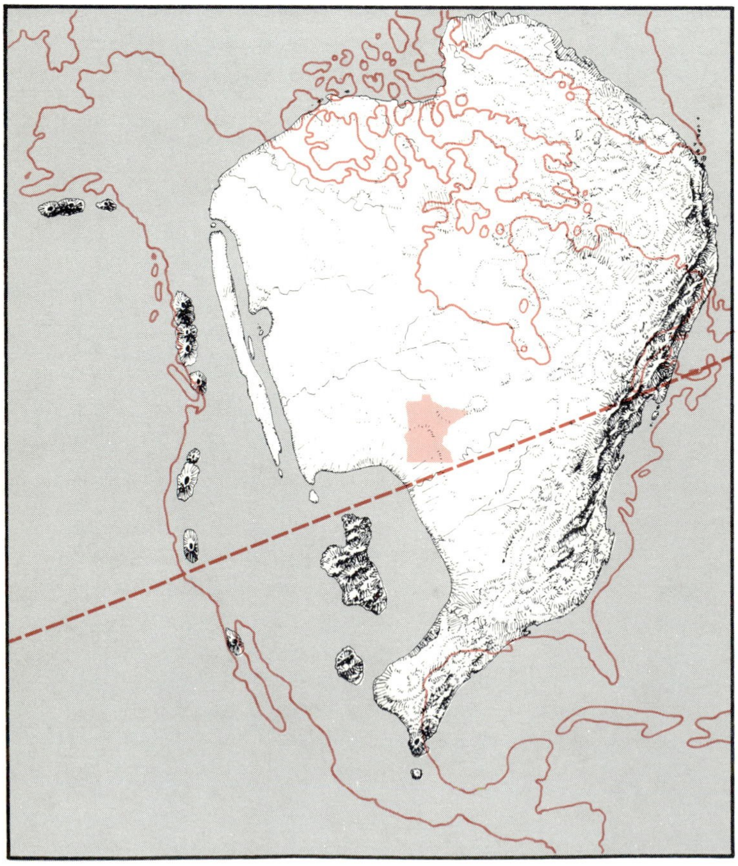

THE MESOZOIC ERA

The Triassic Period
(225 to 190 Million Years Ago)

By Triassic time, what is now Minnesota was located far from the seas, which had withdrawn even farther west than during the Permian (Figure 6-32). Adjacent to the seas were deposits of silt and sand, probably derived from rivers flowing westward off the large landmass that included Minnesota. These Triassic river deposits, as in Wyoming, Utah, and South Dakota, are usually red due to the presence of minor amounts of hematite that formed as the sediments were exposed to an oxidizing atmosphere.

The equator in Early Triassic time still cut through Minnesota, but by Late Triassic time North America had moved northward so that the equator reached from what is now Florida to what is now Texas and beyond. Minnesota was thus 15° to 20° north of the equator and presumably was cooler than in earlier periods.

With no Triassic rocks, Minnesota obviously cannot possess a record of Triassic life forms. However, it is likely

Figure 6-31. Paleogeography of North America during Permian time. The Appalachian Mountains have risen out of the Appalachian Geosyncline, and the seas have withdrawn from most of what is now the Midwest. The dashed line is the possible equator. (After Dott and Batten, 1981.)

that Minnesota was inhabited by an abundance of the dinosaur forms found preserved in both eastern and western North America. The Mesozoic Era, consisting of the Triassic, Jurassic, and Cretaceous periods, is known as the age of dinosaurs. Minnesota, too, probably had its share of the terrible lizards, which were to become even more terrible during the Jurassic and Cretaceous. Flying reptiles were present along with the reptiles on land and may even have soared over Minnesota. The first mammals, small creatures, lived during the Triassic and could also have been Minnesota residents, scurrying through rather lush vegetation, especially in wetter lowlands.

In eastern North America, several downfaulted valleys were depositional sites for Triassic river, lake, and swamp sediments as well as for some volcanic rocks. These valleys appear to have been caused by the initial breakup of the large supercontinent formed by the collision of the continents during Permian time; thus, they may signify the beginning of the present Atlantic Ocean. Dinosaur tracks have been found in these eastern Triassic rocks, and similar

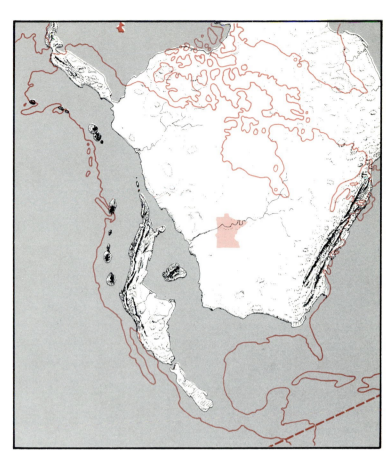

Figure 6-33. Paleogeography of North America during Late Jurassic time. The seas have again transgressed into the continental interior, reaching as far as northwesternmost Minnesota. Except for the Appalachians, the big landmass remained a low continent. The dashed line is the possible equator. (After Dott and Batten, 1981.)

Figure 6-32. Paleogeography of North America during Triassic time. The seas have withdrawn completely from the central and eastern parts of the continent. The dashed line is the possible equator. (After Dott and Batten, 1981.)

tracks may have once recorded the tread of reptilian feet in Minnesota as well.

In western North America, the landmass mentioned earlier was still growing. Streams draining eastward off this land area probably provided an abundance of ecological niches for the reptiles.

The Jurassic Period
(190 to 135 Million Years Ago)

During Early Jurassic time, conditions were much as they were during the Late Triassic. However, Late Jurassic time saw an advance of marine water, called the Sundance Sea, southward from the Arctic into the western interior. The presence of Jurassic marine rocks in the subsurface of the northwestern corner of Minnesota shows that the seas extended quite far eastward (Figure 6-33), probably reaching from the Williston Basin.

The Jurassic rocks of northwestern Minnesota, known

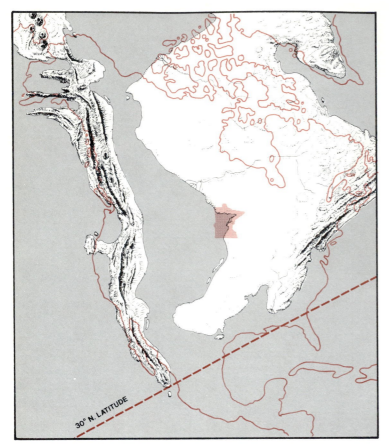

Figure 6-34. Paleogeography of North America during Middle Cretaceous time. This was the time of the last transgression of the seas into the interior of the continent. The eastern shore of the Cretaceous sea crossed through what is now Minnesota. The dashed line is the possible 30°N latitude line, *not* the equator. (After Dott and Batten, 1981.)

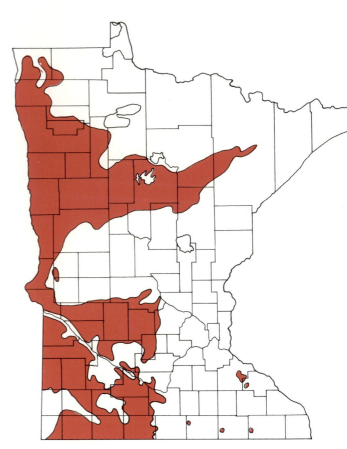

Figure 6-35. Cretaceous deposits (shaded) in Minnesota. In the Cretaceous sea in the southwestern part of the state, islands of Sioux Quartzite probably were present. The eastern edge of the deposits is thin and probably represents the approximate shoreline. (After Austin, 1972c.)

only from drillholes, consist of red mudstone and minor dolomite. These rocks are about 30 m thick where they have been penetrated and are informally called the "Hallock red beds." They contain no fossils, so their age is somewhat uncertain, although Jurassic rocks do exist in North Dakota and Manitoba. Because gypsum is being mined in Manitoba 50 km north of the Minnesota border, it was hoped that the drillholes would penetrate gypsum in Minnesota, too. They did not. Evaporites indicate extensive evaporation, and this requires at least a reasonably hot and semiarid climate. The evaporites in Manitoba may be the northernmost Jurassic evaporites in North America. The equator at this time was 25° to the south of what is today Minnesota, as shown in Figures 6-8 and 6-33.

In latest Jurassic time, the sea retreated northward. Streams draining eastward off the still-growing landmass in western North America formed a broad alluvial plain upon which lakes and swamps were common. Skeletons of dinosaurs are especially abundant in the Morrison Formation

of Colorado, Utah, and Wyoming. It seems likely that Minnesota, too, harbored these dinosaurs, but preservation of their remains was not likely on the low-lying topography.

The Cretaceous Period
(135 to 65 Million Years Ago)
The Seas

About 100 million years ago, the oceans once again invaded the western interior of North America, advancing from both the south and the north. The rising landmass of the west had grown very large, covering an estimated 320,000 km². This landmass formed a rugged *western* shoreline for the Cretaceous sea in what are now Idaho, Wyoming, Utah, and New Mexico, whereas part of the *eastern* shoreline was located in what is now Minnesota (Figure 6-34).

The Minnesota shoreline was not a rugged one, for the landmass dominated by the Canadian Shield was relatively

Figure 6-36. Cretaceous iron-ore conglomerate that was mined in the Enterprize Mine at Virginia in the mid-1950s. Most pebbles are hematite from the underlying Middle Precambrian iron-formation. Similar conglomerates were found in several mines along the Mesabi and probably formed in the vicinity of the shoreline of the Cretaceous sea. (See also Figure 6-37a.)

featureless. Relief was locally about a hundred meters. However, this shoreline was highly irregular, as shown in Figure 6-35. This map depicting the extent of the Cretaceous sea was drawn on the basis of data from innumerable water wells, a few exposures of rock in the western part of the state, and several exposures in iron-ore mines of the Mesabi Range. Most of the sediment deposited in the Cretaceous sea in Minnesota is poorly lithified siltstone and in the Dakotas is mainly black shale. Thus, the streams that carried water and sediment to the sea in Minnesota had sufficient velocity to transport silt-sized material to the shoreline. Locally, along the north edge of the long Cretaceous bay that projected into what is now the Mesabi Range, the shoreline must have been somewhat rugged. Conglomerates made up almost completely of hematite pebbles (Figure 6-36) were present in several open-pit iron-ore mines that were cut through the glacial cover and through Cretaceous rocks before intersecting the Middle Precambrian iron ores. This conglomerate ore, while locally of a high-grade and therefore of some economic importance, is of special interest to geologists for another reason. It shows that high-grade hematite ore must already have been formed in order to be broken up into pebbles, rounded by wave and current action, and deposited as beds. Thus, the weathering that formed the high-grade natural ore out of taconite rock occurred prior to 100 million years ago. Perhaps it happened in earlier Cretaceous time, a point we discuss further in the next section.

Finally, in Late Cretaceous time, the seas withdrew from the interior of North America for the last time. Except for the coastal plains, all of North America has remained high and dry to the present.

Warm Climate and Deep Weathering

The entire world probably had a mild climate owing partly to the widespread, shallow, and warm Cretaceous seas.

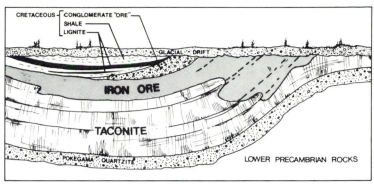

a

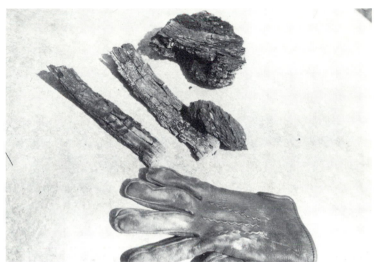

b

Figure 6-37. (a) Cross section of the Enterprise and Alpena mines area at Virginia in the 1940s and 1950s, showing Cretaceous iron-ore conglomerate, shale, and lignite (or low-grade coal). The lignite apparently formed in a swamp at the edge of the sea. (After a sketch by P. A. Lindberg in Gruner, 1946.) (b) Pieces of lignitized Cretaceous pine wood, and a pine-cone (center) from the Enterprise Mine. (Courtesy of P. A. Lindberg.)

Forests were abundant even north of the Arctic Circle. Much of North America probably enjoyed subtropical conditions. That this was true in Minnesota is indicated by the uncovering in the Enterprise Mine at Virginia of a bed of coal as much as 10 m thick and 100 m wide (Figure 6-37). The low-grade coal (lignite) was made up largely of the wood of conifer trees, which must have been growing in abundance in a small swamp at this locality at the eastern end of the long arm of the Cretaceous sea. Other thin coal beds occur elsewhere in Minnesota, and thin coals have been penetrated by numerous water wells. Thus, during Cretaceous time, the surface of low-lying Minnesota probably was dotted with countless swamps, as it is today. Because of a subtropical climate and a lush growth of vegetation, thin coal beds were formed, much as they were further south during Pennsylvanian time.

In southwestern Minnesota the effects of the subtropical

climate can be seen in a deeply weathered zone atop the Precambrian and Paleozoic bedrock. Solution by percolating waters and various chemical reactions through a long period of weathering transformed the granites, gneisses, and other rock types into a soil rich in kaolinite, a soft clay material. The fact that this ancient weathered zone is as much as 60 m thick suggests a long interval of exposure to a humid surface environment (Figure 6-38). Preservation of such an easily excavated material indicates that erosion was not very intense; a thick cover of vegetation might have stabilized the prevailing hills and slopes.

Such well-preserved soil zones give a rare opportunity to reconstruct ancient landscapes. Careful studies by W. E. Parham (1970) of the areal distribution of an iron-rich, hard kaolinite zone with a nodular structure led him to conclude that a peneplain lies buried beneath the Upper Cretaceous marine sediments and Pleistocene glacial drift of western Minnesota. Such topographic surfaces are postulated to be the result of a cycle of weathering and erosion. Their existence had been persistently challenged by students of landscape development, and so the recognition of a peneplain in Minnesota is an important contribution to geomorphology.

The basic requirements necessary for designating an erosion surface as a peneplain are met fairly well by the buried landscape in western Minnesota: (1) it is a surface of very low relief, nearly a plain; (2) a thick zone of deeply weathered rock lies beneath it; (3) the old hilltops are generally of about the same elevation; (4) with the exception of a tough quartzite, all rock types and structures appear to be equally truncated by the erosion surface; (5) a thin cover of sediments is preserved between the weathered hills and

along old floodplains in the major valleys; and (6) the erosion surface is extensive, being at least 275 km long between Blue Earth and Otter Tail Counties and perhaps extending as far north as Winnipeg, Manitoba.

From even a casual observation of the regolith, it is apparent that not all of the minerals in the original rock were equally affected by the weathering processes. In fact, one mineral, quartz, seems to have been almost totally immune to change, judging from its abundant distribution in the residual weathered zone developed on granites and gneisses. Careful chemical analyses of regoliths in southwestern Minnesota and elsewhere by S. S. Goldich (1938) resulted in his recognition of the relative stability of the common rock-forming minerals in the surface environment. Iron- and magnesium-rich minerals, such as olivine and augite, as well as calcium-rich feldspar, are the least stable. These minerals are the major constituents of basalt and gabbro. Increased resistance to weathering is shown by sodium-feldspar and potassium-feldspar and even more so by quartz. Recall that these are the minerals that make up granite. Thus, upon exposure to the same surface conditions, gabbro will decompose more rapidly than granite, and the mineral quartz will survive the longest of any of the common minerals. Here, then, is one reason that quartz-rich sandstone is such an abundant sedimentary rock.

Even though the weathering episode described above produced valuable mineral by-products, especially ceramic clays and perhaps the soft iron ores of the Mesabi and Cuyuna ranges, it is worth pointing out what else might have been. Deep chemical weathering under certain conditions is capable of removing the silica from aluminum silicates, leaving behind a weathering product rich in aluminum. These concentrations of aluminum-rich material, called bauxite, provide most of the world's aluminum ore. Had the chemical alteration of the rocks in western Minnesota been more intense, a major deposit of bauxite would have been formed. How would such an occurrence have influenced the course of events in Minnesota's political and industrial development? We will never know for sure, but the effects would have been profound.

As the soils were forming on the upland surfaces, rivers flowing at lower elevations reworked the weathered material (Figure 6-39a). Finally, the Cretaceous seas flooded the region, burying the old landscape beneath marine muds and silts (Figure 6-39b).

Cretaceous Life

Cretaceous life on the North American continent was varied, including more than 70 genera of dinosaurs, a number of swimming reptiles and flying reptiles, and a few species of small mammals. Remains of Cretaceous dinosaurs are abundant in river, delta, and swamp deposits on the west side of the Cretaceous sea that flooded the western interior of the continent (Figure 6-40). They may have been abundant in Minnesota, too, but such nonmarine sedimentary

Figure 6-38. In-place weathered granite at the base of an exposure near Morton, Minnesota. The granite lies beneath redeposited Cretaceous kaolinitic and lignitic clay. The weathering occurred during the warm Cretaceous Period.

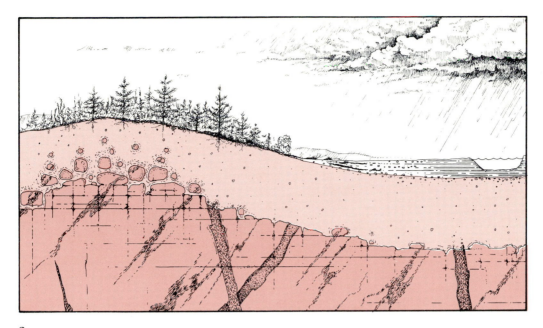

a

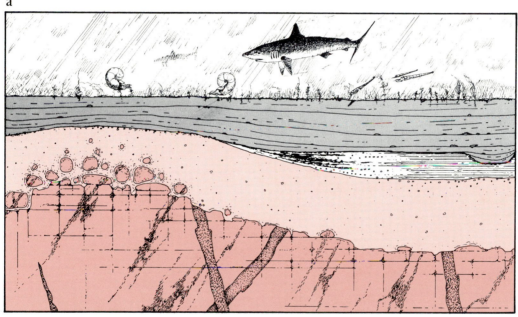

b

Figure 6-39. (a) Diagrammatic illustration of deeply weathered Precambrian rock in the Minnesota River Valley during Cretaceous time. On the right, a Cretaceous river and its floodplain cover the weathered profile. (b) Diagrammatic illustration of the terrane shown in Figure 6-39a, covered by the transgressing Cretaceous sea. Such sequences have been found in the Minnesota River Valley.

rock units, if present in Minnesota, are covered by glacial drift. Yet someone someday may find dinosaur remains in Minnesota.

Remains of other kinds of Cretaceous vertebrates have been found in Minnesota. The skull of a marine crocodile, the largest known from Upper Cretaceous rocks, was found on an open-pit mine dump near Coleraine on the Mesabi Range in 1967. It resembles related species found in Montana and has been named *Teleorhinus mesabiensis* by Bruce

Erickson, of the Minnesota Science Museum. A few mosasaur teeth have also been found, as sparse remains of these reptile swimmers (Figure 6-41).

Shark teeth have been found at several open-pit dumps on the Mesabi Range (Figure 6-42). In Big Stone County, in western Minnesota, some Cretaceous sediments preserved in joints and pockets on granite in granite quarries are made up of small shark teeth. Sharks drop their teeth as new ones grow in beneath them, and such beds are probably the result

Figure 6-40. Cretaceous dinosaurs whose remains have been found in the western states on the west side of the Cretaceous sea shown in Figure 6-34. Duckbilled dinosaurs (left), *Tyrannosaurus* (center), and *Triceratops* could also have roamed Minnesota, but few continental deposits in which their remains might have been preserved have been discovered.

of concentration by waves and currents in areas of slow sedimentation of mud or silt.

Ammonites, coiled cephalopods, were commonly found on several open-pit dumps on the Mesabi Range. Clam and snail shells were once abundant on dumps of the Judson Mine at Buhl. These forms probably lived in the brackish water of lagoons near the shore.

Figure 6-43 is a map of the long Cretaceous bay that extended eastward along the Mesabi Range. Several fossil localities are known, but fossils are now difficult to find at these places because of years of collecting, growth of vegetation, and slumping of original exposures in mines.

It was not until the Jurassic Period that the true flowering plants, and presumably honeybees to pollinate them, came into being. By the end of the Cretaceous Period, this phylum of plants dominated the world's floras, at the same time that mammals came to the fore in the animal kingdom. Beds of Cretaceous coal have already been mentioned as environmental indicators. Leaf impressions, carbonized wood fragments, and pollen grains indicate that a lush floral community thrived upon the landscape of Minnesota.

The Cretaceous rocks of Minnesota contain the remains of a variety of plants, including ferns, conifers, and deciduous trees, with conifers apparently the dominant tree type, especially in the area of the Mesabi Range (Figure 6-44). The reconstructed forests of Minnesota in Late Cretaceous time compare favorably with those of the present-day

Figure 6-41. Swimming reptiles that frequented the Cretaceous sea, as evidenced by their teeth and bones in Cretaceous marine rocks of the western states. There is no reason why they should not have moved about in Minnesota's Cretaceous sea as well. The general lack of exposed Cretaceous marine rocks is probably the main reason such remains are scarce in Minnesota.

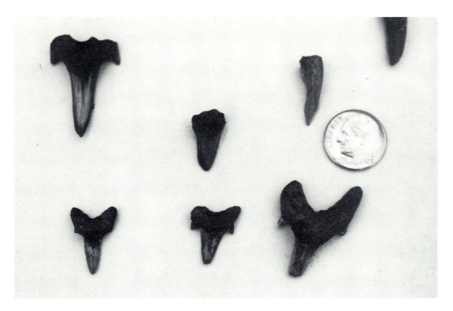

Figure 6-42. Shark teeth found in the vicinity of the Hill-Annex Mine and dumps at Calumet. This collection is housed at the office of the Minnesota Department of Natural Resources at Hibbing.

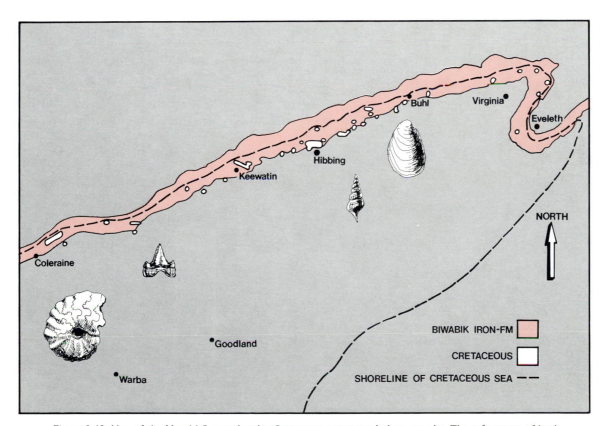

Figure 6-43. Map of the Mesabi Range showing Cretaceous exposures in iron-ore pits. The soft nature of both the Cretaceous sediments and the overlying glacial cover resulted in bank collapse soon after the exposures were made, and little Cretaceous material can be seen today. Snails and clams, perhaps part of a brackish-water assemblage, were prominent in the eastern part of the range (as at the Judson Mine in Buhl), and ammonite cephalopods and shark teeth were abundant at the west end of the range near Keewatin and Coleraine, where the waters were certainly marine. This is part of the long embayment shown in Figure 6-35.

Pacific Northwest, which has a humid climate characterized by mild temperatures the year around. Coastal plain bogs in Minnesota accumulated enough organic material to form thin lignite deposits, but these are neither extensive enough nor thick enough to be economically important today.

Major Rock Formations

On top of the Cretaceous regolith lies an assortment of poorly consolidated, generally horizontally layered sediments, ranging in lithology from bouldery conglomerates to clean, well-sorted sandstone to siltstone and lignite. Two major environments of deposition are indicated: continental, including lake, swamp, and floodplain sediments, and marine, including the beach and offshore sediments of a transgressing sea. The continental sequence, best studied in southeastern Minnesota, is named the Windrow Formation; the marine section of northern Minnesota is called the Coleraine Formation. Deposits in western Minnesota are buried beneath glacial drift and are not easily or abundantly observed, but they are known to be as much as 190 m thick. The deep regolith is covered in many places by sandstone and silt derived by erosion from local sources. This unit changes

upward in the section to marine shale, indicating the advance of a sea from the west. These rocks have been correlated with strata of similar age in the Dakotas and therefore take their names, such as Dakota Sandstone and Pierre Shale, from those formations.

THE CENOZOIC ERA

The Tertiary Period: (65 to 2 Million Years Ago)

As the Cretaceous seas were withdrawing from North America, the west was on the rise (Figure 6-45). Out of the Cordilleran Geosyncline rose the Rocky Mountains. Between the numerous ranges of the Rockies were basins that filled with gravel, sand, and silt carried down from the growing mountains by numerous rivers. East of the Rockies, great fans of river sediments joined together to form a broad plain, today the Great Plains.

As Late Tertiary and modern erosion cut into these river sediments, as in the Badlands of South Dakota, abundant remains of various mammals were uncovered. Oreodonts (sheeplike animals) must have roamed the region in

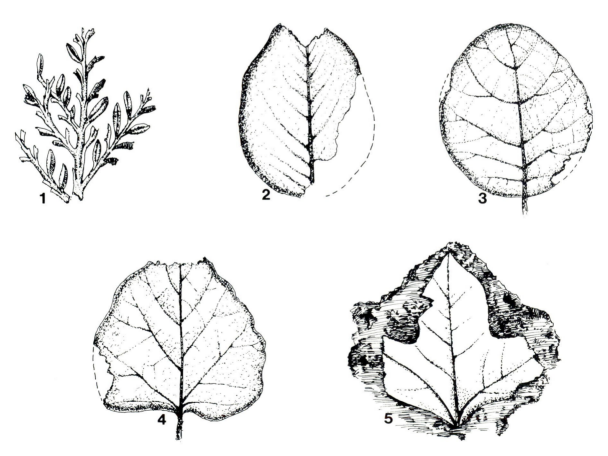

Figure 6-44. Sketches of fossilized Cretaceous leaves discovered in Minnesota: 1 is a sequoia from Austin; 2 is a magnolia from New Ulm; 3 is a protophyllon from Mankato; 4 is *populites* from New Ulm; 5 is a sassafras from New Ulm. (Leaves 1 thorugh 4 are after Lesquereux et al., 1895, and leaf 5 is after Berry, 1939.)

great numbers during the Middle Tertiary along with titanotheres (rhinoceroses) and a host of other mammals, both large and small. The dinosaurs that had reigned supreme through the Mesozoic were now extinct, and mammals filled virtually all ecological niches including the grasslands that first came into existence in about mid-Tertiary time.

Although the widespread river sediments just mentioned never reached Minnesota, it is certainly feasible that the same animals so well preserved in the Badlands were present in what is now Minnesota as well. But Minnesota, far from the Rocky Mountains, was not to experience the rapid accumulation of hundreds or thousands of meters of sediment in which animal remains could be fossilized. Instead, erosion of the already low-lying Minnesota surface continued.

Certainly the rise of the majestic Rockies influenced the climate east of the mountains. The luxuriant subtropical growth of the Cretaceous was no longer present, and a cool, temperate climate prevailed. Minnesota was slowly achieving its reputation as the "icebox of the nation."

It is frustrating to any historian to have to work with incomplete records. What really happened in Minnesota during the Tertiary Period? Even if erosion were the dominant geological process, somewhere there should have accumulated some sediments on a floodplain or in a lake in a depression dissolved into carbonate rocks or at the foot of eroding hills. Yet no rocks of Tertiary age have been positively identified anywhere in the state.

But at least one geologist has marshaled evidence that strongly indicates a Tertiary age for the iron ores of southeastern Minnesota. R. L. Bleifuss (1972), after a careful study of the iron deposits in western Fillmore, southern Olmsted, and eastern Mower counties, concluded that they were formed by weathering of Paleozoic iron carbonates, especially in the Devonian Cedar Valley Formation, under a temperate climate during the Tertiary Period. Such a conclusion is contrary to the long-standing assignment of the same ores to a Cretaceous weathering cycle.

As more subsurface exploration takes place, perhaps rocks of undoubted Tertiary age will be discovered beneath the glacial drift, forming the basis for a longer discourse on the Tertiary history of Minnesota.

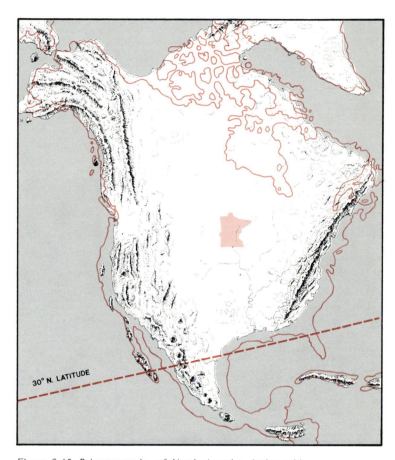

Figure 6-45. Paleogeography of North America during mid-Cenozoic time. Note that the seas are only upon the continents near the coasts. The dashed line is the possible 30°N latitude line, *not* the equator, which was located close to its present position. (After Dott and Batten, 1981.)

A kame complex in western Minnesota indicating glacial activity during the Ice Age. (Courtesy of David F. Reid.)

7

The Quaternary Period (2 Million Years Ago to the Present)

A few kilometers northwest of Montevideo, on the partially wooded floodplain of the Chippewa River, lies a 20-ton block of fine-grained, dark green-gray rock. Many years ago, local observers recognized the fact that this giant boulder contrasted sharply in color and texture with the nearby bedrock, which is a pink, coarse-grained granitic gneiss. They concluded that the dark stranger had traveled to its resting place from a distant source in outer space, and the rock became locally known as the Montevideo Meteorite (Figure 7-1). Centuries earlier, Indians had also been impressed with some of the boulders littering the landscape of the northern portion of the continent. Near the pipestone quarries in Rock County, where the bedrock is a pink quartzite, a cluster of large granite fragments inspired the legend of the three maidens. The village of Red Rock, in Dakota County, is named for a boulder of granite that long before European settlement was kept red-washed with bright ochre pigments to honor its strange presence.

All of these rock fragments have one thing in common. They were plucked from bedrock sources by glaciers, transported by flowing ice, in some instances hundreds of kilometers, and then deposited when the glacier system could no longer carry them. The Montevideo "meteorite" was derived from one of the greenstone belts of northern Minnesota; the "Three Maidens" at Pipestone, from a granite source much farther north; and the red rock of Red Rock, from a granite bedrock source near St. Cloud. The presence of these and innumerable other rock fragments across the length and breadth of Minnesota and adjacent states is one of the most compelling indications that North America at one time in the past experienced an ice age of great proportions.

According to the geological calendar, the last large unit of earth history unfolded in the Cenozoic Era, beginning

about 65 million years ago. The Quaternary Period encompasses just the last 2 million years, and almost all of the familiar features of the present land surface had their greatest development during that time. Volcanism, uplift, faulting, erosion, and deposition all proceeded at such a rapid pace that the face of North America at the end of the Tertiary Period was to be drastically modified. An important contributor to the direction and intensity of that geological facelift was the deterioration of the planetary climatic regime. Replacing the mild, stable climate of the Tertiary was one of cyclical variation, marked by fluctuations between cold and warm periods. The entire surface of the earth

Figure 7-1. Large boulder of greenstone transported by glaciers from northern Minnesota to its resting place near Montevideo, in western Minnesota. Such erratics are excellent indicators of past glacial activity.

was to be modified in response to the changing climate in ways so profound as to justify the definition of a new geologic period, the Quaternary. The first and longest chapter of that period is the Pleistocene Epoch, informally known as the Great Ice Age.

In Minnesota, the events of the Quaternary Period, and especially those of the Ice Age, are only incompletely represented by the geologic record, and even then it is with a midcontinent bias. Therefore, the full sweep of environmental changes and their effects on the land surface, the oceans, the atmosphere, and the biosphere can only be appreciated from a worldwide perspective. Furthermore, because the continental record has been so severely censored by erosion, earth scientists have turned to the ocean basins, where muds and oozes offer continuous recordings of Quaternary climatic fluctuations and some of their geological and biological consequences. Plant and animal communities, both on the continents and in the oceans, reacted to the lowering of world temperatures with massive migrations, selective adaptation, and extinction of species. Shifting weather patterns promoted profound rearrangements of boundaries between arid, temperate, and humid regions. The most impressive change of all was the establishment and growth of glaciers, some of continental proportion, where none had been before (Figure 7-2). Even the outlines of the continents expanded as ocean volumes shrank, starved

Figure 7-3. Striated surfaces, such as this one in northern Minnesota, are produced when dirty glacial ice slides across bedrock. The resulting abrasion wears down the rock, and the scratches indicate the direction of glacier movement.

by the nonreturn of glacier-bound water. Tracing the flow lines and mapping the boundaries of vanished glaciers, as well as recognizing the history of their recurrence, are all made possible by an understanding of glacial processes and the geological features associated with them.

GLACIERS

Glaciers are large masses of moving ice that form wherever snow accumulates faster than it can melt over a long period of time. Snowflakes are first compacted, then partially melted, and eventually refrozen into aggregates of small granular ice crystals called *firn*. This transformation can take place in one season, depending upon the temperature and volume of snowfall. If you dig into an old snowbank in early spring, you will find this frozen material, which skiers call corn snow. After many seasons of continued additions to the pile of cold sediments (ice is a mineral and snow crystals can be called sediment), firn gradually recrystallizes into a tight mosaic of larger ice crystals. When a critical thickness is reached, the mass of ice begins to deform under its own weight. Gravity takes control of the rate and direction of movement of the glacier thus born, just as it controls the velocity of running water and other dynamic systems on the earth's surface.

Once a glacier has been established, its general health is determined by the net accumulation of new snow each year. If more is added than lost, the margins of the ice will be extended. Shrinking margins usually indicate that the glacier is starving, that is, that losses through melting and evaporation are not being compensated by new snow. No change in the position of ice boundaries indicates equal

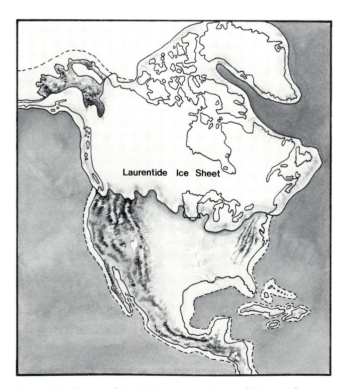

Figure 7-2. Extent of glacial ice during the last (Wisconsin) glaciation of North America, about 16,000 years ago. Glacier boundaries are deduced from mapping the distribution of drift, moraines, and other evidence of glacial activity.

Laurentide Ice Sheet

amounts of gain and loss. Monitoring the boundaries of glaciers thus gives a qualitative knowledge of the mass balance.

As geological agents, glaciers shape landscapes in two different ways. Glacial erosion involves the sculpting of a topographic surface by abrasion and quarrying. Subsequent transport and deposition of the rock debris produced by erosion results in the construction of various landforms both directly from the ice and from glacial meltwaters. The nature of the landforms in a glaciated terrane is ultimately dependent upon the complicated interaction of the geometry of the glacier, its physical state, the nature of the rock and soil across which the ice moves, and the duration of the glacial ordeal. The glacial system is a giant conveyor belt. Erosion produces a load of debris from sources beneath the ice, and sediment is then transported toward the margins, where it is deposited. Regional boundaries can be drawn between zones of erosion and deposition, based on an analysis of the features left behind.

Glacial Erosion

Abrasion is an important erosional process only where the base of the ice is at the melting point, allowing the glacier to slip along a bedrock surface. Lower temperatures freeze the glacier sole to its bed, and no sliding can take place. Even where basal sliding is an important component of flow, clean ice is an ineffective abrasive because it is not hard enough to scratch any of the common rock types. But loaded with rock and mineral fragments, especially such tough varieties as quartzite, granite, and basalt, the sliding base becomes an extremely effective abrasive, like coarse sandpaper.

Bedrock outcrops that have survived contact with sliding glaciers display distinctive scars. *Striations* occur as parallel linear scratches, imparted by the movement of individual mineral fragments or tools embedded in the base of the ice (Figure 7-3). The length of each scratch is a measure of the duration of contact of an individual tool. *Grooves* are features of greater width and depth produced either by larger tools or by concentrated abrasion through time. Regional flow patterns can be reconstructed from careful measurements of these linear tracks. A variety of crescent-shaped marks results from sporadic contact and slip-stick motion with larger rock fragments. The effect of finer abrasion by finely crushed rock "flour" is a smooth, polished finish on striated and grooved bedrock.

Quarrying, or *plucking*, is glacial erosion on a much grander scale. Bedrock weakened by fractures is particularly susceptible. Rock fragments are pulled away by the overriding ice and frozen into the base of the glacier. Generally, this process is most intensive on the downstream side of irregularities on the bedrock surface. One result of differential quarrying accompanied by abrasion is the sculpting of tapered, blunt-nosed hills, called *whalebacks* or *roches moutonnées* (Figure 7-4). Another is the excavation of basins, which eventually may fill with water to become lakes.

Glacial Deposition

The vigor of glacial erosion is also expressed by the volume and character of the sediments that mantle glaciated

Figure 7-4. Streamlined bedrock landforms called whalebacks result from glacial quarrying and abrasion. The glacier that eroded this one in northern Minnesota moved from left to right.

a

b

c

Figure 7-5. (a) Till is debris deposited directly by glaciers. It is characteristically unsorted (that is, without layering) and exhibits a variety of rock compositions and fragment sizes. (b) Outwash displays layering resulting from sorting and deposition by running water. Till and outwash are generally found in close association because the latter is literally washed out of dirty ice margins by meltwater streams. (c) Crossbedding is a common sedimentary structure in outwash. These features were produced by water flowing from left to right.

regions. Typically such deposits display a wide range in size, from boulders to fine rock powder. Angular rock fragments are manufactured by crushing during transportation; contact with other debris in the moving system, or with the bed, results in faceted and scratched surfaces on individual stones. A diverse rock and mineral content reflects the complex composition of the bedrock across which the ice moved.

Till is the name given to unsorted debris deposited directly from glacial ice. Individual rock fragments carried far from their bedrock sources are *erratics*. Meltwater generated along the margins of melting glaciers picks up sediment and redeposits it as sorted beds of sand and gravel called *outwash*. In total aspect, an exposure of till commonly displays an unstratified mixture of large rock fragments held within a matrix of fine sand, silt, and clay (Figure 7-5a). In contrast, outwash is composed of many distinct layers, each containing particles of about the same size smoothed and rounded by transport in running water (Figures 7-5b and 7-5c).

Ensembles of distinctive landforms are constructed of sediments deposited by glacial ice and its meltwater (Figure 7-6). Glacier margins are especially active sites for the accumulation of debris, because there melting is most intense. As a result of meltout at the end of the glacial conveyor belt, considerable volumes of sediment build up to form *end moraines*. These long, narrow bands of till trace the configuration of the ice margin from which the sediment was derived. Till dropped during a glacial advance or deposited by a retreating glacier as a thin blanket marked by low hills and swales is called *ground moraine* (Figure 7-7). Sometimes ground moraine is molded into *drumlins*, streamlined hills with long axes parallel to the direction of ice flow.

Meltwater not only sorts and rounds sediments but also transports glacially derived material far beyond the physical limits of the ice. Streams originating on the glaciers may dump their loads in front of the ice, thereby constructing ramps of sand and gravel in the form of *outwash fans*. Larger systems of braided stream channels deposit ribbons of outwash many kilometers long, filling in old valleys commandeered by the meltwater streams. Streams flowing in tunnels within or beneath the ice eventually plug the openings with sediments. Unmolded by subsequent melting of the confining walls, these tubular outwash bodies remain on the landscape as long, wormlike ridges called *eskers* (Figure 7-8a). *Kames* (Figure 7-8b) are conical hills of outwash originally deposited beneath meltout depressions in stagnant ice.

Accompanying deposition of glacial sediments is the formation of a variety of basins beneath depressions. Such low spots in the land surface are potential sites for the accumulation of water into lakes. Irregular masses of ice buried by debris and subsequently melted produce collapse pits, named *kettles* (Figure 7-8c). Moraines dam natural drainage lines and pond water behind them. Settling and slumping of

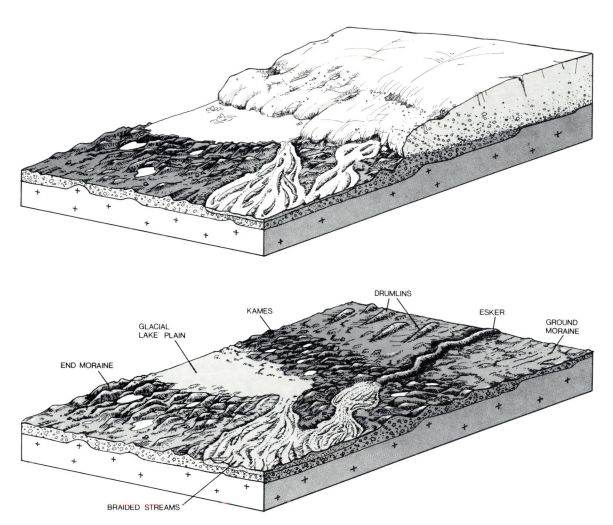

Figure 7-6. These block diagrams portray the various landforms associated with glacier margins. Meltwater produces lakes and braided stream channels. Moraines are built along the margins of glaciers. Eskers are sinuous ridges of sand and gravel marking the former courses of subglacial rivers. Drumlins are streamlined hills formed by actively moving ice.

wet sediments along ice margins also result in a hummocky, pitted topography. Belts of lakes, therefore, typically mark the extent of former glaciers as surely as the sediments themselves. On a highway map of Minnesota, a line can be drawn in the southeastern and southwestern corners of the state that separates areas with and without lakes. That is the approximate boundary of the last glacier to advance across Minnesota.

THE GLACIAL THEORY

It was not until 1837 that the scientific world was presented with the idea that the earth had been gripped in the recent past by a much colder climate, resulting in the expansion of glacial ice. That news came from Switzerland. A young naturalist and teacher at the University of Neuchâtel, Louis Agassiz, first proposed the glacial theory to explain the presence of erratics and striated bedrock in Europe. By 1845, the concept of a great ice age was widely accepted,

Figure 7-7. Low, rolling hills in western Minnesota with intervening shallow lakes and marshland are typical of the ground moraine left behind by the last glacier to cover the area.

101

Figure 7-8. (a) Several landforms, all to be found in or near Glacial Lakes State Park in western Minnesota, are typical of features forming today in the wake of modern retreating glaciers. This is an esker, marking a former subglacial tunnel. (Courtesy of David F. Reid. (b) A kame, unmolded from a filled depression on stagnant ice. (Courtesy of David F. Reid.) (c) A kettle lake, a depression resulting from collapse over a buried, melting ice block. (Courtesy of David F. Reid.)

even by former ardent supporters of the flood theory. Agassiz himself came to the United States in 1846 to accept a professorship at Harvard. He took the opportunity to publicize his ideas to North American geologists, and before long the limits of former ice sheets were being recognized and mapped by the various geological surveys. Agassiz never reached Minnesota, but he did lead an expedition along the Canadian shore of Lake Superior, where he found impressive evidence for his glacial theory.

Many early conclusions relating to the distribution of bouldery sediments had been formed under the theory that the land surface of much of the earth had been submerged and scoured by a great flood, the biblical Deluge. According to that view, the Deluge left a blanket of stony debris, called *diluvium*, in the wake of subsiding currents. Such a Noachian perspective resulted in the consignment of geological features to either antediluvian ("before the flood") processes, to the Deluge itself, or to natural events in the few thousand years since the ark purportedly ran aground. Another theory suggested that icebergs carried by the currents of the great flood, rafted great quantities of bouldery sediment far from their bedrock sources. Subsequent melting of the floating icebergs released the sediment, which accumulated as *drift*, a general term still applied to all glacial deposits, both till and outwash.

At first, all of the glacial sediments and landforms were thought to be the result of one long episode of glaciation, succeeded by the return of the present warm climate. But soon after serious investigations commenced within the framework of the new theory formulated by Agassiz, complicated sequences of more than one deposit of till were discovered. Interlayered with glacial sediment were plant and animal remains, soil, and other evidences of nonglacial conditions. This stratigraphical succession pointed to multiple glaciations. Layers of organic material brought up from water wells dug in the drift were duly noted by most people as a close encounter with the antediluvian world. Only a few geologists recognized the climatic significance of these "remains of Noah's barnyard," as they were commonly called.

A true reading of Minnesota's glacial history did not begin until 1872, when N. H. Winchell became head of the newly founded Minnesota Geological Survey. Given the solid foundation of Agassiz's hypothesis, the task of deciphering the glacial history of Minnesota proceeded briskly. Winchell formed a productive partnership with Warren Upham, a glacial geologist from New England. Together, they delineated the boundaries and documented the activities of several lobes. By 1883, after just a decade of fieldwork, they had mapped the major moraines and integrated them into a grander view of the ice border in North America. Upham's crowning work is a monograph on Glacial Lake Agassiz, published in 1896 by the United States Geological Survey. That lake was an important aftermath of glaciation in the midcontinent. Details of its natural history are given later in this chapter.

Work in other parts of the midcontinent, especially Iowa, Wisconsin, and Illinois, provided more insights into the nature of the climatic changes that dominated the Pleistocene Epoch. Stratigraphic sequences containing multiple till layers indicated four major expansions of glacial ice. Between the tills was evidence of warmer climates in the form of soils, fossils, and sediments of nonglacial origin. Based upon such observations, the Ice Age was subdivided into glaciations and interglacial intervals, each named for a geographic area in the midcontinent. From youngest to oldest, they are:

> Present interglaciation
> Wisconsin glaciation
> Sangamon interglaciation
> Illinoian glaciation
> Yarmouth interglaciation
> Kansan glaciation
> Aftonian interglaciation
> Nebraskan glaciation

Ironically, evidence for a fifth glaciation, the Iowan, originally recognized from deposits in that state, did not survive the scrutiny of modern geologists. And Minnesota, so richly endowed with features of the last advance, received no appellative honors. Although recent studies indicate that the Pleistocene Epoch was even more complicated than suggested by this traditional subdivision, it is still a useful basis for telling the story of the Ice Age.

GLACIAL HISTORY OF MINNESOTA

The dominant effect of climatic change in the midcontinent was the periodic growth and decay of the Laurentide Ice Sheet, a massive glacier centered upon what is now Hudson Bay. Its expansion to maximum size from coalescing ice caps throughout the Canadian Shield resulted in the glaciation not only of what is now Minnesota but of adjacent states as well. At its maximum advance, tongues of ice extended as far south as what are now central Illinois, Missouri, and Kansas (Figure 7-9). Minnesota was almost completely covered by all but the Wisconsin glaciation. A small part of southeastern Minnesota is included in the "driftless area," an area thought by some to have escaped glacial overriding. However, scattered erratics and small patches of drift have been found, indicating that at some time during the Ice Age it too experienced the effects of glaciers.

The diverse bedrock provinces in Minnesota and elsewhere in the Great Lakes region influenced the glacial geology in several important ways. The relative resistance of rock bodies to the effects of weathering and erosion is directly reflected in the patterns of highlands and lowlands that make up the topography developed on exposed bedrock. Topography, in turn, exerted a strong control on the direction of ice flow along the advancing margin of the Laurentide Ice Sheet. Moreover, the contributions of various kinds of bedrock to the glacial load determined the lithologic character of the sediment scattered in the wake of the advancing ice. Following the trail of glacial erratics back to their bedrock sources provides another way, in addition to measuring

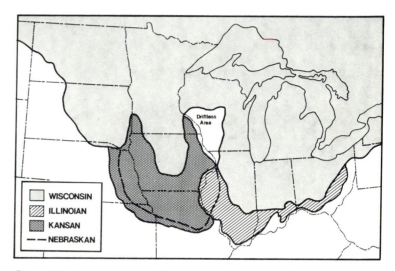

Figure 7-9. The boundaries of the various drift sheets during successive glaciations indicate that glacial ice advanced southward into the midcontinent to roughly the same position four separate times. The most southerly penetration was during the Illinoian.

the direction of striations and grooves, of reconstructing glacier flow paths.

Two long topographic troughs directed the course of ice streams fed by the Laurentide Ice Sheet. The Superior-Minneapolis Lowland lies along the trend of the Late Precambrian rift valley system, floored by lava and later filled with sandstone and shale. Those rather easily erodable sediments were excavated by running water for millions of years prior to the Ice Age. Another lowland straddles the northwestern border of the state and then arches to the southeast to join the other lowland in the Twin Cities. This long topographic depression, now occupied by the Red River and the Minnesota River, is primarily underlain by soft Cretaceous and Paleozoic sediments or by weathered Precambrian rocks. Flanking the lowlands on all sides are high plateaus of more resistant rock types. One plateau, in west-central Minnesota, is capped by up to 150 m of glacial drift, the thickest in the state, accumulated from repeated episodes of deposition by ice flowing out of the bordering lowlands.

The low topographic trends encountered by the Laurentide Ice Sheet along its southern border also exerted important controls on the geometry of the glacier margins. Distinctive ice bulges developed. These bulges or *lobes* in Minnesota are distinguished not only by the kind of drift they left behind but also by such landforms as moraines and drumlins. The Superior and Rainy Lobes advanced periodically from northeastern Minnesota and the Lake Superior Basin. Two glacier tongues from more western sources are the Wadena and Des Moines Lobes.

Because of the striking difference in bedrock lithology between the eastern and western parts of Minnesota and adjacent areas, deposits left by lobes advancing from these regions can be distinguished in the field. The Superior and Rainy Lobes, which crossed the Precambrian terrane of the northeast, produced a bouldery, coarse-textured till of granite, gabbro, basalt, red sandstone, iron-formation, slate, and greenstone. Depending upon the relative amounts of the various rock types, this drift is reddish brown to dark brown or gray-black. The Des Moines Lobe, moving south along the Red River Lowland, was loaded with Paleozoic limestone and Cretaceous shale, along with a large quantity of granite. The resulting till is fine textured, with few boulders, and generally buff to yellow-brown in shallow exposures. In contrast, the till deposited by the Wadena Lobe, though also rich in limestone and granite and of the same color, contains very little Cretaceous shale, indicating a path east of the outcrop area of that rock type. Mixtures of these basic sediment types provide a variety of intermediate drift compositions that require careful analysis before their source can be fixed. Careful mapping and proper interpretation of the various kinds of glacial deposits and landforms are the keys to reconstructing the history of advance and retreat of glaciers in Minnesota during the Pleistocene Epoch. The following summary draws from the work of many glacial geologists, but especially F. Leverett (1932) and H. E. Wright (1972b).

Older Glaciations

The details of glacial activity during the Nebraskan, Kansan, and Illinoian glaciations are largely obscured, either by erosion or burial beneath later deposits. Therefore, studies in Minnesota have not contributed significantly to deciphering the early history of the Ice Age. Erratics of Minnesota bedrock have been found at the very edge of the Kansan drift, near Topeka. With the margins of these older glaciers so far south, ice thicknesses over Minnesota must have been considerable, at least a thousand or so meters.

Both in the southeastern and southwestern corners of the state, beyond the edges of the Wisconsin-age moraines, the topography contains no lakes. Instead, the landscape is one of rolling hills and well-established stream networks. These are the areas of older drift, considerably eroded and displaying numerous bedrock outcrops on the valley sides. Mantling the drift and bedrock is a thin blanket of *loess*, a fine-textured silt, powdery to the touch, deposited by the wind during the last glaciation. Beneath the loess in many

places is a concentration of wind-faceted and polished stones, a lag deposit produced by removal of the finer grain sizes from the older drift sheets by wind. These wind-cut stones are *ventifacts*, and, along with the loess, they indicate that beyond the margins of the last advancing ice sheets, dust-bowl conditions much worse than those of the 1930s ravaged the land surface. In fact, radiocarbon age determinations indicate that such a climatic regime prevailed from approximately 29,000 to 14,500 years ago.

Tills from both northwestern and northeastern sources have been assigned to glacial activities older than the Wisconsin glaciation. However, the bulk of these deposits contain limestone and Cretaceous shale, indicating derivation from bedrock in the area of Winnipeg and the eastern Dakotas. Because of their stratigraphic position beneath younger glacial sediments, these tills were early on referred to as the "old gray drift" and the "old red drift." The former was thought to represent Nebraskan and Kansan glaciations, whereas the latter was assigned to the Illinoian because it lies on top of the "old gray." In fact, it is impossible to make such assignments with any degree of certainty. Some of the old drifts are rich in plant fragments, probably eroded from bogs and forests lying in the path of the advancing glaciers. In recent years, wood samples from several of these deposits have yielded ages greater than 40,000 years. In the jargon of radiocarbon dating techniques, this means that the wood, mainly tamarack and spruce, is too old for the real age to be determined by ordinary methods, yet is at least as old as 40,000 years. Several laboratories have developed techniques for extending the range of radiocarbon age determinations to about 70,000 years, but the reliability of such data is still open to question.

In Iowa, South Dakota, and Nebraska, the discovery of thin layers of volcanic ash interbedded with tills has furnished mineral material that allows the age of the ash to be determined. The oldest deposit of volcanic material gives an age of 1.2 million years, with still older till beneath. All of the tills traditionally assigned to Nebraskan and younger glacial events lie above ash layers that range from 600,000 to 700,000 years old. Therefore, a revised classification is being developed. Unfortunately, no information is available for deposits in Minnesota.

Even though the true ages of these drifts are not known, their antiquity can be appreciated from several other characteristics of their occurrence. Figure 7-10 is a portrayal of the stratigraphic succession along the bluffs of the Minnesota River near Redwood Falls. Two tills lie beneath an old bog deposit dated at greater than 40,000 years. Each represents a glacial advance. The interlayered outwash sands were deposited and soils developed during nonglacial interludes. So intense was the chemical weathering of the second oldest till that even the toughest rock fragments have been rendered to soft clay materials. How long does it take for an ice sheet to form, advance, and then retreat? What length of time is

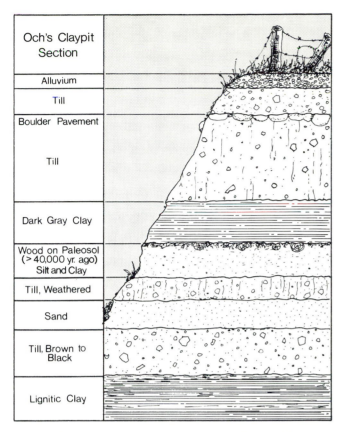

Figure 7-10. This stratigraphic succession near Redwood Falls indicates a complicated history of glacial advances and intervening warmer climates.

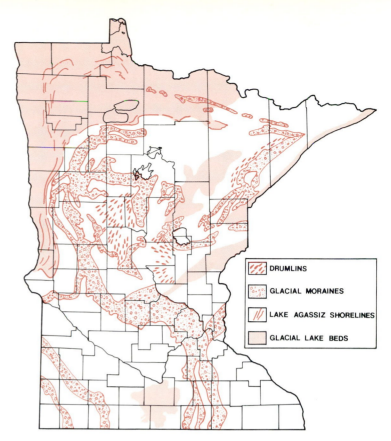

Figure 7-11. Major landforms of the last glaciation. Outside the moraines in the southeast and southwest corners are areas of older drift and bedrock covered by wind-deposited silt (loess).

represented by the establishment of a peat bog and the development of deep soil profiles? How much of the original sedimentary pile has been removed by erosion? No wonder that the early chronology of the Ice Age in Minnesota cannot be stated with certainty.

The Wisconsin Glaciation

Beginning about 75,000 years ago, after a long warm period following the Illinoian glaciation, the climate again deteriorated. The Laurentide Ice Sheet expanded across Canada. During the ensuing millennia, up to about 12,000 years ago, Minnesota was the stage across which tongues of ice advanced and retreated many times. The geological imprint of those episodes during the Wisconsin glaciation remains in the sediments and landforms that make up the surface of the state today (Figure 7-11).

The early history of the Wisconsin glaciation is speculative, because later glacial activity has obscured the record. However, numerous exposures of drift in northwestern and western Minnesota show four tills in superposition (Figure 7-12). They indicate deposition by ice from the Winnipeg Lowland, from the Lake Superior region, and from the eastern Dakotas. The two lowest tills lie beneath horizons dated as older than 40,000 years. The oldest was left by ice

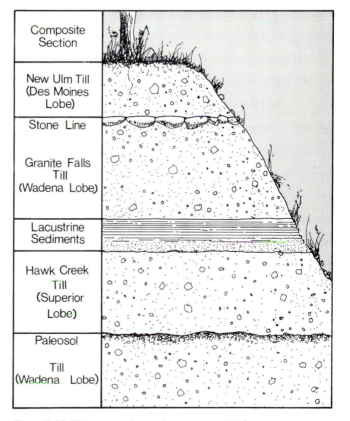

Figure 7-12. This composite section, constructed from exposures along Hawk Creek in Renville County, indicates four glacial advances from different source areas, with intervening episodes of soil formation, lake formation, and erosion.

from the north, the path used by the Wadena Lobe (Figure 7-13, advance 1). The distribution of the second till indicates that a westward advance of ice from northeastern Minnesota reached the Red and Minnesota river valleys and probably the eastern Dakotas (Figure 7-13, advance 2). The timing of this early advance of the Superior and Rainy Lobes is not known for sure, but wood from beneath this till is old beyond the range of radiocarbon dating, greater than 40,000 years. Neither can the exact extent of this advance be fixed. However, from the Lake Superior region to the eastern Dakotas, it left a trail of igneous and metamorphic rock fragments plucked from the Canadian Shield to mark its path. Whatever moraines or other landforms it constructed were largely obliterated by later glacial activity. Small patches of till and gravel in Dakota County near the Twin Cities that lie outside the moraines of Late Wisconsin age may be the only extensive surface exposures of sediments from this advance.

The next important glacial event is represented by a widespread drift sheet composed of sandy till with significant amounts of limestone fragments, apparently derived from Paleozoic bedrock formations in the Winnipeg Lowland. A lack of Cretaceous shale from the eastern Dakotas indicates that the ice entering Minnesota did not originate from that far west. The full sweep of this advance of the Wadena Lobe is still not known, because the drift sheet is buried on all sides by younger glacial sediments. Therefore, the boundaries shown in Figure 7-13, advance 3, are speculative. Till probably deposited at this time is exposed in the open-pit mines of the Iron Range, in the Twin Cities, and along the entire length of the Minnesota River Valley in southern Minnesota. It has been traced across the western borders of the state into the eastern Dakotas. All of the radiocarbon dates indicate that this ice advance occurred sometime before 34,000 years ago. One piece of wood discovered near Mankato enclosed in this till gave an age greater than 72,000 years.

An important readjustment of the margin of the Wadena Lobe is indicated by landforms in west-central Minnesota. Apparently the lobe shrank from its maximum position and then readvanced to construct a massive pile of sediments in the form of the Alexandria Moraine (Figure 7-13, advance 4). Although later overridden by other glaciers, that moraine in large part owes its impressive height and breadth to the activity of the Wadena Lobe. Behind the moraine, ice currents molded the land surface into a swarm of drumlins. About 1,200 individual streamlined hills of glacial sediment fan across Otter Tail, Wadena, and Todd counties. Ranging from a southerly trend to a southwesterly one to a westerly one, the areal pattern represents the radiating flow of the Wadena Lobe in its waning stages. Eventually, flow stopped altogether, and the dirty ice margin slowly collapsed to form a broad, hummocky, lake-dotted moraine.

Some time later, vigorous renewed growth of the Rainy

and Superior Lobes, fed by the shifting centers of snow accumulation upon the Laurentide Ice Sheet in Canada, sent glacial ice once again from northeastern Minnesota and the Lake Superior Basin deep into the heartland of the state. In a remarkable sequence of events, beginning about 30,000 years ago and spanning the next 15,000 years, that large mass of ice advanced, halted, and disintegrated to produce many of the elements still seen in the landscape of central and northeastern Minnesota. The lobes, having coalesced into what was, from a surface perspective, a continuous ice cover, consisted of a thin part covering the highlands west of Lake Superior and a much thicker segment that filled the Lake Superior Basin. Far to the west, this ice field was joined to the more weakly advancing Wadena Lobe (Figure 7-13, advance 5).

The maximum stand of this glacial advance is now marked by a system of looping moraines. The Wadena Lobe segment constructed the Itasca Moraine. The remaining ice margin in Minnesota and Wisconsin left behind the St. Croix Moraine, a broad garland of thick, hummocky, bouldery sediment more than 450 km long. Much of this moraine displays a kame and kettle topography typical of deposition associated with stagnant, dirty glacial ice. Behind the margins of the moraine, the flowing ice constructed several drumlin fields, notably in the vicinity of Brainerd and Pierz. Farther northeast, the Toimi Drumlins mark the direction of the upland-based Rainy Lobe.

Sometime after 20,000 years ago, the glacier-making processes again slowed and, cut off from the supply of ice necessary to replenish that lost to melting, the Rainy-Superior Glacier thinned and stalled in its forward movement. Beneath the starving glacier in its course south of Lake Superior between Duluth and the Twin Cities, a vast plumbing system of tunnels developed to drain away the large volumes of meltwater (see Chapter 12). At first, hydrostatic pressure was so high in the water-filled tunnels that the subglacial streams had sufficient velocity to erode deep valleys into the floor. Eventually the tunnels collapsed, and the waning meltwaters deposited ribbons of sand and gravel in the previously formed valleys. Today, an ensemble of eskers within valleys is all that remains of that old subglacial drainage system.

Oddly out of step with the demise of the Superior Lobe was the development of a vigorously flowing protrusion of the ice sheet into the Red River-Minnesota River Lowland. By 14,000 years ago the leading edge of that glacier had reached a position in central Iowa, earning for it the name Des Moines Lobe. Several offshoots of this long, broad ice mass in the form of sublobes invaded territory just abandoned by the Superior Lobe (Figure 7-13, advance 6). The Grantsburg Sublobe moved in a northeasterly direction downhill into the Twin Cities Basin, where it breached a long segment of the St. Croix Moraine. Its northern margin blocked drainage flowing southward, and Glacial Lake

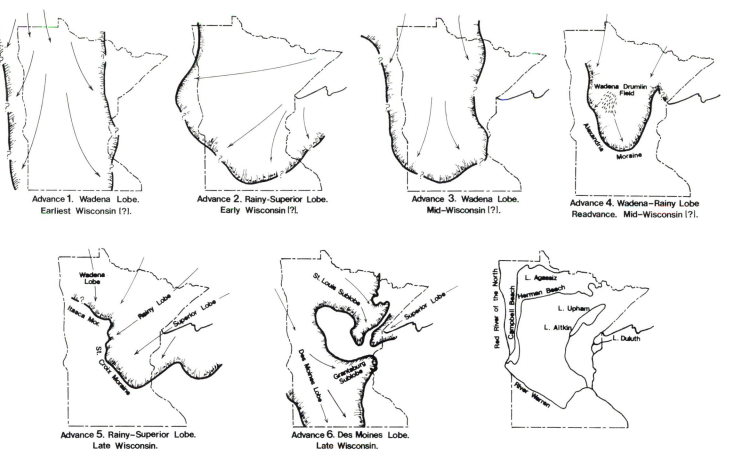

Advance 1. Wadena Lobe. Earliest Wisconsin (?).

Advance 2. Rainy-Superior Lobe. Early Wisconsin (?).

Advance 3. Wadena Lobe. Mid-Wisconsin (?).

Advance 4. Wadena-Rainy Lobe Readvance. Mid-Wisconsin (?).

Advance 5. Rainy-Superior Lobe. Late Wisconsin.

Advance 6. Des Moines Lobe. Late Wisconsin.

Figure 7-13. Summary of ice activity during the Wisconsin glaciation as reconstructed from the distribution of glacial sediments and landforms. Dates are approximate. (After Wright, 1972b.)

Grantsburg rose behind the ice dam (Figure 7-14). Somewhat later, perhaps 12,000 years ago, the St. Louis Sublobe probed eastward almost into the Lake Superior Basin. In front of that retreating ice margin, about 11,600 years ago, water became ponded to form Glacial Lakes Upham and Aitkin.

At 14,000 years ago, active flow in the Superior Basin was primarily restricted to the main glacier mass north of the Canadian border, with perhaps a reservoir of ice in the Lake Superior Lowland. That basin, because of its great depth and therefore large storage capacity, was to become the staging ground for several more ice invasions of short duration and extent by the Superior Lobe. In a series of pulses, ice overflowed the rims of the basin, first to a stand at the Highland and Mille Lacs Moraines (Figure 7-15). A weaker advance of the Rainy Lobe on the upland brought that glacier to what is now the Vermilion Moraine. Several final pulses followed before full retreat.

The disintegration of the Des Moines Lobe left important imprints upon the land surface of Minnesota as well. Its former extent is everywhere marked by a distinctive till, fine-textured and containing a high volume of Paleozoic limestone and Cretaceous shale fragments. This drift, along

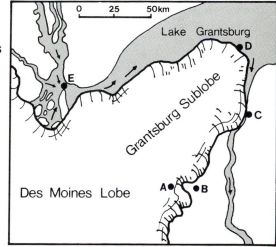

A● Minneapolis
B● St. Paul
C● St. Croix Falls
D● Grantsburg
E● St. Cloud

Figure 7-14. Glacial Lake Grantsburg resulted when the Grantsburg Sublobe blocked the southward flow of drainage into the Mississippi and St. Croix river systems, about 16,000 years ago. (After Cooper, 1935.)

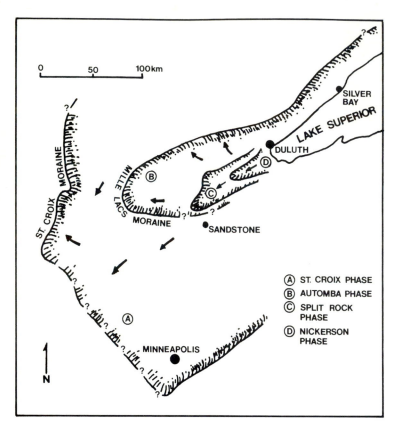

Figure 7-15. Late Wisconsin phases of the Superior Lobe are marked by nested moraines, showing waning vigor in a sequence of advances from the Lake Superior region. (After Wright, 1973.)

with the loess swept from its surface by wind, is the parent material for most of the soils of the western and southern parts of the state. Several moraines were constructed by the lobe (Figure 7-16). The Bemis Moraine marks a standstill near its maximum development, and the Altamont Moraine, an important halt during its general retreat. By 12,000 years ago the Des Moines Lobe had melted back into the Red River Lowland. With its final retreat, the glacial ordeal was over, and a new landscape emerged from the cover of ice to become part of the familiar face of present-day Minnesota.

AFTERMATH

The destruction of the Laurentide Ice Sheet was decreed by a simple meteorological change, a rise in the mean annual temperature of the earth's atmosphere beginning about 13,000 years ago. In response to that warming trend, more glacial ice was converted to water than was replenished by snowfall. At first, the end of the Ice Age was most apparent in eastern North America. There, rapidly shrinking glacier margins along the continental shelf were lapped by rising ocean waters. Worldwide, the return of water to the ocean reservoirs from the glacial deep freeze would eventually

cause sea level to rise about 130 m. In Minnesota, as well as in other areas in the middle of the continent, the retreat of the Laurentide Ice Sheet was accompanied by another kind of flooding. Meltwater, coursing from the rapidly downwasting glacier surfaces, flowed across the land in an intricate system of streams and rivers. Thousands of natural basins, ranging in size from a few hectares to tens of thousands of square kilometers, and uncovered in the wake of the melting ice, filled to overflowing as well. Slowly, a new landscape evolved as the warmer climatic regime persisted in imprinting its own control on the surface environment.

Glacial Retreat

Ice margins retreat in two different ways. Neither involves a reversal of flow, despite the implications of the word "retreat." In one situation, glacial flow continues to deliver

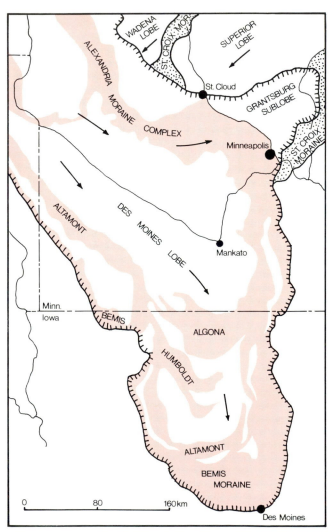

Figure 7-16. Moraines of the Late Wisconsin Des Moines Lobe. Drift from this ice sheet contains limestone fragments from Winnipeg and shale from the eastern Dakotas.

ice and rock debris to the terminus. However, since more is lost to evaporation and melting than is replenished by forward movement, the boundaries of the glacier recede. Any temporary halt in the general deterioration of the ice margin may be marked by the construction of a significant pile of sediments melted out from the glacial conveyor belt. These *recessional moraines* are useful in documenting the history of deglaciation in a region.

A second style of glacial waning involves the stalling of flow to the point where movement ceases throughout a variable width of ice in the terminal zone. Such wholesale stagnation is followed by the slow melting of the dead ice, a process that produces a distinctive suite of landforms. Sediments that melt out and accumulate on the surface of the downwasting glacial ice provide an insulating blanket, and the ice beneath can persist for thousands of years. These ice-cored terranes evolve underground drainage and cavern systems, similar to those produced by the solution of limestone. Eventually, a landscape of hills and kettle lakes emerges as the buried ice is finally melted and the collapse is complete.

The retreat of the Superior and Des Moines Lobes across Minnesota was swift. From its maximum expansion to a terminal position in central Iowa about 14,000 years ago, active ice of the Des Moines Lobe disappeared from the state in just 2,000 years or so. From a geological point of view that is a very short time. By 10,000 years ago the Superior Lobe had vacated its basin and the Great Lakes region was entirely free of ice. The thick midsection of the Laurentide Ice Sheet was gutted by iceberg *calving* in Hudson Bay by 8,000 years ago. In that process, large volumes of ice were detached along the floating margin and drifted out to sea. A small ice cap on Baffin Island is the largest remnant of that once formidable glacier. Greenland has always supported a separate ice sheet.

Glacial Lakes

As the glacial ice thinned and melted back, the deployment of meltwater and runoff from precipitation was influenced more and more by the emerging drift-covered bedrock topography. At first the retreating ice margins acted as dams, and in conjunction with moraines and other topographic barriers they helped to pond considerable amounts of water. None of these glacial lakes lasted for more than a few thousand years; however, their vast extent influenced the early postglacial environment profoundly.

The largest of the glacial lakes, Lake Agassiz, came into existence when the Des Moines Lobe melted back across a topographic divide near Browns Valley, which now separates north-flowing drainage in the Red River Basin from the south-flowing Minnesota River. Even before that, water became ponded in back of the Bemis Moraine as the southwestern side of the Des Moines Lobe melted down into the Minnesota River Lowland. Other lakes, collectively called

Lake Minnesota, formed in the lowland itself as the ice retreated northward. North of the Twin Cities, Lake Grantsburg received water and sediments from south-flowing meltwater streams blocked by an eastern extension of the Des Moines Lobe. In the northeast, Lakes Aitkin and Upham rose in front of the deteriorating St. Louis Sublobe, and Glacial Lake Duluth was trapped between the rim of the Superior Basin and the Superior Lobe itself. Dozens of smaller glacial lakes contributed to the general flooding of the landscape.

Evidence for Glacial Lake Agassiz in the form of shoreline features and sediments accumulated beneath its waters has been found throughout an area of about 320,000 km^2 (Figure 7-17). Expansion of the lake began by at least 12,000 years ago, when the Des Moines Lobe retreated northward into the Red River Basin. Meltwater was trapped between the ice margin and a higher topographic rim, a recessional moraine, forming the southern boundary of the Red River Lowland. Various stages of lake level are recorded in a series of beaches, some of which can be traced continuously for hundreds of kilometers. They indicate that Lake Agassiz fluctuated significantly in size and depth, with a

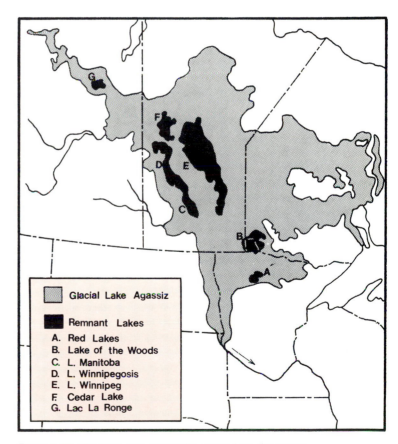

Glacial Lake Agassiz

Remnant Lakes

A. Red Lakes
B. Lake of the Woods
C. L. Manitoba
D. L. Winnipegosis
E. L. Winnipeg
F. Cedar Lake
G. Lac La Ronge

Figure 7-17. The total area eventually covered by the waters of Lake Agassiz. The lake was never this large at any one time. Darker patterns indicate remnant lakes still occupying parts of the basin. (After Elson, 1967.)

surface area at any one time not exceeding 128,000 km². In places it was more than 120 m deep.

At its highest level, the lake overtopped a moraine dam near what is now Browns Valley, and an outlet river became established. This Glacial River Warren eventually excavated the valley now occupied by the Minnesota River. For several thousand years the southern outlet was an important control on the level of Lake Agassiz. Eventually, the retreating ice margin at the north uncovered lower outlets, and the lake shrank to such a low elevation that River Warren ceased to flow. In its place, the Minnesota River became established.

One interesting aspect of the Lake Agassiz beaches as they lie on the modern landscape is that they rise in elevation to the north. Originally each continuous beach ridge bounded the flat surface of a particular lake level. Since then, they have been tilted as much as 0.6 m per kilometer. This vertical uplift is thought to represent a slow rebound of the earth's crust in response to glacial unloading. The amount of postglacial rebound is directly related to the original thickness of the ice. High ice loads resulted in a greater depression of the crust and therefore a larger vertical recovery after the ice was removed.

In the Lake Superior Basin, water became ponded in front of the retreating Superior Lobe, and a series of glacial lakes preceded the final development of Lake Superior. The western end of the basin was uncovered first, about 12,000 years ago, and Glacial Lake Duluth is the name given to a body of water that stood variously at elevations between 323 and 335 m. That lake found outlets in Minnesota through the Moose River and in Wisconsin through the Brule River, eventually to discharge via the St. Croix River Valley into the Mississippi drainage basin (Figure 7-18). From that high level, the ancestral Lake Superior dropped to a stage more than 60 m lower than the present lake, as further retreat of the ice uncovered lower outlets in the vicinity of Sault Ste. Marie. At that time the bedrock sill controlling the level of the outlet river was depressed more than 60 m

below its present altitude by the weight of glacial ice. Slow rebound of the crust was accompanied by a rise in the level of the lake to near its present height by about 4,000 years ago. All of the present shoreline features date from that time.

Rivers

Accompanying the filling of lake basins was the establishment of drainage lines to accommodate surface runoff generated by the melting ice. Sediment-rich meltwater streams constructed broad plains of sand and gravel in front of the retreating ice, filling older valleys with thick deposits of outwash. Spillover from the larger glacial lakes, sometimes of catastrophic proportion, eroded deep gorges along the courses of the outlet streams. Some of those valleys today are occupied by rivers of much smaller discharge, so-called *underfit streams*. The Minnesota River is a classic example.

Rivers are dynamic physical systems that extend themselves across the landscape into upland areas, sometimes capturing drainage from other basins (Figure 7-19). They have a natural tendency to continue developing into longer and more intricate networks of branching open channels until the entire landscape is most effectively served by their drainage service. The plumbing system that carried water away from the margins of the last glaciers was modified to suit the new hydrologic cycle, one in which glacial meltwater was not a source of discharge. Eventually, thousands of kilometers of new stream courses were imprinted upon the land surface to await human recognition with such names as Mississippi, Rum, St. Croix, and Crow.

The response of rivers to changing conditions is clearly documented in the valleys of the Minnesota River and the Mississippi River in the vicinity of the Twin Cities. The broad valley of the Mississippi downstream from Fort Snelling is in sharp contrast to the narrow gorge that separates Minneapolis from St. Paul upstream from that point. Disregarding the present volumes of water flowing in each river, the natural extension of the Mississippi River Valley is along the course of the Minnesota River, the path of spillover from Lake Agassiz. In both valleys, the modern floodplains are enclosed by steplike surfaces at various elevations above the valley floors. These river terraces are former floodplains and channel bottoms constructed when the rivers flowed at those higher levels. They were left high and dry as the result of active downcutting, during episodes of vigorous erosion of channel bottoms.

A long time ago, geologists recognized that rivers tended to develop stability over the long run, in response to the many variables that make up a stream system. J. H. Mackin defined the tendency for balance in this way:

A graded river is one in which, over a period of years, slope is delicately adjusted to provide, with available discharge and prevailing channel characteristics, just the velocity required for the transportation of the load supplied from the drainage basin. The graded stream is a

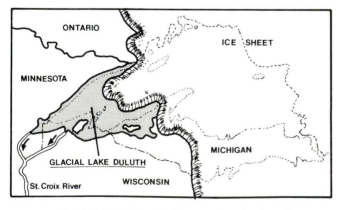

Figure 7-18. Glacial Lake Duluth formed when the Superior Lobe melted back to expose the Lake Superior Basin. It lasted only a few thousand years. (After Green, 1978.)

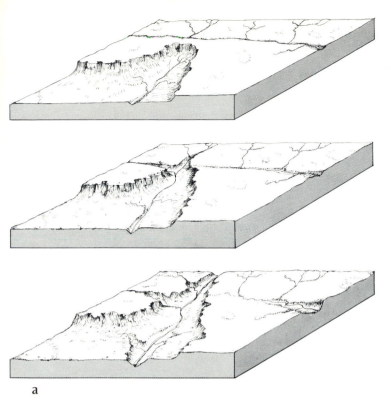

a

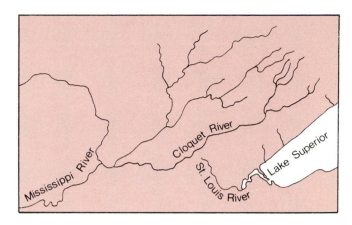

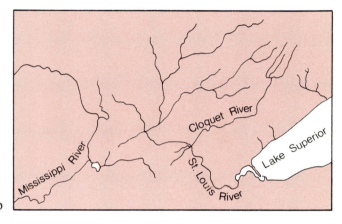

b

Figure 7-19. (a) The extension of a stream valley upriver by *headward erosion* lengthens the stream's course and sometimes results in the capture of drainage from other systems. (b) *Piracy* of a considerable system of tributaries from the Mississippi River by the St. Louis River resulted in significant drainage changes. (After Martin, 1911.)

system in equilibrium; its diagnostic characteristic is that any change in any of the controlling factors will cause a displacement of the equilibrium in a direction that will tend to absorb the effect of the change (1948: 471).

Major disruptions in the balance of a river can be perpetrated by such things as large-scale changes in water volume, sediment load, and base level, that lowest level to which the river can flow. Ultimately, base level for most large rivers is the sea level of the time. But over the short run, base level for every tributary stream is the level of the river into which it flows.

Water volume, sediment load, geological characteristics of the drainage basin, and base level are independent variables; the stream has no direct control over their magnitude or character. It must simply respond to the runoff furnished by the hydrologic cycle in its basin and the amount of sediment delivered to it by slope processes on the valley sides. Eventually, the river is braked by the level of the ocean, below which it cannot erode, nor above which it cannot significantly deposit at its mouth. Dependent variables are such things as the size and shape of the channel, the average drop in elevation along a given stretch of river, called the *gradient* or *slope*, and the mean velocity of the running water. These characteristics are determined by the independent variables.

The volume of water coursing through a particular section of stream channel in a given period of time is called *discharge*. Fluctuations in discharge are readily discernible in a stream channel because they are reflected in the level of the water above the channel bottom. These levels, or stages, generally fluctuate seasonally in response to an irregular supply of runoff from the drainage basin. A simple summary of the volume of water flowing through a channel at a particular place during a specific period of time is a *hydrograph*. Figure 7-20 shows the seasonal changes in discharge for three small Minnesota streams, indicating climatic variations in their drainage basins.

Besides carrying away runoff, natural channels also flush away rock and mineral debris in the form of sediment. A key control on the amount and size of particles that a river can transport is the average velocity of the running water. The maximum volume of a particular size of particle that a stream can carry at any one time is termed *capacity*, and the largest size, expressed as a diameter, is the stream's *competence*. Simply stated, the faster a river flows, the more work it can accomplish in terms of sediment transport. This is because faster flow means more energy. What determines the velocity? Two factors are the amount of discharge and the steepness of the slope. Given a volume of water, the steeper the slope, the faster the velocity. But what determines the stream's slope? The answer is the river itself, and

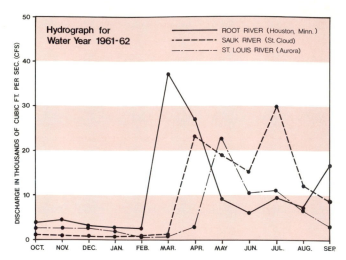

Figure 7-20. These hydrographs show changes in discharge through one year for three streams in Minnesota. Notice the lag in peak runoff from south to north as winter gives way to spring.

that is the key to understanding Mackin's concept of a graded stream.

When a river is presented a volume of sediment that is too large for it to carry away, the material is deposited. Eventually, deposition raises the bottom of the channel, giving the river a higher elevation at that point and thus increasing the gradient. The higher the slope, the greater the velocity generated on that slope. The faster the water flows, the larger the competence of the stream. As deposition continues to elevate the streambed, the time finally comes when sufficient velocity is generated to carry even the largest particles away. At that point the river is graded. Deposition ceases, and the slope remains constant. Thus, the stream is an adjusted conveyor belt continually carrying debris washed into it from the valley sides while maintaining a stable channel gradient.

The river will remain graded until something changes, and then the whole system will respond in such a way as to overcome the effects of change. Evidence for profound changes in the natural history of the Mississippi and Minnesota rivers is found all along their courses. Some of the details already told, such as the floods of Glacial River Warren, are based on observations of the sediments and valley characteristics of those rivers. The following story links a series of events, from about 20,000 years ago to the present, into an explanation of the features found in the valleys today. It is presented as an unusually clear example of the response of rivers to change.

A few miles downstream from St. Paul, near Pine Bend, prominent river terraces lie at elevations of about 250 and 265 m above sea level. The modern floodplain occurs here at about 214 m. Beneath the terrace surfaces, a little digging reveals that thick sequences of sand and gravel lie on top of

the bedrock sides and floor of an older, probably preglacial, valley. Borings for Lock and Dam No. 2 at Hastings revealed further details of the valley there, and Figure 7-21, stage 5, is a summary of the composition and configuration of the various deposits. It is obvious that the river has entrenched, filled, and then cut down again several times. These cycles of cutting and filling are a reflection of the river's response to change, beginning with the formation of a bedrock valley in preglacial times (Figure 7-21, stage 1).

The bouldery gravel underlying the highest terrace and making up a significant volume of the total valley fill is directly traceable to the St. Croix Moraine, a deposit of Late Wisconsin age marking the edge of the last advance of the Superior Lobe into the Twin Cities area (see Figure 7-13, advance 5). Rock types in the terrace gravel match closely those found in the moraine. Figure 7-21, stage 2, recreates the probable origin of this deposit. Meltwaters from the wasting Superior Lobe found their way into the deep bedrock-walled valley of the Mississippi River. Unable to carry the enormous volume of coarse debris washing in from the dirty ice margin, the river began to deposit sand and gravel. Eventually the upper part of the valley became completely clogged with sediment, and the river developed a braided network of channels. Finer sediment was deposited downstream. The net effect on the entire upper Mississippi River Valley was to increase the slope.

Retreat of the Superior Lobe and expansion of the drainage basin resulted in the concentration of runoff into one or two major channels. Much of the water was probably from the overflow of glacial lakes and swamps and therefore barren of sediments, since most were trapped in the lake basins. In response to these new conditions, the river had sufficient energy to excavate the coarse debris previously dumped. That interval of downcutting left remnants of the previous valley fill as terraces (Figure 7-21, stage 3).

Even before all of the outwash from the Superior Lobe had been excavated from the valley, renewed glacial activity in the form of an advance of the Des Moines Lobe from the west had begun. Although ice from this lobe did not reach the Mississippi River Valley at what is today Pine Bend, it effected great changes in the river's behavior. At its maximum stage, a tongue of ice from this advance, the Grantsburg Sublobe, extended as far north as what is now Pine City and eastward to what is now Grantsburg, Wisconsin, blocking the drainage of the Mississippi and St. Croix rivers to produce Glacial Lake Grantsburg, whose outlet was the St. Croix River (See Figure 7-14). The sequence of events following the melting of this glacier was carefully worked out by W. S. Cooper (1935) in his explanation of the formation of the extensive Anoka Sand Plain.

At Pine Bend, the Mississippi, cut off from the surface runoff of its drainage basin by the intrusion of the Grantsburg Sublobe, lost its competence to excavate further, and downcutting ceased. Again, sediment-charged meltwater found its way into the valley, but this time the debris was not so coarse grained and the rock types were different. Deposition began and braided channels again developed (Figure 7-21, stage 3). In time, the river's gradient again was constructed to such a position that it was able to carry the outwash from the margin of the Des Moines Lobe far downstream. But the fill was not thick enough to overtop the high terrace associated with the activity of the Superior Lobe.

Deposition ceased in response to the retreat of the Des Moines Lobe shortly after 14,000 years ago, and another period of slow trenching occurred. An extensive terrace at about 245 m was stranded at that time. Final clearance of ice from the Minnesota River Valley set the stage for the development of Lake Agassiz and the drainage of that large glacial lake by River Warren.

River Warren was a high-volume stream competent enough to erode a deep trench across the entire state. The waters of that flood entered the Mississippi River Valley near Fort Snelling, cutting deeply into the glacial gravels and the Paleozoic bedrock as well (Figure 7-21, stage 4). The river segment above Fort Snelling was left hanging, and a waterfall developed. That waterfall eventually eroded the deep gorge of the Mississippi River in the Twin Cities as it retreated upstream to its present position as St. Anthony Falls in downtown Minneapolis (Figure 7-22). A calculation based on the rate of retreat of that falls was the first determination of absolute time of the end of the Ice Age in Minnesota.

The excavation by River Warren produced a deep, low-gradient valley sufficient to carry the high volume of spillover as long as Lake Agassiz was in existence. But that large body of water was eventually drained away, and by 9,000 years ago River Warren had been reduced to a trickle. Into the wide, deep valley, tributaries continued to bring sediment that the major stream, now the Minnesota and Mississippi rivers, could not carry away. At each point where major tributaries entered, natural sediment dams built up to pond the upstream segments of the valley. In this way a series of river lakes was formed, the largest being Lake Pepin (Figure 7-23). Thus began another episode of valley filling.

Lake Pepin once extended as far upstream as St. Paul, a reflection of the large volume of sediment washed into the valley by the Chippewa River near Wabasha and the low gradient of River Warren. A delta at the head of that early Lake Pepin received sediment from the Minnesota and upper Mississippi rivers. In the 10,000 or so years of its existence, the delta has advance downstream by continuous addition of sediment to its present position near Red Wing, thereby shortening the length of the lake by almost 80 km. In time,

Figure 7-21. This sequence of cross sections, reconstructed from topography and exposures on the Mississippi River near Pine Bend, outlines a sequence of events from about 16,000 years ago to the present.

113

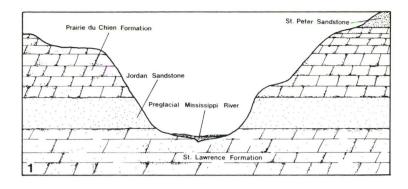

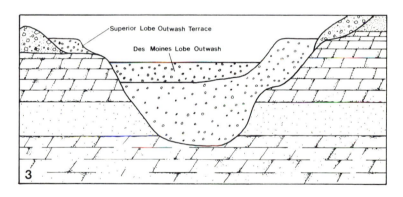

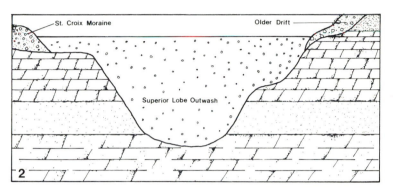

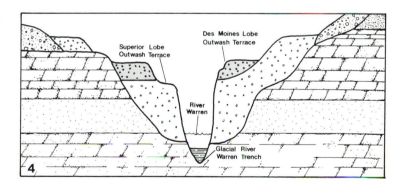

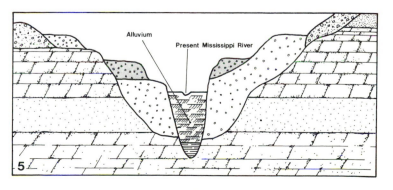

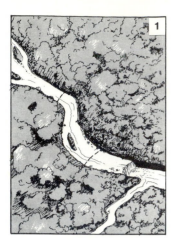
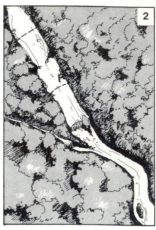

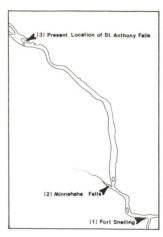

a

b

Figure 7-22. (a) St. Anthony Falls originated near Fort Snelling when River Warren undercut the Mississippi River at that point. Its gradual retreat approximately 11 km upstream left an impressive gorge. (b) This engraving of the falls was made from a 1766 sketch by Jonathan Carver. (Courtesy of the Minnesota Historical Society.)

Lake Pepin will be completely filled, and the effect of the Chippewa River fan dam will be negated. The broad floodplain of the Mississippi River above Red Wing is underlain by a thick fill of deltaic sediments (see Figure 7-21, stage 5).

One side effect of all the episodes of valley filling was the damming of tributary valleys. Many of these became temporary lakes, but rapid infilling quickly destroyed them. One prominent survivor is Lake St. Croix, whose deep bedrock basin is the lower segment of the St. Croix River, excavated in part by spillover from Glacial Lakes Duluth and Grantsburg. As the early Lake Pepin delta migrated past the mouth of that valley, sediment partially dammed the river. Point Douglas is the name of that barrier today. Valley filling continues to the present, some of it the result of dams constructed by human geologic agents.

Vegetation

As the climate cooled during major glaciations, vegetation was particularly affected. Long before glacier lobes advanced into what is now Minnesota, profound changes in the distribution of plant communities had already been accomplished. A southward shift in the tree line was accompanied by the

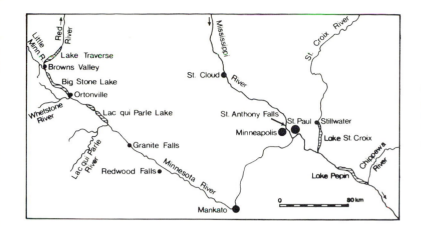

Figure 7-23. A series of natural dams was constructed by the tributaries along the course abandoned by River Warren, resulting in the formation of a series of river lakes in the valley. Lake St. Croix backed up behind sediment deposited as a delta in early Lake Pepin, which had extended to St. Paul.

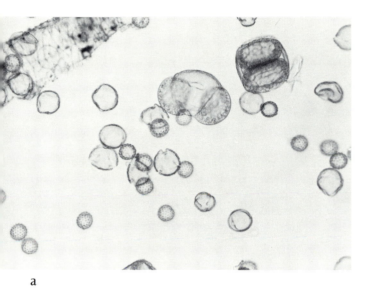

a

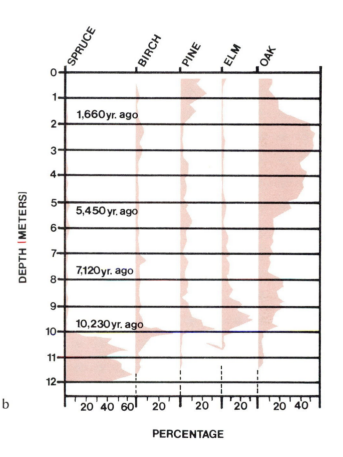

b

PERCENTAGE

Figure 7-24. (a) Pine, oak, ragweed, and sour dock pollen, typical of material accumulating today in small lakes in southern Minnesota. These spherical, spiny ragweed grains are about 19 μm in diameter. (Photo by L. J. Maher, Jr.) (b) This partial pollen diagram shows the abundance of pollen for several tree species in bog sediments near St. Paul. The decrease in spruce pollen and increase in oak pollen through time is interpreted as a change from cool and moist to warmer and drier climate. (After Winter, 1961.)

establishment of tundra conditions, including permanently frozen ground and intense frost action north of the tree line. Advancing ice eventually buried the already frozen and barren landscape.

In response to climatic warming, plants reinvaded the newly deglaciated land surface, slowly establishing prairies and forests in equilibrium with the climatic conditions and landscape features of the time. The march of vegetation was accompanied by animal migrations as well. The fossil record contains a wonderful history of climatic change.

Sediments that accumulate in bogs and lakes generally contain a history of plant succession in the form of pollen and other plant parts, including seeds, woody fragments, and leaves. Most ponds and lakes in glaciated areas originated as kettles or basins resulting from direct glacial erosion or deposition. Therefore, they were on hand to act as traps for fossils during the passage of successive waves of plant immigrants in the wake of retreating ice. Cores of sediment are taken from lakes and bogs, and the pollen record is then transcribed into a history of climatic change (Figure 7-24).

Studies in Minnesota instigated by H. E. Wright (1969) indicate that tundra conditions prevailed in central Minnesota from about 20,000 to 14,000 years ago. Tundra persisted in northeastern Minnesota a few thousand years longer. Although no direct evidence of such conditions has yet been found in the southern half of the state, presumably there also cold-climate vegetation dominated the land surface periodically. Pollen diagrams from southern Minnesota show that spruce was the dominant tree type for at least 3,000 years following the demise of the Des Moines Lobe; then birch and alder began to replace the spruce. In the Twin Cities area that change took place about 11,000 years ago. About 10,000 years ago jack pine and red pine established a foothold in the state, and from that time on pine quickly became the dominant forest type in the north.

As the climate grew still warmer, southern Minnesota's birch forests were replaced by trees typical of a more temperate climate, such as elm and oak. Eventually, the trees declined altogether and prairie succeeded the forest. By 7,200 years ago, some prairie aspects reached almost as far as Duluth. At about this same time, white pine had reached northern Minnesota from the east.

The pollen record clearly shows a reversal in climate, beginning about 7,000 years ago, that resulted in a slow retreat of prairie and deciduous forest from the north. As noted by Wright, "We now enjoy a climate that is probably not unlike that of 9,000 years ago, only a short time after the ice sheet left the Lake Superior region" (1969: 405). This return to cooler conditions was accompanied by a prolific development of bog vegetation, which has resulted in the accumulation of enormous volumes of peat.

Ecological disturbance of another kind is also contained in the pollen record. The influence of human occupation, especially since European settlement of Minnesota, is marked by a decline of tree pollen in response to deforestation and an increase in grass and weed pollen. The clearing of the forests and expansion of agriculture has significantly changed the composition of pollen presently accumulating in Minnesota's lakes. Especially noticeable is the increasing abundance of *Ambrosia* or ragweed (Figure 7-25). In fact, palynologists can fairly well trace the European migration from Maine to Minnesota by dating the rising percentage of this pollen type, which is generally very dramatic in the uppermost sediment of most lake cores. Such increases are noted at about 1775 in Maine, about 1830 in lower Michigan, and approximately 1890 in Ely and Duluth. Accompanying the increase in ragweed is the decline in abundance of pollen of pine, widely used for lumber or cleared for farming.

Animal Life

Evolution, migration, and extinction of species all contributed to a pattern of constant change in the composition of land animals in North America during the Pleistocene Epoch. Certain evolutionary trends seem to have been triggered by the colder climate. Because larger animals lose body heat more slowly—a condition resulting from the smaller ratio between surface area and volume—big became more than just beautiful as the Ice Age intensified. Giant-sized beavers, bears, elephants, and rhinoceroses flourished. Other adaptations included the development of wool for thermal insulation in such animals as the woolly mammoth

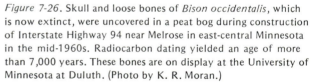

Figure 7-26. Skull and loose bones of *Bison occidentalis*, which is now extinct, were uncovered in a peat bog during construction of Interstate Highway 94 near Melrose in east-central Minnesota in the mid-1960s. Radiocarbon dating yielded an age of more than 7,000 years. These bones are on display at the University of Minnesota at Duluth. (Photo by K. R. Moran.)

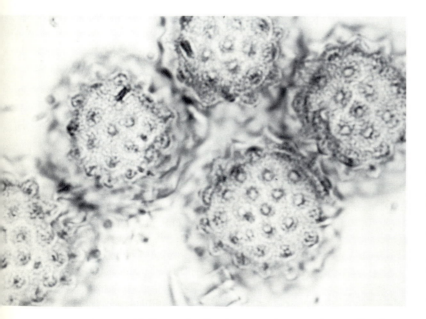

Figure 7-25. Ragweed pollen (*Ambrosia*) is a hay fever sufferer's bane but a significant marker in the pollen record, indicating human disturbance of the environment. (Photo by L. J. Maher, Jr.)

Figure 7-27. Browsing herbivores along the margins of the retreating glaciers in what is now Minnesota, about 13,000 years ago, included the mammoth, mastodon, musk-ox, and giant beaver. Also abundant were bison, elk, moose, and reindeer.

and the woolly rhinoceros and the growth of blubber in polar bears as a source of energy in winter.

Compared to that in other areas, the fossil record in Minnesota is extremely poor for reconstructing the life of the past few million years: glaciation has severely limited the chances for animal remains to be preserved and left undisturbed. But fossil finds elsewhere, especially in areas that were never glaciated, reveal that North America supported a wonderful menagerie of animals, many of them now extinct. From the famous asphalt deposits in Los Angeles, called Rancho La Brea, more than 4,000 individual mammals have so far been excavated. Certain caves in the Grand Canyon contain mummified remains of giant sloths and dung so well preserved that it still has a manurelike smell. Bog deposits in the Great Lakes states indicate that many large animals, including bison, mammoths, and mastodons, migrated northward as the ice sheets retreated (Figure 7-26).

At the beginning of the Nebraskan glaciation, the larger mammals included at least three varieties of mastodon, several species of saber-toothed cat, horses, camels, ground sloths, llamas, bear, pronghorn antelope, and jaguars. Many smaller animals, such as the mouse, fox, skunk, otter, raccoon, and mink are still common today.

The Middle Pleistocene was a time of intercontinental migration. Low sea levels during the Kansan glaciation allowed free interchange across land bridges into North America from both Eurasia and South America. Lumbering mammoths found their way into Alaska across the Bering Strait land bridge, along with antelope, musk-ox, and some carnivorous predators, including several species of saber-toothed cat. Among the returning immigrants was the horse *Equus*, whose ancestors had evolved during the Tertiary in North America.

The Illinoian and Wisconsin faunas included the ground sloth, mastodon, mammoth, llama, camel, horse, giant beaver, giant short-faced bear, giant armadillo, saber-toothed cat, woodland musk-ox, and bison and many smaller animals. Figure 7-27 is a portrayal of a scene about 13,00 years ago in Minnesota. Among the remains of some of these animals are stone and bone tools, indicating that human beings were also a part of the environment.

The disappearance of animal species through extinction is a common event according to the recorded histories of the past several hundred years. Human predation and environmental disturbance directly relatable to human activity are the most often cited causes. Yet the decimation of wild animal populations during the last thousand or so years does not compare in magnitude to the extinctions that accompanied the end of the last glaciation. Beginning about 11,000

Figure 7-28. Hunters drive a herd of bison over a steep quartzite cliff in what is now Blue Mounds State Park, near Luverne. This hypothetical scene illustrates hunting skills that may have contributed to the extinction of many large animals at the end of the last glaciation.

years ago, major extinctions occurred among the large land-based animals of North America. Within a short period of time, perhaps as little as 2,000 years, all of the elephants, horses, ground sloths, saber-toothed cats, and camels and various species of bison had disappeared. More than 200 genera were affected. This zoological catastrophe is one of the great mysteries of the Ice Age.

Two conditions have been cited as possible contributors to the demise of so many large species: climatic change with accompanying habitat disturbance and the development of hunting skills by a burgeoning population of human predators. The coincidence of the appearance in North America of skilled hunters with the disappearance of big-game animals supports the latter hypothesis. Evidence exists that hunting techniques included stampede and fire drive (Figure 7-28). Many radiocarbon dates of kill sites in the Great Plains indicate the presence of big-game hunters with sophisticated stone weaponry between 13,000 and 11,000 years ago. Evidence for early hunters in what is now Minnesota is found mainly as stone tools associated with bison bones. One site near Lake Itasca contains the remains of at least 16 bison that were killed and butchered between 8,000 and 7,000 years ago.

Whatever the cause of the many extinctions, the result, in the words of A. L. Wallace, is "a zoologically impover-ished world, from which all the hugest, and fiercest, and strangest forms have recently disappeared" (1876: 400).

Lakes and Bogs

A lake is a body of relatively still, generally fresh water that has been trapped in a natural basin or impounded behind some kind of barrier or dam. By its very nature, it is one of the most fragile elements of the surface environment. For a lake to persist, there must be a sufficient flow of water into the basin to replace losses from evaporation, spillover, and seepage. Furthermore, the basin containing it must have the capacity to accommodate the storage of whatever sediments accumulate beneath its waters, without completely filling. Finally, the lake basin must maintain its closure against the downcutting of outlet streams. The fact is that not many lakes are long-lived, geologically speaking, and Minnesota's landscape will eventually dry out as sediment infilling and stream development continue to deplete the total volume of lake basins.

The importance of past glacial activities in the origin of most of the state's lakes has already been outlined. Erosion by abrasion and quarrying left hundreds of low spots on the bedrock surface. Dams in the form of moraines constructed along glacier margins ponded water, too, as did more irregu-lar piles of glacial sediment left behind by the melting of

Plate 1. Minnehaha Falls in Minnehaha Park, Minneapolis. Minnehaha Creek, a tributary to the Mississippi River, flows over a ledge of Platteville Formation, and the turbulent water erodes the weakly cemented St. Peter Sandstone below. (Photograph by Jerry Stransky)

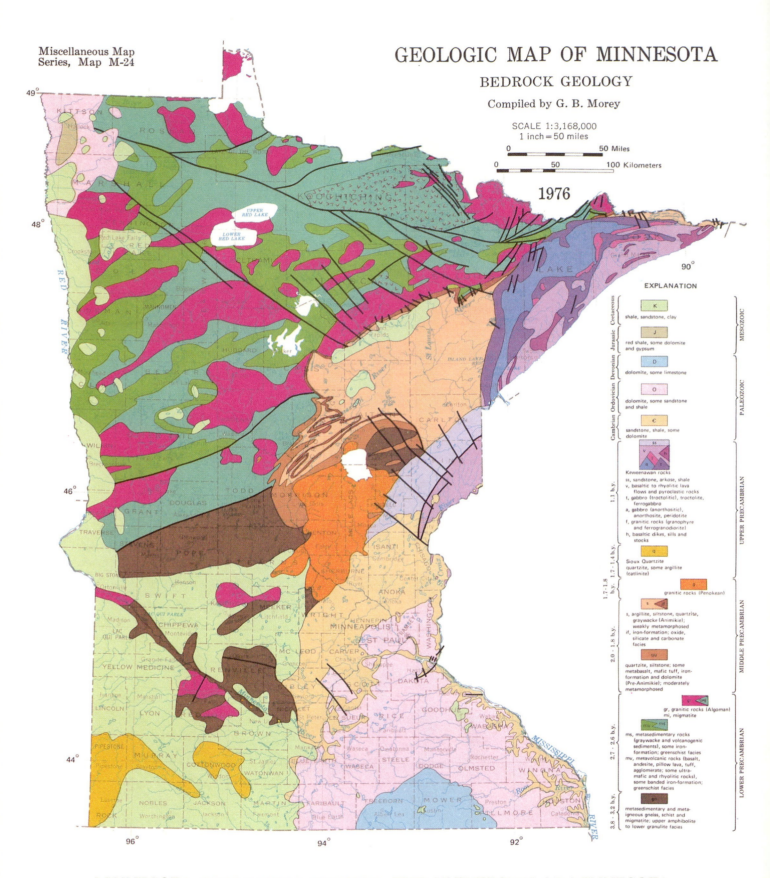

GEOLOGIC MAP OF MINNESOTA

BEDROCK GEOLOGY

Compiled by G. B. Morey

SCALE 1:3,168,000
1 inch = 50 miles

0   50 Miles

0   50   100 Kilometers

1976

EXPLANATION

K
shale, sandstone, clay

J
red shale, some dolomite and gypsum

D
dolomite, some limestone

O
dolomite, some sandstone and shale

€
sandstone, shale, some dolomite

ss / v / t / a / f / h
Keweenawan rocks
ss, sandstone, arkose, shale
v, basaltic to rhyolitic lava flows and pyroclastic rocks
t, gabbro (troctolitic), troctolite, ferrogabbro
a, gabbro (anorthositic), anorthosite, peridotite
f, granitic rocks (granophyre and ferrogranodiorite)
h, basaltic dikes, sills and stocks

q
Sioux Quartzite
quartzite, some argillite (catlinite)

g
granitic rocks (Penokean)

s / if
s, argillite, siltstone, quartzite, graywacke (Animikie); weakly metamorphosed
if, iron-formation; oxide, silicate and carbonate facies

qu
quartzite, siltstone; some metabasalt, mafic tuff, iron-formation and dolomite (Pre-Animikie); moderately metamorphosed

gr / mi
gr, granitic rocks (Algoman)
mi, migmatite

ms / mv
ms, metasedimentary rocks (graywacke and volcanogenic sediments), some iron-formation; greenschist facies
mv, metavolcanic rocks (basalt, andesite, pillow lava, tuff, agglomerate; some ultramafic and rhyolitic rocks), some banded iron-formation; greenschist facies

gn
metasedimentary and metaigneous gneiss, schist and migmatite; upper amphibolite to lower granulite facies

MESOZOIC — Cretaceous / Jurassic
PALEOZOIC — Devonian / Ordovician / Cambrian
UPPER PRECAMBRIAN — 1.1 b.y. / 1.7 - 1.4 b.y.
MIDDLE PRECAMBRIAN — 1.7 - 1.8 b.y. / 2.0 - 1.8 b.y.
LOWER PRECAMBRIAN — 2.7 - 2.6 b.y. / 3.8 - 3.2 b.y.

MINNESOTA GEOLOGICAL SURVEY - THE UNIVERSITY OF MINNESOTA

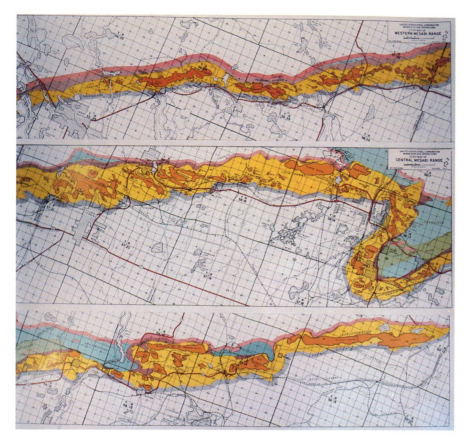

Plate 3. Strip map of the Mesabi Range. Unfortunately, this book isn't long enough to accommodate the map in a single strip. The top map covers the western one-third, the middle map the central one-third, and the lower map the eastern one-third. The low-grade iron-formation is yellow, and the open pit mines are orange. The large pit just north of Hibbing in the middle map is the famous Hull-Rust-Mahoning pit, the largest in the world. The two big pits just to the left of the bend in the range are U.S. Steel's Minntac taconite pits. The two big pits near the center of the lower map are Erie Mining Company taconite pits and the long pit at the right end of the lower map is the Reserve Mining taconite pit. (Courtesy of the United States Steel Corporation, 1977).

Plate 4. The Pillsbury high-grade open pit west of Chisholm was active from 1898 to 1961. The Monroe pit is in the background. Minnesota Highway 169 now crosses this mine, resulting in an excellent observation post. The Iron Range Interpretative Center is located just to the left of the photo near the mine. Note the elongate nature of the pit which followed the naturally enriched ore.

Plate 5. Polished slab of algal structures about 2000 million years old from the Mary Ellen Mine at Biwabik. The white and red finger-like columns grew upward and trapped sediment, which shows as specks, between them. They were probably made of calcium carbonate and have been replaced by white chert and red jasper. The dark gray material is hematite.

Plate 6. The most abundant rocks in the earth's outer shell are granite, (lower left) and basalt (upper right). Granite makes up the bulk of the continental crust. Its fine-textured chemical equivalent is rhyolite (upper left). Basalt makes up the bulk of the oceanic crust. Its coarse-textured relative is gabbro (lower right). The granite is from the St. Cloud area, the rhyolite is from Grand Marais, the basalt with amygdules is from the North Shore of Lake Superior, and the gabbro is from Duluth.

Plate 8. Pillowed greenstone about 2700 million years old at Gilbert, Minnesota, behind the High School. These pillows, exposed in two dimensions on a horizontal surface, have been tilted to a vertical position. The tops of ideal undeformed pillows are curved upward in a convex form and the bottoms have points that point downward. The tops in this photo are towards the top of the photo. Note the glacial striations parallel to the hammer handle.

Plate 7. A piece of copper-nickel ore from the Duluth Complex. The yellow blebs are chalco-pyrite; the nickel minerals are too fine-grained to see. The bulk of the rock consists of plagioclase (light gray), augite (dark), and olivine (dark). The sample came from a test pit south of Ely. (See Figure 8-23).

Plate 9. These minerals are the most important constituents of the Earth's crust. Quartz, the feldspars, augite, hornblende and the micas make up the bulk of igneous and metamorphic rocks. Calcite precipitates out of water to form limestone. Clays form at the Earth's surface by the weathering of the feldspars and the micas. All these minerals can be found in sedimentary rocks. The top row, from left to right, includes plagioclase feldspar, potassium feldspar, quartz, muscovite mica, and calcite. The bottom row, from left to right, includes augite, hornblende, olivine, biotite mica, and clay.

Plate 10. This large banded Lake Superior agate on a Lake Superior beach was once an amygdule in a lava flow. The colors are characteristic of Lake Superior agates. Note that several of the adjacent pebbles are volcanic rocks with smaller amygdules of other materials.

Plate 11. Thin lava flows northeast of Shovel Point by Illgen City (see Figure 5-10). Note that the angle of dip of the flows is gently off to the right, toward the center of Lake Superior.

Plate 12. The Sawtooth Range on the North Shore, formed by eroded lava flows tilted 8 to 20 degrees to the left towards Lake Superior. The view is to the southwest from the Grand Marais Coast Guard Station.

Plate 13. Reconstructed post at Grand Portage. Hat Point is in the distance at the left and Grand Portage Island is at the right.

Plate 14. Gooseberry Falls just below the Highway 61 bridge at Gooseberry Falls State Park. Three flows make up the falls; one makes up the upper half of the falls, another the middle to lower part, and a third is at the very base of the falls. (Courtesy of J. C. Green.)

Plate 15. The Duluth Harbor Basin of the Duluth-Superior Harbor. Included in this picture is the Duluth Entry of the harbor which is maintained by the Corps of Engineers, and numerous docks and slips used by vessels for loading and unloading shipments in the Duluth-Superior Harbor. In the background is the cross-channel and the upper channels of the Harbor. (Courtesy of U.S. Army Corps of Engineers.)

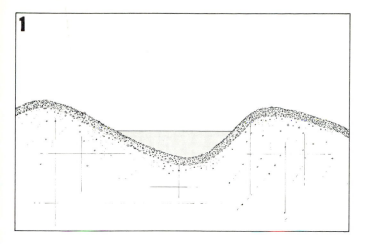

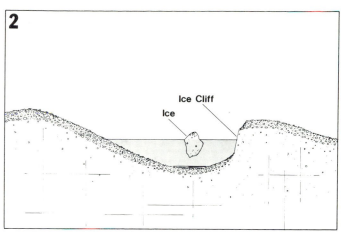

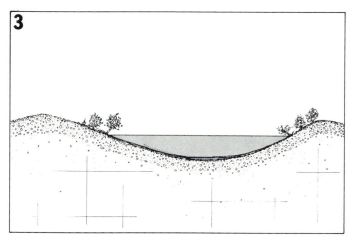

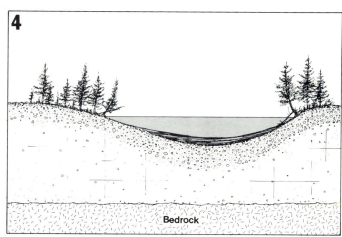

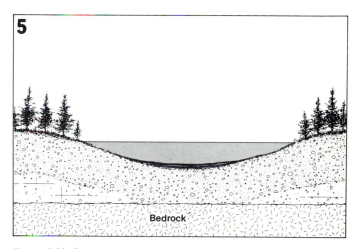

Figure 7-29. Stages in the development of a kettle lake on stagnant glacial ice. Final stability comes only after all the buried ice has melted out. (After Driscoll, 1976.)

debris-rich ice. Many lakes owe their existence to the irregular melting of stagnant or dead ice, accompanied by shifting patterns of sedimentation on the wet, dirty, unstable ice-cored topography. In order to better understand the natural history of Minnesota's kettle lakes, F. G. Driscoll (1976), studied lakes forming today along the stagnant margin of the Klutlan Glacier, Yukon Territory. He recognized five stages in their development (Figure 7-29), beginning with the accumulation of meltwater in low areas. Areal expansion and deepening by iceberg calving and melting are accompanied by infilling and the growth of vegetation.

Sediment infilling at the margins along with the accumulation of mineral and organic debris in the deeper parts of a basin begin as soon as the lake makes its debut upon a landscape. Kettle lakes, even before glacial ice is completely melted, become loaded with sediments dumped from the unstable slopes that border them, including both glacial debris and plant remains. Core samples taken from many Minnesota lakes include a zone near the base loaded with wood fragments and other plant trash, the remains of the earliest vegetation cover to establish itself in the postglacial environment. Eventually, stable slopes are attained as areal expansion ceases and the last ice melts.

The total sediment column is an archive documenting successive changes in the evolution of a lake's natural history,

including environmental disturbances in the vicinity of the basin. Changing water chemistry through time is generally reflected in the fossil content of the lake sediments. Certain aquatic plants and animals are extremely sensitive to changes in water quality. The addition of nutrients generally triggers the increased growth or "bloom" of algae, accompanied by an upward shift of total organic productivity with a decline

Figure 7-30. This small lake in northern Minnesota will soon disappear as bog vegetation establishes a complete cover.

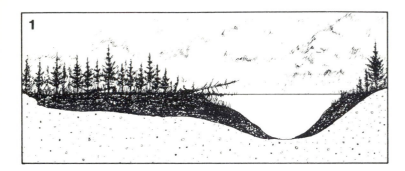

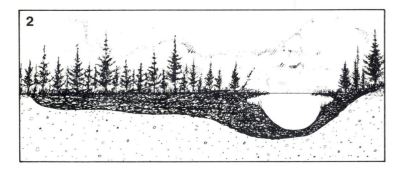

Figure 7-31. A lake is converted to a peat bog by the progressive growth of plants from its margins toward the center.

in recreational value. The consequences of human disturbance, including deforestation, agriculture, mining, sewage disposal, and water supply development, are all reflected in both the present condition of Minnesota's lakes and in the

upper few centimeters of sediment accumulated since European settlement in the 19th century.

In just the last 12,000 years, since the disappearance of glacial ice, many basins have already been completely or partially filled by natural processes; some have become peat bogs (Figure 7-30). Hundreds, probably even thousands, of others have been drained by human endeavor and converted into dry land for agricultural and other uses.

Peat is an organic deposit composed of partially decayed plant material that has accumulated over a long period of time in an environment conducive to its preservation. Generally, such preservation is enhanced by an absence of oxygen, which is vital to the existence of bacteria and other organisms that normally destroy plant remains. Cool temperatures also retard the decomposition of plant matter; therefore, the formation of peat is encouraged by the persistence of oxygen-poor stagnant waters under cool climatic conditions, just the environment prevailing in the lowlands of northern Minnesota for the last 5,000 years.

Peat began to accumulate almost immediately after vegetation had established itself in the wake of retreating ice. The oldest deposits began to form about 12,000 years ago by a process known as lake filling (Figure 7-31). Progressive growth of plants from the margins of a lake toward its center in the form of a floating mat is accompanied by the deposition of dead plant and animal debris on the lake bottom. As the lakefill proceeds, a succession of plant communities invades and colonizes the floating mat, which thickens and extends itself. Aquatic plants and sedges are the pioneers, establishing a frontier settlement along the shallow margins of the lake. As infilling changes water depth, semiaquatic plants and mosses become established, followed by shrubs such as Labrador-tea and willow, herbaceous plants, and finally trees, especially black spruce and tamarack. After complete filling, the bog that results may become raised as vegetation continues to accumulate, especially when sphagnum moss becomes established.

Peat bogs resulting from lakefill are the predominant type found over much of the Canadian Shield, at least where it is not covered by the sediments of the larger glacial lakes but rather is pockmarked with smaller depressions in morainic belts and areas of glacial scour. A typical succession produced by lakefill grades upward from a basal aquatic peat through sedge to woody forest peat and finally to an upper sphagnum moss horizon.

Large deposits of peat in northern Minnesota and elsewhere can be attributed to the spread of wet, peat-favoring environments by a process known as paludification or swamping. Climatic changes toward a cooler, wetter regime beginning about 5,000 years ago instigated the development of most of the peat now found in the northern part of the state. The extension of wetlands was especially prevalent across the low-relief, poorly drained lowlands formerly occupied by Glacial Lakes Agassiz, Aitken, and Upham. As a consequence of the buildup of plant remains at favorable sites, the water table rises beneath the thickening mat, thus encouraging the growth of even more plants and ensuring the proper conditions for preservation of dead material. Expansion of the living bog, even up slopes, ensues, along with an accelerating rate of peat accumulation.

The extensive peatlands now covering much of the eastern arm of the former Glacial Lake Agassiz have most recently been studied by M. L. Heinselman (1963). There, bog expansion and rising water tables in many places resulted in the deterioration of growing conditions for trees. As evidence, he cites the superposition of as much as 3 m of sphagnum or sedge peat atop woody forest remains in areas that today are almost totally treeless. Figure 7-32 is a summary of the apparent history of environmental change as a direct result of gradual swamp extension and water table rise in a Lake Agassiz bog.

Modern peatlands fall easily into two categories based on vegetation—bog and fen. Bogs are peat-filled areas dominated by sphagnum moss, commonly with stands of black spruce and tamarack. The water of bogs is extremely acid and mineral-poor because it is replenished mainly by precipitation. Fens, on the other hand, support a more luxuriant cover of grasses, sedges, and reeds. Their waters are richer in minerals and less acid, replenished by circulation from the groundwater reservoir. Both bog and fen typically display a variety of vegetation patterns. Some contain teardrop-shaped forest islands separated by water tracks and some (string bogs) have alternating bog ridges and swales with contrasting vegetation types (Figure 7-33). Their origin is not clear, but they are certainly related to water movement patterns through the bogs and to the availability of nutrients.

The formation of organic-rich deposits in wet environments during the last 12,000 years in what is now Minnesota is not a unique event. Coal and lignite deposits of Cretaceous age had a similar origin, in a slightly different geologic setting. And even more ancient concentrations of dead plant material are mined in the great coal belts of Mississippian and Pennsylvanian age in Appalachia and the Interior Lowland area of North America. Left to "ripen," the great peat deposits of Minnesota might eventually be compressed into a coal field, if the necessary sedimentary cover is deposited on top of it.

Figure 7-33. Vertical air photograph of a portion of the vast peatland north of Upper Red Lake, in Beltrami County. The straight lines indicate drainage ditches several kilometers apart constructed for agricultural development in the early 1900s. The large dark areas are mostly tree-covered "raised bogs," which are nourished only by rain and snowfall. The light areas are largely sedge-covered water tracks, some of which carry nutrient-rich waters from the mineral uplands bordering the peatland. (Courtesy of H. E. Wright.)

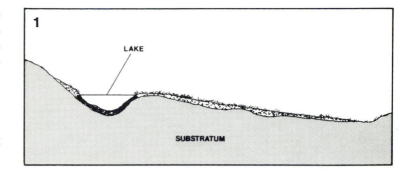

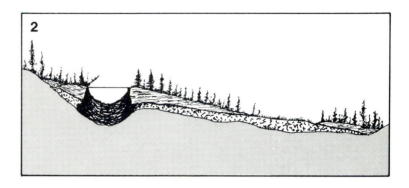

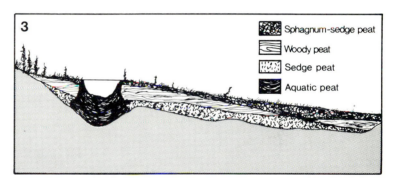

Figure 7-32. The progressive extension of swamp environment is promoted by the rise of the water table beneath a thickening vegetation mat. Eventually a peat deposit will result. (After Heinselman, 1963.)

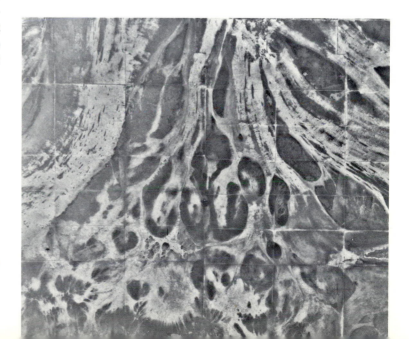

PART III

Mineral Resources

If someone from another state were to ask what kinds of minerals were mined in Minnesota, the answer would probably be a short one—iron ore. Indeed, iron ore accounts for the lion's share of Minnesota's mineral production, which is now $1.75 billion per year. This makes Minnesota the number one state in nonfuel mineral production. But such an answer would be somewhat incomplete, for different rocks and clays have been quarried at several places in the state, and sand and gravel have been removed from thousands of pits.

And, although iron ore would have been the best answer for a century or so, it's possible that iron ore will be only part of the picture in the not-too-distant future. The search by companies for the ores of copper, nickel, zinc, uranium, and other metals is an ongoing process. Some of the geological studies done by the Minnesota Geological Survey, the Minnesota Department of Natural Resources, and, to a lesser extent, the United States Geological Survey are aimed at providing base-level data that will be of value to the many companies engaged in mineral exploration. Already, great reserves of copper and nickel have been proven out by drilling in the gabbros of the Duluth Complex near Ely and Babbitt and farther south toward Duluth. Some of the gabbroic rocks are rich in titanium, a possible future resource. Rarer elements such as gold and vanadium are also occasionally mentioned in discussions of Minnesota's mineral potential.

If the phrase "mineral resources" is defined in a broad and nontechnical manner, then resources such as water and soil are also logical topics for inclusion in this part of the book. In one view, agriculture is the largest open-pit mining operation in Minnesota! Chapter 8 deals with the metallic minerals, and Chapter 9 with the nonmetallic resources, including fuels and water.

The first train load of Mesabi ore, in 1892, on the Duluth, Missabi and Northern dock at Duluth. As C. K. Leith (1903) stated, "Reports of the discoveries of wonderful deposits of iron ore in the Mesabi district, the soft nature of the ore, its flat attitude, and the strange methods of mining it had aroused great curiosity and had started much discussion. There was a general disposition to be skeptical as to the value of the ore found in such unusual conditions and so cheaply mined. The arrival of the first train load of ore was thus a matter of much interest." (From Leith, 1903: facing p. 28.)

8
Metallic Minerals

IRON ORE

Discovery of Iron Ore in Minnesota

Iron ore in the Lake Superior region was first discovered near Negaunee, Michigan, in 1844 by a surveying party when great variations were noted in the behavior of compass needles. The canal and locks at Sault Ste. Marie at the east end of Lake Superior were completed in 1855, and that same year the first iron ore from Negaunee, on the Marquette Range, moved through the locks, destined for Cleveland. The Menominee Range began producing in the 1870s and the Gogebic Range was opened up in the 1880s (see Figure 4-7).

Meanwhile, what was going on in Minnesota? In 1855, the year the first Michigan ore passed through the Sault locks, a total of only a few dozen people lived in the settlements at Fond du Lac, Oneota, and a trading post on Minnesota Point, all of which became parts of Duluth in 1857. Iron ore was discovered by George Stuntz near Lake Vermilion; he had gone up, along with hundreds of others from Duluth, to check out a report of gold in 1865. J. G. Norwood, a geologist, may already have sighted the iron-formation in 1848. Nevertheless, Stuntz noted the iron ore and was commissioned by the Minnesota legislature to construct the Vermilion Trail in 1868-1869. In 1884, Minnesota's first iron ore was shipped from open pits at the discovery site, the Soudan Mine of the Vermilion district, via the new Duluth and Iron Range Railroad to Two Harbors. Four years later, ore was being shipped from the underground Chandler Mine at Ely, on the eastern end of the Vermilion Range, 32 km from Tower and Soudan.

Interestingly, "gold rushers" had traveled to Lake Vermilion on the Vermilion Indian Trail, which crossed the hills of the Mesabi Range. Iron had been reported on the Mesabi at least as early as 1866 by Minnesota's first state geologist, Henry H. Eames, who was sent by the state to check over the mineral potential of the northern half of the young state. Reportedly, Christopher Weiland, of Beaver Bay, pointed out the iron ore near Babbitt, only to have Eames reply, "To hell with iron. It's gold we're after." Geologists who visited the few outcrops of iron-formation pronounced it low grade and uninteresting, but, unknown to them, glacial debris covered the good ore. In 1879, after years of searching for iron ore that their father had mentioned, four Merritt brothers and four nephews of theirs organized another exploration program. Test pits were put down and on November 16, 1890, J. A. Nichols, in the employ of the Merritts, discovered iron ore that contained 65% iron at a site that was immediately named Mountain Iron (Figure 8-1).

During the next decade, Duluth was the entry point for the many people pouring into the Mesabi Range. Ore was quickly discovered at Biwabik, Virginia, Eveleth, and Hibbing and then at various locations as far to the west as Coleraine. The first Mesabi Range iron ore was shipped in 1892 from Mountain Iron to Duluth via the Duluth, Missabi and Northern Railroad, which connected with the Duluth and Winnipeg Railroad for the last leg. By 1893, 10 mines were already in operation. E. J. Longyear, who had drilled the first holes on the range, put down more than 7,000 holes and pits in the next two decades and proceeded to build the worldwide Longyear Company (Nute, 1951).

The Merritt brothers, who were so instrumental in the discoveries and the early development of the Mesabi, were to lose their fortunes before they really had them, because of overextension of financial obligations and the financial panic of 1893. Paul DeKruif's (1929) book about the Merritts, entitled *Seven Iron Men*, details this interesting

Figure 8-1. The Mesabi Range discovery site at Mountain Iron. The x marks the spot where the first ore was removed. The photo was taken in 1890. (Courtesy of the Northeast Minnesota Historical Center, University of Minnesota, Duluth.)

story. The history and development of the Mesabi is a colorful tale that is best told with visual aids. The Iron Range Interpretative Center at Chisholm depicts much of this story focusing on the geology, the technology, and the people who came from all over the world to find and to mine the ore, and ultimately to homestead, in northern Minnesota.

The Cuyuna Range has its own special history. It is an area of low relief, not a topographic range held up by hard rock, as are the Vermilion and Mesabi ranges. In central Minnesota, as far back as 1859, surveyors and timber cruisers were having trouble with their compass needles in certain areas. In 1882, Henry Pajari looked for magnetic iron-formation, the suspected cause of the problems, without a discovery. Later, careful surveys revealed the belts of magnetite-bearing iron-formation, the Trommald Iron Formation. Investors obviously hoped that the buried iron-formation would prove minable, but high-grade ore pockets, which would be nonmagnetic hematite rather than magnetic magnetite, were difficult to locate. An old Cornish mining adage was applicable: "Where it is, there it is. Where it ain't, there I is." After many fruitless drilling ventures on buried anomalies, for there are no rock outcrops, ore was hit in 1904 by Cuyler Adams, the surveyor for whom the Cuyuna was named. He had spent many years of searching with a special Swedish dip needle that measured the amount of magnetic attraction. Actually, the minable high-grade ore was in oxidized nonmagnetic pockets, in the strips of magnetic iron-formation, as on the Mesabi. In 1909 the first ore from the Cuyuna Range was shipped to the Superior, Wisconsin, docks over 160 km of railroad that crossed numerous bogs. It took nearly a month for the initial trainload of ore to arrive from the Kennedy Mine, but at last the Cuyuna, too, was on its way.

What of the Gunflint Range, along which J. G. Norwood discovered iron at Gunflint Lake in 1850 and which he thought was an eastern extension of the Vermilion Range? Perhaps Norwood just rediscovered the ore, for it was supposedly noted at Gunflint Lake in 1780 by Frenchmen as they traveled the canoe route up the Pigeon River (Zapffe, 1938). No ore has been mined from this range, either over its 32 km length in Minnesota or over its much longer length in Ontario, because the iron-formation contains too much chert and has no naturally enriched ores. However, the Paulson "mine" just west of Gunflint Lake was reportedly being prepared for production when the richer Mesabi iron ores were discovered, and the present Gunflint Trail uses much of the old trail built from Grand Marais to the "mine." The Port Arthur, Duluth, and Western Railroad (known to some as the Poverty, Disease, and Want because of its poor financial situation) reached the mine in 1893, but the mine was abandoned and the railroad was sold to a logging company.

The last iron-ore area to be explored in Minnesota was an area in Fillmore, Olmsted, and Mower counties in southeastern Minnesota. Although the presence of limonite-hematite ore there was known by 1882, the possibilities were not seriously explored until the 1930s. Two companies mined ore there from 1942 to 1968.

The Vermilion Range and part of the Mesabi Range could easily have been part of Canada, except for a stroke of luck, as related by Zapffe (1938), who paraphrased Williams (1907). Negotiators at the Treaty of Paris in 1783 (Benjamin Franklin among them) had an inaccurate map that showed Lake of the Woods draining into Lake Superior through the Pigeon River. Had they known that the Pigeon River was only 55 km long, the negotiators might have selected the

126

St. Louis River as the international boundary. (At the same session, Benjamin Franklin reportedly deflected a pencil line a bit so as to include the Upper Peninsula of Michigan in United States territory.)

Origin and Production of Minnesota's Iron Ores

So far in this book, iron ore and iron-formation have been mentioned several times—in the section of Chapter 3 on Early Precambrian greenstone belts, in the section of Chapter 4 on Middle Precambrian iron-formations, and in the section of Chapter 6 on the Cretaceous seas over Minnesota. The iron ranges of Minnesota are shown in Figure 4-1. Three Minnesota ranges have produced iron ore—the Vermilion Range of Early Precambrian age, and the Mesabi and Cuyuna Ranges of Middle Precambrian age. In addition, minor but interesting deposits of ore of probable Tertiary age have been mined in southeastern Minnesota. The iron-formation and iron ores of these three different ages have different origins, as we shall see.

More than 400 Minnesota mines (Table 8-1) produced about 3 billion metric tons (t) of iron ore between 1884 and 1980. (By contrast, Michigan and Wisconsin together have produced a total of 1 billion metric tons.) About 94% of that total came from the Mesabi Range. In 1942, during World War II, 64 million metric tons of high-grade ore were mined from the Mesabi alone! In 1978, Minnesota produced 51 million metric tons of iron ore, with more than 75% of that being taconite pellets which are produced from taconite rock. (Taconite rock was not called iron ore until quite recently, for it was too low grade to process at a profit.) That the Lake Superior district remains important to the nation's steel industry is shown by the fact that in 1978 nearly 90% of the domestic ore used by the steel industry was Lake Superior ore.

The Vermilion Range

The old ores of the Vermilion Range have a volcanic origin. As volcanism was going on in the area, iron, silicon (in the form of silica), and other elements were given off through submarine volcanic vents and fissures, probably in warm or hot waters. When the right chemical conditions were reached, the silica was precipitated as thin bands of chert, and the iron was precipitated as thin bands of one or more iron minerals, either hematite, magnetite, siderite, pyrite, or iron silicate. (Geologists term such rocks *exhalites*, because their components were literally exhaled by volcanism.) The resulting interbedded sequence of the Vermilion Range, which now consists of chert, magnetite, and hematite, is normally not rich enough to be ore (Figure 8-2); further movement and concentration of iron minerals within the iron-formation is necessary. Although the exact nature of this enrichment is subject to interpretation, it is probably the result of hot waters that moved through the iron-formation (Figures 8-3 and 8-4).

The largest of the 11 mines of the range, the Soudan Mine, was closed in 1962, but the United States Steel Corporation sold it and 400 surrounding hectares (ha) to the state for the magnificent sum of $1 on the condition that it be made into a state park (Figure 8-5). (The park is further described in the section on the western Vermilion district in Chapter 10.) A total of about 14 million metric tons of ore containing 63% to 66% iron were removed from the Soudan. Several small mines in the Tower-Soudan area produced some ore in the early days of mining on the Vermilion Range. There were 5 mines in Ely—the Chandler, the Pioneer, the Zenith, the Sibley, and the Savoy. The last of these was closed in 1964. A total of 70 million metric

Figure 8-2. Banded iron-formation near the Soudan Mine in Tower-Soudan State Park. White bands are white chert, some of the darker bands are red jasper, and the darkest bands are metallic steely gray hematite with some magnetite. Each band of rock lithology consists of thin microbands on the order of 1 mm or less in thickness. This alternation of composition is related to the solubility of iron and silica, for the iron-formation is a chemical precipitate related to Early Precambrian volcanism.

Table 8.1. Minnesota's Iron Mines, as of 1979

	Mesabi Range	Cuyuna Range	Vermilion Range
Exhausted	148	21	9
Inactive	173	33	2
Operating	43	1	. . . . . .
Total number of mines	364	55	11
Potential mines (Reserves)	83	29	. . . . .

Source: Data from Trethewey, 1979.

Note: Not included here are the numerous small "brown" iron-ore (limonite-hematite) mines of southeastern Minnesota that are no longer operative.

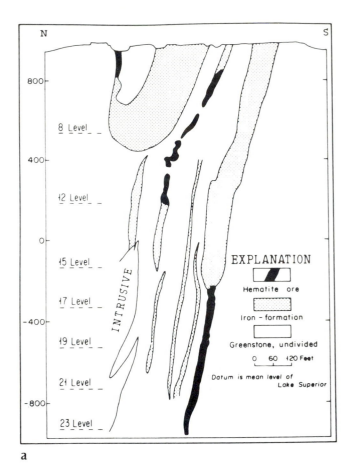

a

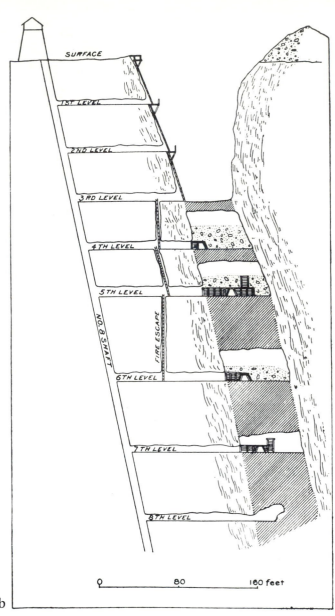

b

Figure 8-3. (a) Generalized geologic cross section of the ore in the Soudan Mine. The view is toward the east. The high-grade iron ore is shown by the darker shade and iron-formation by the lighter shade. The white portions are various types of greenstone. The ore extends below the 27th level (700 m deep), not shown on the figure. (After Klinger, 1956.) (b) Diagram showing mining methods in the Soudan Mine. (From Clements, 1903.)

Figure 8-4. Hematite ore in the Soudan Mine at a depth of 700 m. Hematite is slightly reflective in this photo.

tons was produced from hematite in the Ely Trough, a synclinal fold (Figure 8-6). Total production from the 11 mines of the Vermilion Range was nearly 95 million metric tons.

The Mesabi Range

The origin of the Middle Precambrian iron-formation of the Mesabi Range was presented in Chapter 4. The sedimentary rock type of this formation, taconite, usually contains 25% to 30% iron. Fortunately, at some time in the past natural processes enriched the low-grade iron-formation to form natural iron ores. Oxygen-rich alkaline groundwaters moving through faults and other cracks in the iron-formation dissolved out much of the silica (SiO_2), leaving oxidized iron minerals (hematite and limonite) and thus increasing the iron percentage from that characteristic of taconite (25% to 30%) to 50% to 55% and even higher (Figure 8-7 and Plate 4). The removal of silica from the rock units resulted

a

b

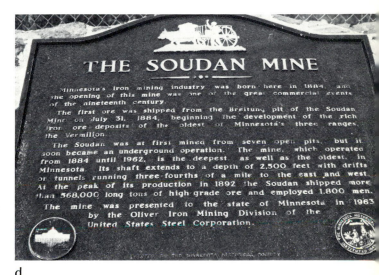

d

c

Figure 8-5. (a) Early open-pit iron mine at Soudan. Note the sledgehammers. Miners used ropes to move from one level to another. The photo was taken in the 1880s. (Courtesy of the Tower-Soudan Historical Society.) (b) Early open-pit iron mine at Soudan. Note the precipitous pit walls. The photo was taken in the 1880s. (Courtesy of the Tower-Soudan Historical Society.) (c) Present-day headframe at the Soudan Mine. Built in the 1920s, it is still in use but transports tourists, rather than iron ore, down (and up). (d) Plaque at Tower-Soudan State Park.

in a decrease in volume, which is commonly expressed as sagged and folded rock layers (Figure 8-8). When this enrichment actually occurred is a good question. It could have happened as early as 1,700 to 1,600 million years ago or as late as the Cretaceous Period, between 130 and 65 million years ago. Cretaceous sedimentary rocks on the Mesabi include conglomerates made of hematite-ore pebbles, so the enrichment occurred prior to the advance of the Cretaceous sea into northeastern Minnesota (see Figure 6-36).

Because of the gentle dip of the iron-formation on the Mesabi, a broad zone intersects the ground surface, and the alkaline groundwaters have come into contact with much iron-bearing rock (see Figure 4-9). The iron-formation there

a

b

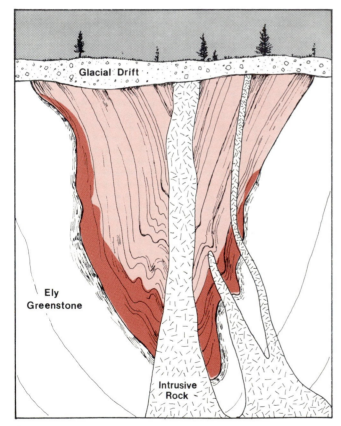

c

Figure 8-6. (a) Ely in the early days of iron mining. The Chandler Mine is at the left. (From Clements, 1903.) (b) Early open pit and entrance to the underground workings at the Chandler Mine in Ely. (From Winchell and Winchell, 1891.) (c) Generalized geologic cross section of an iron-ore body in a syncline at Ely. The ore is dark and the iron-formation is a lighter shade. The maximum depth of mines at Ely was about 225 m. (After a diagram from the United States Steel Corporation.)

a

b

Figure 8-7. (a) The McKinley Mine at McKinley, one of the last high-grade natural ore pits in operation in 1980. The ore is treated at the mine site to remove fine clay and quartz silt and sand impurities; this raises the ore grade. Note the elongate nature of the pit, which followed the zones in which the iron content of the original taconite rock was increased naturally by the solution of silica by alkaline waters. The photo was taken in 1979. (b) Mine in the Virginia area. The photo was taken in 1960. Note the thin glacial drift layer on the left and the elongate nature of the ore zones.

Figure 8-8. The Hull-Rust-Mahoning open pit, in Hibbing.

Figure 8-9. High-grade natural ore beneath glacial till. Easy removal of the unconsolidated till enhanced the mining operation. Similar relationships existed throughout the Mesabi Range. The photo was taken in 1960 in the Harrison Mine at Nashwauk.

Figure 8-10. Large shovel from the early days of mining in the Mountain Iron Mine. (From Winchell et al., 1899.)

has few natural surface exposures because of a cover of much younger glacial sediments, but the unconsolidated glacial debris was easily removed by large shovels (Figures 8-9 and 8-10). Furthermore, the high-grade ore was quite soft and was easily mined in large open pits. The hundreds of mines on the Mesabi Range are shown in Plate 3, and a smaller portion of the Mesabi Range is shown in greater detail in Figure 8-11). Whereas the early years of mining on the Mesabi saw several shallow underground mines, the development of larger equipment saw them transformed into open pits (Figure 8-12a).

Ownership of the numerous iron-bearing properties was and is complex, with property boundaries commonly cutting across ore bodies. Thus, various pits merged into single larger ones, the most famous being the Hull-Rust-Mahoning pit at Hibbing, which encompasses dozens of mines. Open pit mining started here (Figure 8-12b). In all, about 1.2 billion metric tons of material (half of it iron ore) were removed from the pit, which has a maximum depth of about 150 m and covers an area of nearly 8 km². (By comparison, only about 40% as much rock was moved during the building of the Panama Canal.) The Hull-Rust-Mahoning pit, from which high-grade ore was removed, will be surpassed in size by some of the taconite pits. The Reserve Mining Company pit at Babbit is already 15.5 km long and covers an area of

about 15 km². The other 364 high-grade iron-ore mines of the Mesabi were obviously smaller than the Hull-Rust-Mahoning; some were rather small pits (Figure 8-13).

The property boundaries on the Mesabi have created some interesting situations. A case in point is the history of the Sliver Mine at Virginia (see Figure 8-11). A lawyer dealing with legal matters on property boundaries and ownership realized that two adjacent properties did not quite abut each other; that is, a narrow piece of ground was not owned by either company. He obtained ownership of this land, and the resultant Sliver Mine produced nearly 2.8 million metric tons of high-grade ore for which he received a royalty of about $.50 per ton. The making of this millionaire took from 1908 to 1926, the life of the mine, but it was still a relatively short process.

The Cuyuna Range

The origin of the iron-formation and the iron ore of the Cuyuna Range is essentially the same as that of the Mesabi Range. One mine on the Cuyuna Range is shown in Figure 8-14. Partially because of the steep dips of the folded iron-formation (see Figure 4-8), some mines on the Cuyuna were of the underground type even though large machinery had already been developed for open-pit operations. The open pits are now completely flooded. At the peak of World War I, in 1918, 32 mines were operating. The villages of Trommald, Riverton, Manganese, Cuyuna, Crosby, and Ironton were founded near mines, but only the latter two were destined to survive as towns of moderate size. New mines continued to be opened until the 1950s, but currently only one mine (the Algoma-Zeno) is being worked to obtain as much as 90,000 to 136,000 t of manganiferous iron ore annually. About 96 million metric tons of ore have been produced from the range. In all, 167 land and mining companies searched for or produced ore on the Cuyuna. All ore came from Crow Wing County, although the range extends

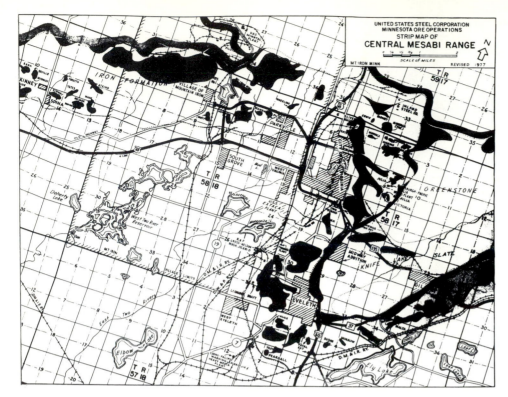

Figure 8-11. Enlarged portion of the strip map of Plate 3 in the "Virginia Horn" area. The large pits in the upper left are United States Steel's Minntac taconite pits at Mountain Iron. Some pre-iron-formation rock units appear as long black lines on this photograph and should not be confused with pits. (Courtesy of the United States Steel Corporation.)

a

b

to the northeast and southwest into Aitkin and Morrison counties.

Southeastern Minnesota

The "brown ore" of southeastern Minnesota (in Fillmore, Olmsted, and Mower counties, near the towns of Chatfield, Spring Valley, Etna, and Ostrander), has a unique origin quite different from that of the important Precambrian iron ores. However, concentration or enrichment of iron from rocks with low iron content was still necessary, as it was with the Precambrian iron-formations. R. L. Bleifuss (1972) has studied these ores in detail and concluded that they are the result of the oxidation of a siderite (iron carbonate) zone of the Devonian Cedar Valley Formation to goethite (limonite) and hematite (Figure 8-15). He thinks this weathering and residual accumulation occurred under temperate climatic conditions during a portion of Tertiary time, an idea that contrasts with earlier opinions that the weathering occurred during the Cretaceous. (Refer back to Chapter 6 for additional details on Cretaceous and Tertiary weathering.)

The ore in southeastern Minnesota was mined from 1942

Figure 8-12. (a) The Adams Mine, of Eveleth, showing old underground workings exposed by an early open pit. The photo was taken in 1895 or 1896. (Courtesy of the Northeast Minnesota Historical Center, University of Minnesota, Duluth.) (b) The Hull-Rust-Mahoning high-grade open pit at Hibbing, the largest iron mine in the world. It is an amalgamation of several mining properties. Some high-grade natural ore remains. The Hibbing Taconite Company is now mining taconite from the big pit area, and some will be produced from the floor of the old open pit as well. The photo was taken in 1960.

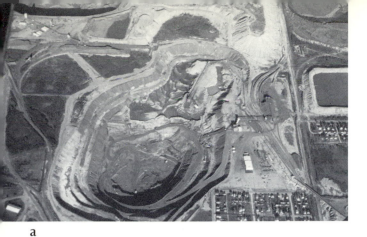

a

b

Figure 8-13. (a) Aerial view of high-grade iron mining from small pits on the western Mesabi Range in 1955. (b) The Lind Mine between Grand Rapids and Coleraine, one of the westernmost high-grade iron mines on the Mesabi Range. Note the thick glacial drift. The photo was taken in 1976.

Figure 8-14. The Mahnomen Mine at Ironton on the Cuyuna Range. The U-shaped fold on the far wall of the pit contains manganiferous iron ore in the center and iron ore on the sides and bottom. The photo was probably taken in the early 1950s.

to 1968. More than 7 million metric tons of ore were removed from more than 50 mines that ranged in size from small, thin pods in farmers' fields, as shown in Figure 8-15b, to deposits of a few hundred thousand metric tons. The ore was trucked to a railroad, and then moved to steel mills in Missouri. Many of the fields were then restored to agricultural use.

Iron-Ore Reserves

What does the future hold for Minnesota's iron-ore industry? How much ore remains to be mined? This question, for all practical purposes, can be reformulated in the following way: how much ore is left on the Mesabi Range?

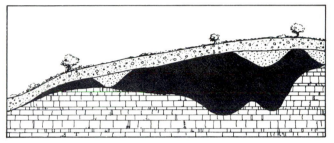

a

b

Figure 8-15. (a) Generalized cross section of a representative small iron-ore body (dark) in southeastern Minnesota. Note that it rests on carbonate rock. (After Bleifuss, 1972.) (b) Small iron-ore "mine" in southeastern Minnesota. The photo was taken during the 1950s. (Courtesy of the Minnesota Geological Survey. From Bleifuss, 1972.)

Figure 8-16. United States Steel Corporation's Minntac taconite plant at Mountain Iron. The west pit is in the left background. The view is to the southwest. (Courtesy of the United States Steel Corporation.)

Ironically, there is still considerable ore on the Vermilion Range, especially in the Soudan Mine, but the deep and expensive underground mines may never again be economically feasible. Although parts of the Cuyuna could be redeveloped if necessary, that range is not now considered a potential iron-ore source. However, the Cuyuna Range includes some iron ore with 5% to 10% manganese; and this constitutes the nation's greatest reserve of manganese. (Manganese is used by the steel industry; 6 kilograms [kg] per metric ton of iron ore removes the oxygen and sulfur and produces a clean, sound metal.)

High-grade (natural) iron-ore reserves on the Mesabi in 1980 are pretty well exhausted. They are estimated at only 145 million metric tons, and even part of that may ultimately be left in the ground. Total Mesabi taconite-ore reserves in 1978 were estimated by R. W. Marsden (1978) at 43 billion metric tons. Of this total, 25 billion metric tons are good taconite ore; the balance includes 3.6 billion metric tons of leaner ore, 3.9 billion metric tons of ore that yields a concentrate product containing more than the desired amount of silica, 10.5 billion metric tons of iron-formation that has been oxidized to hematite, and some miscellaneous types. Some of the lean ore is removed as

necessary when better taconite underlies it and may then be processed rather than hauled to the waste dumps. The hematite ore is similar to some of Michigan's concentrating-grade ore, which is processed by flotation methods rather than by magnetic separation as in Minnesota. Because flotation may be used in Minnesota some day, this 10.5 billion metric tons can be classed as potential ore. At the present rate of use, about 180 million metric tons of crude ore a year, it will take about 240 years to use up the 43-billion-metric-ton reserve! Furthermore, this figure of 43 billion metric tons was calculated for only the Biwabik Iron Formation to the point of "zero profit" at 1974 costs and values; that is, the calculation assumed that ore could be mined profitably only where the overlying Virginia Formation was thin. In the future it may be possible to remove more of the overlying rock than is economically feasible today, and this would make additional ore available.

Thus, Minnesota has great reserves of iron ore, with taconite plants and labor force already on site. However, worldwide factors also affect Minnesota's iron industry. In 1980, about 35% of the iron ore used by our nation's steel industry was imported from 10 foreign countries, with

Canada and Venezuela supplying most of this total. Whereas Minnesota once produced about 25% of the total world annual production of iron ore, it has produced less than 8% of the world's total since 1968. Minnesota's iron ore production has not markedly decreased, but the world's ore supply has increased greatly with discoveries in Australia and elsewhere.

The Taconite Process

Back in 1918, Edward W. Davis, of the University of Minnesota's Mines Experiment Station, was already concerned about the eventual depletion of the big Mesabi Range's high-grade iron ore. He compared the range to a long slice of raisin cake. The "raisins" or high-grade ore were being eaten up, leaving only the taconite "cake." Davis spent decades studying ways to commercially concentrate the magnetic taconites of the Mesabi. This research, funded by the Minnesota legislature, was of fundamental importance in assuring the continuation of mining in Minnesota.

Mining of the very hard taconite rock was delayed until the perfection of a drilling technique known as jet piercing, which uses a kerosene and oxygen flame to heat the rock and water to spall it in the drillhole. Explosives can then be set off in carefully spaced holes to break the rock into pieces of manageable size. Taconite has been called the world's hardest rock. Some minerals are harder, but taconite is indeed tough, a difficult material to mine. Recent advances in drilling technology include the use of hard, tungsten-carbide "buttons" on bits that are rotated under weights of about 25 t. These cause the taconite to flake or spall.

With the advent of the taconite process, new plants had to be built (Figure 8-16). As of 1982, eight plants existed in Minnesota, built at a total cost of more than $2 billion (Figure 8-17). Once mined, the magnetic taconite is ground to a fine powder, with the small pieces about 0.044 mm in diameter. This grinding frees the small magnetite grains, which can then be separated from the bulk of the chert by strong magnets. Finally, the powdered magnetite is formed into marble-sized pellets and baked for use in making iron. Bentonite, a clay-rich material that forms from the alteration of volcanic ash in the western states, is used as the "binder" for the pellets, which are composed of hematite when they cool. Details of the process are shown in Figure 8-18.

The small, uniform size and spherical shape of the pellets permits air to permeate the mixed charge of coke, limestone, and pellets in the blast furnace, rapidly reducing and melting the pellets. This greatly increases the capacity of the blast furnaces, roughly doubling the amount of steel a furnace can produce as compared to what it can produce using natural ore. Thus, the marketability of the natural ore that remains unmined has greatly diminished.

In Minnesota, only magnetic taconite is processed into pellets. Methods are being developed whereby nonmagnetic taconite, which contains hematite and limonite rather than magnetite, can be used. At present, a froth flotation process used in northern Michigan seems to offer the most potential.

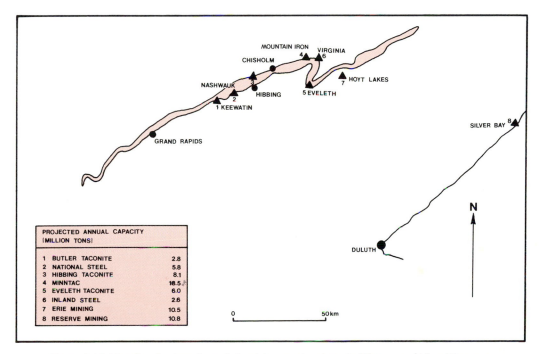

PROJECTED ANNUAL CAPACITY (MILLION TONS)	
1 BUTLER TACONITE	2.8
2 NATIONAL STEEL	5.8
3 HIBBING TACONITE	8.1
4 MINNTAC	16.5
5 EVELETH TACONITE	6.0
6 INLAND STEEL	2.6
7 ERIE MINING	10.5
8 RESERVE MINING	10.8

Figure 8-17. Map showing locations of the eight taconite plants in Minnesota. (After Minnesota Department of Economic Development, 1975.)

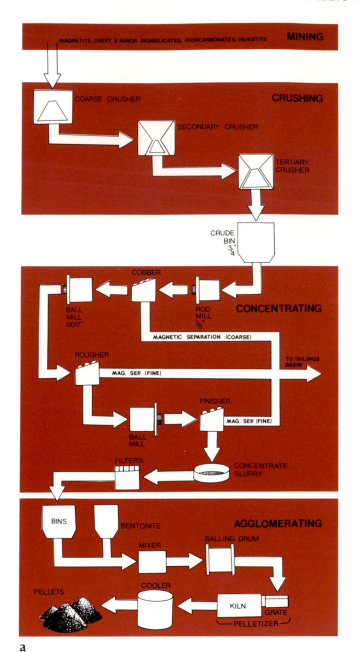

MINING

MAGNETITE, CHERT, & MINOR IRONSILICATES, IRONCARBONATES, HEMATITE

CRUSHING

COARSE CRUSHER

SECONDARY CRUSHER

TERTIARY CRUSHER

CRUDE
BIN
3/4"

CONCENTRATING

COBBER

ROD
MILL
1/8"

BALL
MILL
.0017"

MAGNETIC SEPARATION (COARSE)

ROUGHER

MAG. SEP. (FINE)

TO TAILINGS
BASIN

FINISHER

MAG. SEP (FINE)

BALL
MILL

FILTERS

CONCENTRATE
SLURRY

AGGLOMERATING

BINS

BENTONITE

BALLING DRUM

MIXER

PELLETS

COOLER

KILN

GRATE

PELLETIZER

a

Figure 8-18. (a) The taconite process. (b) The finished product taconite pellets, with about 63% iron and 6% silica.

b

In this process, finely ground hematite is floated on bubbles generated by a chemical mixture, thus separating the hematite (Fe_2O_3) particles from the chert (SiO_2) particles.

Iron Mining and the Environment

Iron mining on the Mesabi Range in Minnesota has created 16,000 direct jobs and countless other indirect jobs, making it a leading employer in northern Minnesota. The taxes paid to state and local governments on iron ore (both natural ore and taconite) totaled about $120 million in 1979.

Yet with the good comes the not-so-good. Mining means that the land surface must be altered in order to accommodate the open pits, the waste rock piles, the ore-processing plants, and the tailings sites. It does not really matter that much of the land surface had already been changed from its original virgin state by logging and the forest fires that commonly followed close on the slashings of the logging process in the late 1800s and early 1900s. Some alteration of the land is necessary if a mining industry is to exist. Trade-offs must be made.

Not too many years ago, few voices raised concerns about the preservation of the environment. Today, the people, the government, and the mining industry, too, are all acutely aware of environmental problems. While mining will disturb a given area for the duration of the operation, the area can be returned to a useful condition. Open-pit mines can become lakes and tailings areas and waste piles, when properly prepared, can become grassy or forested areas. With proper planning, even the wildlife habitat can be as good as or better than it was before mining. One thing must be kept in mind, however. The cost of reclaiming much of the land used by mining may be prohibitive, and it is eventually paid by you, the consumer, in higher product cost.

Whereas open-pit mining operations leave obvious holes on the surface, underground mining can mean less surface disturbance. Shallow underground mines, however, may eventually result in collapse of the overlying surface, as has occurred extensively in Michigan and on Minnesota's Vermilion Range, where the depth of the ore made underground mining necessary (Figure 8-19).

The fine tailings that are a by-product of the taconite process have about twice the volume of the separated magnetite, and their disposal has been of great concern. Reserve Mining Company's plant at Silver Bay, the first taconite plant in the world, was built on the shore of Lake Superior in the mid-1950s, when the process required great amounts of water. The tailings were dumped into the lake, where, it was shown, they would generate turbidity currents and be carried to a deep trough on the lake bottom a few kilometers offshore. And more than 99% of the tailings were carried to the bottom, based on geophysical measurements of the thickness of the underwater sediment. However, a fraction of a percentage of the tailings, composed of the

a

b

Figure 8-19. (a) Area of collapse in the village of Ely, over old underground mines. A small portion of the depression is attributable to open-pit mining in the early days of mining operations and also during World War II. The photo was taken in 1977. (b) Old photo taken underground in the Chandler Mine at Ely showing mine collapse. Unlike the Soudan Mine, which is in very hard and solid rock, the Ely mines needed constant shoring up with timbers. Dozens of miners were killed by mine collapse. (From Clements, 1903.)

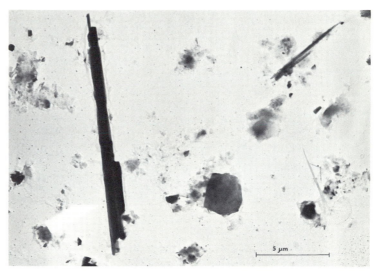

Figure 8-20. Electron microscope photograph of cummingtonite needles removed from the water of Lake Superior. At lower right is a skeleton of a diatom, a microscopic silica-secreting plant. Other fragments are flakes of clay. The bar scale is 0.005 mm long, or about 0.002 in. (Courtesy of P. M. Cook, United States Environmental Research Laboratory, Duluth.)

Figure 8-21. Reserve Mining Company's on-land tailings disposal site 11 km inland from the lakeshore plant. The photo was taken in the summer of 1980, shortly after the site began receiving tailings. Tailings are transported in a slurry through a pipeline, which is visible in the foreground. (Courtesy of Reserve Mining Company and Basgen Photography.)

finest particles, was agitated by waves and moved by currents to surrounding areas in Lake Superior. Much of this fraction was composed of microscopic cleavage fragments of an amphibole mineral, cummingtonite, which was formed where the iron-formation at the eastern end of the Biwabik Iron Formation was metamorphosed by the intrusion of the magmas of the Duluth Complex 1,100 million years ago (Figure 8-20).

The reactions that transformed the original iron minerals into cummingtonite, which is stable under the increased

Figure 8-22. Revegetated taconite tailings area on the Mesabi Range.

Figure 8-23. International Nickel Company's prospect pit, which exposed copper-nickel mineralization in the Duluth Complex south of Ely in 1968. The presence of mineralization has long been known because of the rusting of pyrite at this locality. The pit was filled in shortly after the photo was taken.

temperature and pressure caused by the gabbroic intrusion, can be expressed as follows:

$$\text{Iron carbonate} + SiO_2 + H_2O \rightarrow$$
$$\text{cummingtonite} + H_2O + CO_2$$

or

$$\text{Iron silicates} \rightarrow \text{cummingtonite} + H_2O$$

Though not technically asbestos fibers, as they were called in the legal proceedings during the long controversy, cummingtonite cleavage fragments have a similar elongate form and it seemed possible that the human body would react to cummingtonite fragments much as it does to asbestos fibers, which are known to be carcinogenic when inhaled in sufficient quantities. Fortunately, an on-land disposal site has been built a few kilometers inland at a cost of about $370 million (Figure 8-21).

The taconite mined by most other companies farther west along the Mesbi Range, away from the metamorphosing effects of the Duluth Complex of gabbroic rocks, does not contain cummingtonite. However, stabilization of the waste tailings is desirable in order to eliminate any problems that might result from the fine dust. This has been accomplished largely by the planting of vegetation, mainly grasses and legumes (Figure 8-22).

Recall that the Biwabik Iron Formation on the Mesabi Range dips toward Lake Superior at a gentle angle of about 10°. As the taconite close to the surface is mined away, future mining may require moving down the dip, toward Lake Superior, and hence to greater depths. Thus, underground mining might again be necessary, since it would not be feasible to move all the overlying rock. It has been suggested that 50 to 100 years from now we may see ore-hauling trucks being driven in tunnels dug down the 10° incline of the beds. It has even been suggested that the taconite-processing plants of the future may be located underground.

COPPER AND RELATED DEPOSITS

Copper and Nickel in the Duluth Complex and Related Rocks

The presence of copper and nickel in a few rusty gabbro outcrops south of Ely has been known for several decades. During the 1960s, International Nickel Company (INCO) dug a pit into the rock at one location on the Spruce Road and shipped out several tons of material for metallurgical studies (Figure 8-23). Since then, much exploration has gone on between Ely and Duluth, with an estimated $22 million spent on exploration by 1980. The result is that an estimated 4.5 billion metric tons of copper-nickel ore are thought to be present in a narrow 5-km-wide and 80-km-long strip along the northwestern edge of the Duluth Complex. At 1977 prices for copper and nickel, the ore is worth about $5 billion. Even more important than the market value of the ore is the fact that it makes up 25% of the total United

States reserve of copper and 12% of the world's supply of nickel, increasing the nation's known reserves of nickel by 50 times. Also present in the ore are minor but retrievable quantities of cobalt, gold, silver, platinum, and titanium.

The ore is generally low grade, averaging only 0.5% to 1.0% total copper and nickel (Plate 7) with higher grades present at a few places. However, the tonnages are so great that it becomes a major resource. Copper is mainly present in the mineral chalcopyrite ($CuFeS_2$); and nickel, in the mineral pentlandite [$(Fe,Ni)_9S_8$].

The ore is found in the basal zone of the Duluth Complex, which was described in Chapter 5 as a thick sill (see Figures 5-9 and 5-17). The rock type here is troctolitic gabbro, a rock that consists mainly of plagioclase, pyroxene, and olivine. Prior to detailed mapping and study of the Duluth Complex in the 1960s and 1970s by several geologists of the Minnesota Geological Survey (B. Bonnichsen [1972a], D. M. Davidson, Jr. [1972], H. D. Nathan [1969], W. C. Phinney [1972], R. B. Taylor [1963], and P. W. Weiblen [1965]) it was generally thought that the chalcopyrite and pentlandite had crystallized first, before the more volumi- nous rock-forming minerals, and then because of their higher density compared to the magma had settled to the bottom of the still-liquid sill. Tilting and erosion later exposed the bottom ore-bearing zone. However, the mapping revealed the presence of several individual intrusions rather than one huge one. Thus the world-renowned Duluth

Gabbro became the Duluth Complex, and the ideas about the genesis of the ore had to change too. P. W. Weiblen and G. B. Morey (1976) suggested that the sulfur in the copper and nickel minerals was derived from the middle Precambrian Thomson and Virginia Formations as magma incorporated and melted chunks of the black slates and graywackes, and that the source of the metals was the magma itself.

The ore in INCO's holdings south of Ely (Figure 8-24) would probably have to be mined by open-pit methods, for the basal contact of the gabbro with older rocks here dips southward at a shallow angle. The Minnamax property of American Metals Climax Exploration, Inc. (AMAX), at Babbitt (see Figure 8-24) would be largely mined by under- ground methods because the contact of the gabbro with older rocks dips southward at a steep 30° angle that places the ore at a greater depth than at Ely. AMAX sank a 4.2-m-diameter test shaft 527 m deep and then moved laterally as far as 350 m to test the ore zone (Figure 8-25). Figure 8-26 shows the ore. In early 1982, AMAX decided

Figure 8-25. AMAX's test shaft near Babbitt in 1979. (Courtesy of AMAX, Inc.)

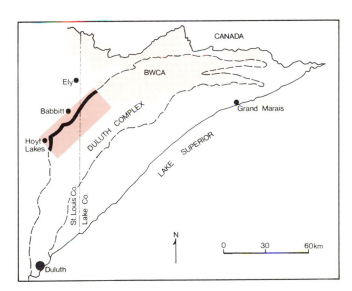

Figure 8-24. Generalized map showing the area in which 13 companies have drilled for copper-nickel mineralization along the base of the Duluth Complex. The narrow, dark zone contains numerous bodies of ore. Much of the ore is at a depth of 600 m, for the base of the gabbro dips to the southeast at 20° to 35°. The western part of the Boundary Waters Canoe Area is not shown. INCO's area is southeast of Ely and AMAX's area is near Babbitt. (After Bonnichsen, 1974, and Listerud and Meineke, 1977.)

Figure 8-26. A piece of drill core from AMAX's Minnamax Project near Babbitt. The main copper mineral, chalcopyrite ($CuFeS_2$), shows as light-colored blebs. (Courtesy of AMAX, Inc.)

not to develop the property in spite of their expenditure of $21 million, citing the low price of copper and adverse economic conditions.

A three-year moratorium on copper-nickel development was imposed in 1976 in order to allow a detailed study of the environmental, social, and economic impact that mining of the gabbro would have on northeastern Minnesota. The study was completed in September, 1979. Although the report made no recommendations about mining, it did indicate that mining, milling, and smelting seem feasible if modern technology is employed and if pollutants from other industries are virtually nonexistent. One point that must be considered is the impact on the Boundary Waters Canoe Area of an open-pit mine with possible metal-bearing waters and a sulfur dioxide-emitting smelter near Ely. Another is the extra growth stress on the two nearest towns, Ely and Babbitt, which could double in size. On the positive side, mining of the present reserves over a 50-year period could create 2,500 direct permanent jobs, another 5,000 indirect jobs, as much as $300 million in tax revenues, and large royalties to the state.

Other copper and nickel occurrences are present in Upper Precambrian mafic dikes and sills near the northeastern tip of Minnesota. Although 36,000 t of ore containing 0.4% copper, 0.2% nickel, and low values of platinum, paladium, and gold have been found in one deposit, the minerals are not present in large enough amounts to mine (Mudrey, 1972).

The possibility that a smelter will be located in Minnesota brings up the problem of "acid rain." "Acid rain" is of world-wide concern, mainly because coal-burning power plants and industrial smelters place sulfur dioxide into the air where it combines with water to form sulfuric acid. Automobile exhaust in the air combines with water to produce nitric acid. When it rains or snows, the acidic precipitation can end up in lakes where the increased acidity affects the plant and animal life at all levels of the aquatic food web from bacteria to fish. The total numbers of organisms and the numbers of species decrease in acidic waters, and, ultimately, reproduction is affected and even fish can vanish completely. The complex natural ecosystem is fragile.

In early 1982, new data showed that the ability of many BWCA lakes to withstand acidification is decreasing. The buffering capacity of a lake (the capability of neutralizing acidic components), which is estimated by its alkalinity, is largely a function of the amount of carbonate and bicarbonate in the water. The chemistry can be complicated, but carbonates in solution are largely the products of weathering of calcium-rich rocks such as limestone, and much of the igneous-metamorphic bedrock of the BWCA is a low producer of buffering ions. Glacial drift contains some limestone, but there was not much glacial deposition in northeastern Minnesota in general and in the BWCA in particular.

Therefore, the runoff waters from rain and snow that reach the lakes are acidic. Thus, BWCA lakes are highly susceptible to acidification and could become acid rather easily as their original alkaline contents are utilized in buffering added acidic components. Of the 1338 lakes in the BWCA, 1218 are sensitive to acid deposition; of these, 308 are extremely sensitive, 575 are moderately sensitive, and 334 are potentially sensitive (Minnesota Pollution Control Agency and others, 1981). Of the 2561 lakes present in Cook, Lake, and St. Louis Counties, 2024 are sensitive. Some areas in North America and in Europe already contain acidified lakes without fish. If a smelter is ever built in Minnesota, it would have to be of a type that would not add to the "acid rain" problem.

Copper in Upper Precambrian Volcanic Rocks

Paleo-Indians mined native (metallic) copper from the Upper Precambrian volcanic rocks of the Keweenaw Peninsula of Michigan and Isle Royale, perhaps as much as 5,000 years ago. In fact a copper culture developed there, with the copper traded as far away as what are now Montana, New York, and Georgia. A much newer "copper culture" was started in Michigan in the 1840s when mines were opened on the Keweenaw Peninsula, removing copper from what were once vesicular tops of lava flows and porous conglomerates associated with the flows.

Because the Michigan copper was in the volcanic sequence, it was only natural that prospectors should look for copper in the volcanics of Minnesota as well. Small amounts of native copper and other copper minerals have been noted at many places on the North Shore of Lake Superior and especially along the French River near Duluth and the Stewart River near Two Harbors. About 20 known occurrences are located between Duluth and Two Harbors and between Tofte and Grand Marais, with the copper present as either metallic (native) copper or as copper sulfides. Masses of native copper weighing more than 6 kg were recovered, and locally exploration was quite extensive but without success.

Along the French River, the French River Mining Company mined about a ton of copper over a period of several years, starting in 1863. The only copper "mine" is located near the Little Knife River; 20 holes were drilled and a 33-m-deep shaft with 60 m of drifting (tunnels) was put down by the Mining Corporation of Canada in 1929 and 1930. Three buried flow tops contained copper, but not in great enough quantities to be commercial. Other mining attempts were made farther south along the volcanic belt (see Plate 2) where the lavas are locally exposed. Just east of Hinckley on the east bank of the Kettle River, the Great Northern Copper Company reportedly mined native copper from

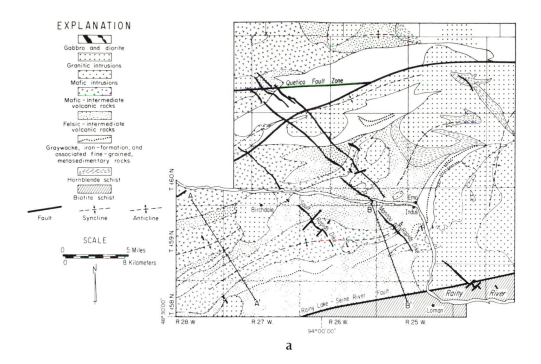

a

1897 to 1902, and there are several test sites just east of Pine City.

Why is there no copper on the North Shore when it is so abundant in the Keweenaw Peninsula and even on Isle Royale? W. S. White (1957) has suggested that the steep dips of the lava flows in the latter two areas allowed hot copper-bearing fluids generated at depth beneath the thick pile of lava flows to move upward along porous flows and conglomerates, whereas the gentle dips (10° to 25°) of Minnesota's lava flows (see Figure 5-10) may have hindered such movement. Boulders of native copper are occasionally found in glacial drift of Minnesota, and, though their source is difficult to pinpoint, Isle Royale is a likely place.

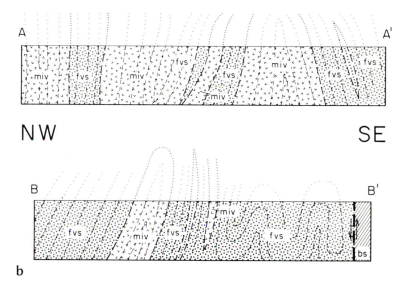

b

Figure 8-27. (a) Geologic map of an area on the Minnesota-Ontario border that has been interpreted as an explosive volcanic center. The Canadian portion of the map is from Ontario geologic survey maps. (From Ojakangas, Meineke, and Listerud, 1977.) (b) Geologic cross sections across the same map trending in a northwest-southeast direction. The folds are drawn on the basis of top indications given by pillowed greenstones and on a few graded beds in drill cores. The abbreviation miv represents mafic intermediate volcanic rocks; fvs represents felsic intermediate volcanic rocks; and bs represents biotite schist. The thin lenses with a single line of dots represent iron-formation. (From Ojakangas, Meineke, and Listerud, 1977.) (c) Interpretive geologic model of same area as shown in a and b during volcanic activity in the sea, prior to uplift and folding. A sequence of explosive felsic volcanic rock units and minor precipitated ("exhaled") iron-formation (dark) rests on the greenstone base. Compare with the model of Figure 3-9. (After Ojakangas, Meineke, and Listerud, 1977.)

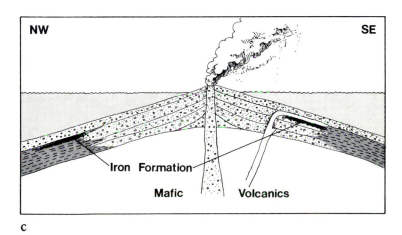

c

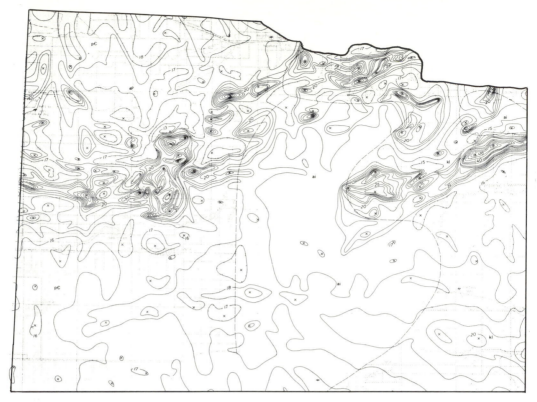

Figure 8-28. Aeromagnetic map of a portion of northern Minnesota (Canada is the white area in the upper right corner) showing differences in magnetic susceptibility of the bedrock. Closely spaced contours with higher numbers indicate greater differences than do widely spaced contours with lower numbers. Closed lines with tick marks on the inside indicate areas of lower magnetic intensity. Compare with the geologic map of Figure 11-10 and with the gravity map of Figure 8-29. (From Bath, Schwartz, and Gilbert, 1964.)

Figure 8-29. Gravity map of a portion of northern Minnesota showing differences in gravitational attraction of the bedrock. The contours are in milligals and all are given as negative numbers. The lower the number (e.g., −30), the lower the gravitational attraction of the underlying rock. The higher the number (e.g., −10), the higher the gravitational attraction. Compare with the geologic map of Figure 11-10 and with the aeromagnetic map of Figure 8-28. (From McGinnis, Durfee, and Ikola, 1973.)

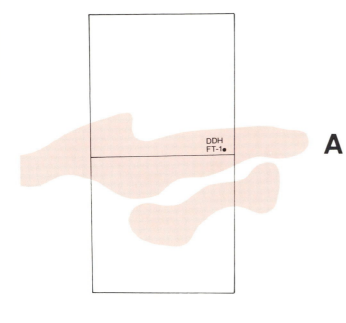

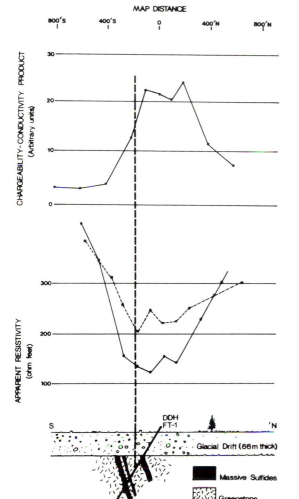

Lower Precambrian Copper-Zinc Deposits

The volcanic-sedimentary (greenstone) belts described in Chapter 3 have received much study because they contain most of the copper, zinc, gold, and silver deposits of Canada. The evidence is very strong that the deposits are of volcanogenic origin. Most of such deposits are located near explosive felsic volcanic centers, so a main objective of exploration in any such area is to attempt to locate such centers. Both the felsic chemical composition of the rocks and the large size of the explosive fragments are suggestive of proximity to a vent area. Several such vent areas have been approximately located in northern Minnesota (Figure 8-27 shows one example), but associated ore deposits, if present, have not yet been found.

Earlier in this chapter, iron-formations of Early Precambrian age were described as exhalites because the ore-generating fluids were literally exhaled by volcanism. The same can be said of the sulfide ore minerals.

During the late 1960s, more than $10 million was expended by companies on mineral exploration in the volcanic-sedimentary belts of northern Minnesota. Although exploration decreased during the 1970s, some companies have nevertheless continued their efforts. Because of the thick glacial deposits that cover most of the rocks and because deposits may be beneath other rocks, geophysical techniques have to be employed. Whereas aeromagnetic and ground gravity surveys are used to outline the broad distributions of covered rock types (Figures 8-28 and 8-29), other techniques measure the electrical conductivity of buried rocks and thus locate smaller highly conductive rock

Figure 8-30. In 1972, the Ridge Mining Company drilled a 215-m-long inclined drillhole near Birchdale, in Beltrami County, to test an electromagnetic anomaly. The drillhole intersected three zones of massive sulfide—27 m of 20% to 90% sulfide, 10 m of 30% to 70% sulfide, and 6 m of 60% sulfide. Unfortunately, the sulfides are all pyrite and pyrrhotite, which are iron sulfides; only traces of copper-bearing sulfide (chalcopyrite) were detected in the hole. The remainder of the core consists of greenstone (metabasalt and metaandesite) with some possible rhyolite. The anomaly was initially located during airborne electromagnetic surveys, as shown in part A; with the drillhole indicated by the numbered dot; the squares are 1 mi on a side, and two areas of higher conductivity are outlined. The drillhole is not shown on Figure 8-32 but is located near hole T 25 A 1, 55 km southwest of Baudette. Then, a ground survey along a measured grid produced a magnetic map and electromagnetic and induced polarization data. The vertical line that extends through parts B, C, and D of the diagram shows the location of the drillhole relative to two profiles derived from the field measurements. Part B shows the zone of high electrical conductivity, part C shows the zone of low resistance to the passage of an electrical current, and part D shows the inclined drillhole and the massive sulfides encountered in the volcanic sequence. The data on which the illustration is based were submitted by the Ridge Mining Company and are on file at the Minnesota Department of Natural Resources at Hibbing.

143

Figure 8-31. Part of a drill core from northwestern Minnesota that contains 54 m of 5% to 100% sulfide mineralization in a sequence of explosive volcanic rocks. However, the sulfide (reflective in the photo) is all pyrite and pyrrhotite, both rather useless iron sulfide minerals, without any copper or zinc mineralization. The sulfide portion of the core has been split for sampling purposes. This is another example of a technical geophysical success, for sulfides were located, but it was not an exploration success. This is Exxon drillhole No. R-4-1, 45 km southeast of Baudette.

Figure 8-32. Map showing locations of exploration drillholes on state lands in a portion of northwestern Minnesota. Another 27 drillholes in the shaded Birchdale-Indus area are not shown. The small crosses represent a single drillhole, and circled crosses represent more than one drillhole. (From Ojakangas, Meineke, and Listerud, 1977.)

units such as massive sulfide bodies (Figure 8-30). Drilling is then necessary to test the anomalies. Several massive sulfide zones were drilled in northern Minnesota but were found to contain only pyrite and pyrrhotite, two minerals composed of iron and sulfur (Figure 8-31). They lacked the copper, zinc, and lead found in minerals such as chalcopyrite ($CuFeS_2$), sphalerite (ZnS), and galena (PbS). Although the finding of such massive pyrite-pyrrhotite bodies is a technical success, that is not much consolation to the geologists and the companies expending the great amounts of time and money on the search for ore. Figure 8-32 shows some of the drillholes on state-owned land; drilling on some private land is not shown. Geochemical studies of soil, water, and lake sediments are also used to zero in on areas with, for example, high trace amounts of copper or nickel.

Although no volcanogenic ore deposits have yet been found in Minnesota, the prospects have to be judged as fairly good if for no other reason than that the ore-bearing Canadian belts extend into Minnesota. A base-metal mine in Ontario, Sturgeon Lake, is located only 160 km north of the international border in the same volcanic-sedimentary belt that extends into Minnesota in the Rainy Lake and Baudette areas, and many others are in the same belt but farther away. A few Minnesota localities do contain some zinc or copper; the best to date is a 6-m-thickness of 1% copper (including 1 m of 3.38% copper) in a drillhole near Soudan. Four ore deposits, one of which ranks among the world's largest, were also found in volcanic-sedimentary accumulations in north-central Wisconsin in the 1970s; these rocks may be Middle Precambrian rather than Early Precambrian as in northern Minnesota. It is probably only a matter of time before an ore deposit will be found in one of Minnesota's volcanic-sedimentary belts.

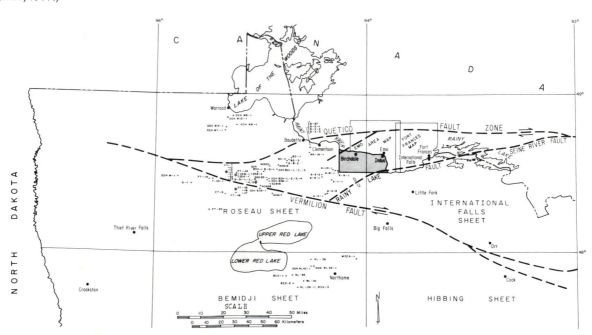

URANIUM DEPOSITS

The energy crisis of the 1970s emphasized the dependence of the United States upon foreign energy sources. This, along with other factors, helped intensify the search for domestic energy reserves, including uranium. This search brought uranium exploration companies into Minnesota in the late 1970s. Why Minnesota? One rule of mineral exploration is to search areas that have geologic characteristics similar to those in areas that are known to contain ore deposits. In the 1970s, several exceedingly rich uranium deposits were found near the buried erosional surfaces (unconformities) present beneath the Athabasca Sandstone of northern Saskatchewan and the Kombolgie Sandstone of northern Australia, both of Late Precambrian age. Ore is found at the unconformity, in the rock beneath it, and in the overlying sandstones. Geologists differ in their theories on the origin of such deposits: some think that they were deposited by rising warm waters and others think that they were deposited by descending surface or near-surface waters. But most agree that the unconformities played a key role in the genesis of the deposits and all agree that the ore is near unconformities.

There are several unconformities present in Minnesota's rock column, as described in preceding chapters. The major ones are located at the bases of Middle Precambrian rocks (Chapter 4), Upper Precambrian rocks (Chapter 5), Cretaceous rocks (Chapter 6), and Pleistocene sediments (Chapter 7). Though any unconformity may be of some interest to uranium seekers, the two within the Precambrian column, about the same age as those in Australia and Saskatchewan, are the most tantalizing. In addition, radioactivity levels 10 to 20 times higher than background have been found near Denham and at the old Arrowhead gold-graphite mine near Mahtowa (Figure 8-33).

The Upper Precambrian Sioux Quartzite of southwestern Minnesota (see Figure 5-2) is similar to both the Athabasca Sandstone of northern Saskatchewan and Kombolgie Sandstone of northern Australia and rests unconformably upon older rocks as do the other two units. Consequently, it is being explored for uranium deposits. The Athabasca, Kombolgie, and Sioux all appear to have been deposited on broad alluvial plains, where systems of sediment-laden streams dropped their loads. Oxygen-rich surface waters in the higher reaches of the drainage systems could have dissolved small amounts of uranium from source rocks that may have contained only a few parts per million of uranium and then carried it in dilute solution until reducing chemical conditions caused it to be precipitated as new uranium minerals, typically pitchblende (UO_2). Such chemical conditions may well have existed at the unconformities; great quantities of water could have flowed past chemically favorable sites. The minerals could have been deposited by

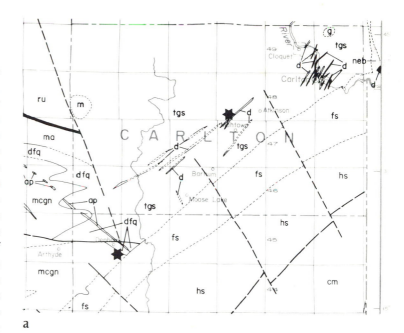

a

b

Figure 8-33. (a) Geologic map of a portion of east-central Minnesota that was actively being explored for uranium during the late 1970s and early 1980s. Note especially the Fond du Lac Formation (symbolized by fs), which rests unconformably upon the Thomson Formation (symbolized by tgs), and the Denham Formation (symbolized by dfq), which rests unconformably upon the McGrath Gneiss (symbolized by mcgn). The stars indicate locations of two minor radioactive "hot spots" that have been known since the early 1950s. The geology has been changed somewhat in a new map by Morey, Olsen, and Southwick, 1981. (From Morey, 1978.) (b) A piece of slightly radioactive pyritized and fractured graphitic slate from the old Arrowhead gold-graphite mine just north of Mahtowa. The radioactivity resides with the pyrite between pieces of slate.

waters flowing over the eroded surfaces before, during, or even after deposition of the sands. If the correct chemical conditions were present near shear or fault zones where broken rock made space available for the precipitation of the new minerals, the deposition of uranium would have been even more likely. Deposition of uranium by warm waters rising from depth would occur at the same spots, with the shear and fault zones serving as the channels for the waters as well as the depositional sites.

The Fond du Lac Formation, younger than the Sioux Quartzite, appears to be very similar to feldspathic sedimentary rocks of the Martin Formation just north of the Athabasca Sandstone of northern Saskatchewan. Although they are of a different composition than the Sioux Quartzite and the Athabasca Sandstone, the Fond du Lac and the Martin formations nevertheless share many characteristics with them. They were deposited by river systems upon major erosional surfaces. Where the Fond du Lac rests upon the faulted, folded, and eroded black organic and pyrite-bearing Thomson Formation, the chemical environment and the space necessary for precipitation of uranium minerals appear to have been readily available (Figure 8-34). Again, theories involving either warm rising waters or cool, descending waters can be applied. Of special interest in this area is the presence of some natural gas in a water well near the village of Kettle River. This gas, which appears to have been generated within the Thomson Formation, would be an excellent reducing agent and could have caused the precipitation of uranium from water. The gases could have moved into both the Thomson and the Fond du Lac formations.

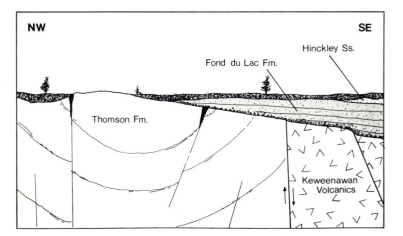

Figure 8-34. Diagrammatic northwest-southeast cross section across the Thomson-Fond du Lac boundary of the map of Figure 8-33. Details of relationships are unknown because of cover by rock units and glacial till. However, this illustrates the potential for uranium mineralization (solid black spots) where faults intersect the unconformity or intersected the unconformity before erosion removed the Fond du Lac and the unconformity itself. The fault between the volcanics and the Thomson Formation is hypothetical.

How do companies search for uranium deposits in such rock units? Obviously, a uranium deposit situated at the ground surface is the easiest type to locate. Consequently, searching is done along the edges of younger rock units that may have been eroded off uranium deposits along the unconformity, thereby exposing the deposits. Locating a deposit beneath the younger rock unit is like trying to find a needle in a haystack but more difficult. If you lost a needle in a haystack, you could carefully remove the haystack, piece by piece, until you found the needle. When they search for uranium in the "rock stack," geologists don't even know whether uranium was ever present, and obviously they cannot remove the rock stack. Expensive drilling is necessary, and a small but rich uranium deposit could easily be missed by a few meters. Early in a search, aerial radiometric surveys are often used to measure radioactivity levels over a region. Water, sediment, soil, and even vegetation may be analyzed for uranium and radioactive daughter products of natural decay to further narrow the search to specific areas. Surveys by hand-carried scintillometers, the successors to Geiger counters, are used to measure radioactivity of rocks; however, just 30 cm of soil can mask a high-grade uranium deposit. Whereas a scintillometer measures total gamma radioactivity and does not distinguish between uranium, thorium, or radioactive potassium, a portable device called a spectrometer does distinguish uranium. Other geological techniques, especially electrical conductivity, are used to locate buried fault zones, fault zones containing abundant pyrite, and graphitic layers in the rock, all of which would have been favorable sites for the deposition of uranium. Although there are scattered rock outcrops in southwestern Minnesota where the Sioux is exposed and in east-central Minnesota where the Fond du Lac overlies the Thomson, the thick glacial cover (see Figure 2-5) makes exploration for minerals, especially uranium, difficult. All rocks contain some uranium—usually less than 4 parts per million (ppm)—and will register on a scintillometer, but to find uranium in minable concentrations is quite a task.

This discussion has centered on the potential for unconformity-type uranium deposits in Minnesota. Other types of deposits could also exist. It is at least theoretically possible that uranium-bearing conglomerates of early Middle Precambrian age, such as those present north of Lake Huron in Ontario (see Chapter 4), could exist deep in the subsurface of east-central Minnesota beneath upper Middle Precambrian rock units. Pegmatites in the Lower Precambrian units of northern Minnesota could contain uranium minerals; some with abnormally high radioactivity are present in Minnesota's Northwest Angle. The metamorphic migmatite terrane (see Chapter 3) west of the Vermilion Batholith in Northern Minnesota (see Plate 2) appears to be similar to, but older than, rocks in Southwest Africa that contain low-grade but large uranium deposits; some rock exposures in the Big Falls area have radioactivity levels many times greater than background. There is even a remote possibility that the Upper Precambrian sandstones of Minnesota (the Fond du Lac and the Hinckley) and other rock units of the Lake Superior region contain sandstone-type uranium deposits similar to those in much younger rocks of Wyoming, Colorado, and New Mexico.

What if a uranium discovery were to be made? Concern has been expressed that if an exploration drill penetrated an ore body it might contaminate the groundwater with radioactivity. Actually, ore bodies are likely to have been deposited in porous rock in the first place, so the system has probably always been open to contact with groundwater. This principle is, in fact, used in exploration. Geochemical study of water samples from both water wells and streams can delineate areas of higher radioactivity. In the east-central Minnesota area, samples collected by the Minnesota Geological Survey and analyzed by the United States Department of Energy have shown 43 water wells to be so anomalously radioactive as to exceed safety levels established by the government. This is a natural contamination. If uranium deposits were discovered and mined in Minnesota, both federal and state nuclear regulatory agencies would strive to make certain that the mining operations had little harmful environmental impact. The experience gained by mining elsewhere over the past few decades is invaluable and should help assure safe mining operations. When mining is completed, the land can be restored as closely as possible to its original state.

GOLD AND SILVER DEPOSITS

California had its Gold Rush in 1849, and Minnesota had its Lake Vermilion Gold Rush in 1865 and 1866. State Geologist Henry Eames published, in September, 1865, an encouraging assay on a vein quartz specimen from Lake Vermilion that indicated a gold value of about $23 per metric ton and a silver value of about $4 per metric ton. Eames reportedly wrote that the slates around the lake were traversed by numerous veins of quartz, nearly all of them showing the presence of the precious metals—gold and silver (1866). However, a Scot by the name of McEwen claimed to have found gold there 10 years earlier, in 1855. He may be the one who, according to legend, plunked a nugget down on a bar one night and, when properly plied, answered that he found it "up on the lake."

People hurried to the area. Ossian Dodge, a reporter with the pen name Oro Fino ("fine gold"), wrote letters back to the St. Paul Pioneer Press. One of these stated:

The majority of these points (on Lake Vermilion) are composed of talcose slate sprinkled with iron- and gold-bearing quartzes. As a general thing, they are heavily timbered and are covered with thick moss. In the latter being removed, the eye beholds bright veins of quartz, lavishly enriched with gold (1865).

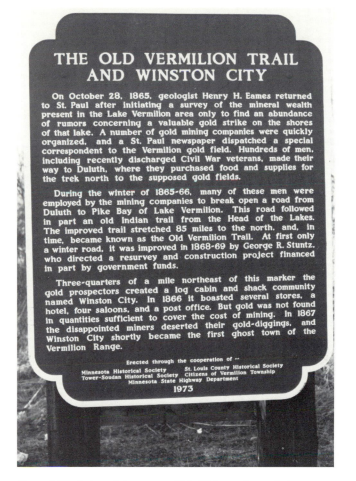

Figure 8-35. A sign of the past on U.S. Highway 2 near Lake Vermilion.

Specimens with values as high as $165 per metric ton for gold and $8 per metric ton for silver were reported. The winter of 1865 and 66 saw the formation of several gold-mining companies in St. Paul. With spring came trainloads of supplies and machinery heading up to Duluth and Superior, there to be transferred to wagons for the remaining overland route via a newly hacked out "road" to Lake Vermilion. Included in the shipments was machinery for an ice cream parlor; the milk was to come from a Major Newsom's oxen. Mills were built to process the ore, and townsites were planned (Figure 8-35).

Enthusiasm was high and the speculators were busy. Then a Mr. Walker sent some ore samples to a Professor Chase for assay. Chase replied that the gold content was almost nil and he was attacked as inept. But by the end of the summer of 1866 people began to realize that fool's gold (FeS_2) was abundant and that no one had actually seen gold. Few miners remained after that. Perhaps fortunes had been made by the freight haulers and the outfitters, but none were made in the gold fields. The prospectors and miners must have tried hard, for prospect pits and shafts

a

b

Figure 8-36. (a) Rainy Lake City in May, 1894, when it had a population of 350 people. It is not even a ghost town, for nothing remains today. (Courtesy of the Koochiching County Historical Society.) (b) Stamp mill at Rainy Lake City about 1.5 km east of the Little American Mine. (From Winchell et al., 1899.)

dot the numerous quartz veins of the area and there is an old adit (tunnel) on Gold Island in Lake Vermilion. On the Canadian side of the border, however, gold was eventually found at Lake of the Woods and on the North Shore of Lake Superior.

A Duluthian who had made a fortune in gold on the Canadian side of Lake of the Woods sent prospector George W. Davis to the American side of the border area. In July, 1893, Davis stopped on Little American Island (Figure 8-36) on Rainy Lake, near the entrance to Black Bay, and found gold in a 2-m-wide quartz vein. Word of the Little American Mine spread quickly, and by 1894 pits and shafts dotted the islands and the mainland of the American side of Rainy Lake. The Little American remained the largest mine with its 60-m-deep shaft. Getting to Rainy Lake usually involved taking the train from Duluth to Tower, where iron ore was already being mined in the "gold fields" of the 1865 gold rush, boating across Lake Vermilion to the Vermilion River on the *Libby*, traveling by stage up the Crane Lake Portage Road to the town of Harding on Crane Lake, taking the steamer *May Carter* to Kettle Falls, walking a 200-m portage to Rainy Lake, and finally boarding another steamer (the *Walter S. Lloyd*) for Rainy Lake City, which was located at the entrance to Black Bay on Rainy Lake.

At the new Rainy Lake City site, 14 saloons, three hotels, a bank, a newspaper office, a barber shop, a post office, several stores (including a wild game butcher shop where one could buy a deer for a dollar), a sawmill, and a school were built (Figure 8-36). A stamp mill was erected for crushing the quartz in order to free the gold, and a thousand people attended its opening on July 4, 1894. Less than 30 t of ore yielded $988.50 worth of gold in the first crushing, which took 11 days. However, the quantity of ore was small and the operation folded after producing an estimated $4,500 to $5,600 worth of gold in 1894 and 1895. The mill had run for 52 days and had crushed 450 t of ore. In 1895, N. H. Winchell, the state geologist, described the 13-m-deep shaft of Little American Mine as follows:

With no timbering, no pump, nothing but a hand
windlass for hoisting ore, rock and water and the mill
a mile away, it is a good example of the folly of robbing
a mine in order to provide ore for present purposes
without proper development for future mining
(1895:79)

By 1897, Rainy Lake City was just another ghost mining town. The prospectors and miners had left for richer fields, including the Klondike area in the Yukon, which experienced a big gold rush in 1897 and 1898. In 1906, only one person still called Rainy Lake City home. A few summer cabins have been built on the old townsite.

The quartz veins on Little American Island contained pyrite, chalcopyrite, and tourmaline in addition to the gold. The vein is one of many veins and pods of quartz found along a major fault zone in schists (Figure 8-37). Most of the other pits and shafts are also on this fault, which extends for 200 km to the east in Canada and 80 km to the west. It has been named the Rainy Lake Fault in Minnesota and the Seine River Fault in Ontario. Chemical analyses on rocks from several of the pits show small amounts of gold and silver. Near the fault in Canada, three gold mines have operated in the past.

Gold has always been the favorite of prospectors, and "mines," all too commonly parts of fraudulent stock schemes, were developed at several places in Minnesota. Among these are the Arrowhead gold-graphite mine at Mahtowa 65 km south of Duluth and a probably fraudulent mine near Delhi in the Minnesota River Valley. The Buyck Gold Mine, or Crane Lake Gold Mine, on the Vermilion River about 32 km northeast of Orr was both a mining and milling operation during the 1920s. Many of its buildings are still in use as a resort. Legend has it that this was a profit-making stock venture, with the mine "salted" by means of a shotgun blast of gold. However, some say gold was actually recovered from the rock, which is a large pyrite-bearing inclusion of biotite schist in granite. Gold has also been reported in small quartz veinlets in granitic rock near Virginia. Gold has even been rumored to be present in the St. Peter Sandstone near Fort Snelling, in Minneapolis, a virtual impossibility.

The richest silver mine in the Lake Superior region was Silver Islet, a small island in Ontario, 65 km northeast of the northeastern tip of Minnesota. Silver worth $3,350,000 was mined there between 1868 and 1884 with special procedures necessary to keep Lake Superior out of the 70-m-deep shaft. Another $1,885,000 worth of silver was mined on the Ontario mainland to the west. The ores were associated with mafic dikes intrusive in the Middle Precambrian Rove slate. These rocks are also present in adjacent Minnesota; prospecting has resulted in the location of ore minerals in quantities too small to be of economic value.

Silver was explored for intermittently between 1880 and 1929 in downtown Duluth at the "Point of Rocks" in gabbro of the Duluth Complex. The old shaft is rumored to be beneath a present-day house and an adit purportedly intersects a railroad tunnel. There is a story that $30,000 worth of silver and copper ore was bagged up, awaiting shipment to the east, but the sulfide minerals oxidized, thereby rotting the bags, and the ore was discarded into the bay.

Minor exploration for silver and gold is still going on in northern Minnesota. Research by state agencies is conducted on Minnesota rocks from time to time. For example, a 1979 study revealed traces of gold in chemical analyses from the old McComber iron mine on Armstrong Lake east of Soudan; this mine produced 7,624 t of hematite ore from 1917 to 1919.

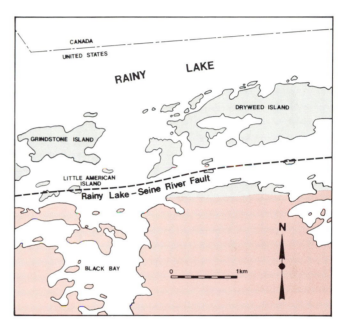

Figure 8-37. Geologic map of part of Rainy Lake showing the location of the Little American Mine on the Rainy Lake Fault. (After the Minnesota Geological Survey, 1969.)

A small peat bog in northern Minnesota.

9

Nonmetals, Fuels, and Water

Metals are the glamour elements when it comes to mineral exploration and exploitation, and the iron deposits of Minnesota deserve premier status in an inventory of the state's mineral wealth. Or do they? If mineral resources include water, then a price must be set on lakes, streams, swamps, and groundwater reserves. But how does one determine in dollars and cents the value of rivers and lakes as effluent disposal systems, transportation routes, and water supplies? It is difficult, and even impossible when biological, recreational, and aesthetic uses are added to the picture. Consider also the exploitation of the soil in the extraction of nutrients by agriculture. In one justifiable view, plowing and crop harvesting combine to become the largest open-pit mining operation on earth. What is the value of minerals withdrawn each year by farmers from the state's soil bank without being totally replaced by fertilizers? Less glamorous and less important resources range from clay to building stone to sand and gravel. Even agates and other pretty stones have a value and give thousands of "prospectors" happy hours of exploration time.

NONMETALLICS

Sand and Gravel

In 1978 Minnesota's total production of sand and gravel was valued at $61 million (Briggs and Rafn, 1979). Every one of the state's 87 counties has active gravel pits, although most of the production occurs around the larger metropolitan areas. The quality of natural aggregate varies considerably. Gravel and sand derived from glacial deposits of the Des Moines Lobe contain high volumes of shale fragments, which degrade the usefulness of the material for many purposes. Generally, the largest bodies of high-grade aggregate are found in the moraines of the Superior and Wadena lobes and as outwash fans and terraces constructed by

meltwater streams beyond their margins. Eskers, those long, sinuous, ice-tunnel deposits, are easy targets for sand and gravel prospectors. Unfortunately, there is reason to fear that they will disappear from the landscape, consumed by power shovels and front-end loaders. Still, Minnesota contains a vast reserve of this important resource. Tens of millions of dollars each year in transportation costs are saved because glaciers so richly endowed the state with high-quality, naturally washed sand and gravel, conveniently located for easy extraction. Gravel pits, in fact, have become one of our most common "landforms" (Figure 9-1).

Clays

Clays have been quarried at several locations in the state. Some were formed by Cretaceous weathering (see Chapter 6) and others accumulated in glacial lakes, such as Glacial Lake Duluth, or along glacial rivers. The clays have been used to make brick, tile, and pottery.

Dimension Stone and Crushed Rock

Perhaps the first mineral industry in the state involved the quarrying of rock as *dimension stone*, that is, cut for building purposes. Limestone was quarried for Fort Snelling as early as 1820. Early homes, many still standing, were partially or totally constructed from the rock formations found nearby. Even glacial boulders were piled into foundations, walls, and hearths, as they still are. Sandstone, limestone, and granite are the most commonly quarried rocks.

The granite quarries of the Minnesota River Valley in the vicinity of Morton and Ortonville, as well as the more extensive developments in the St. Cloud region, produce a great variety of beautiful and structurally sound dimension stone (Figure 9-2). "Diamond Pink," "St. Cloud Red," "Opalescent," "Rainbow," and "Dark Pearl," some of the trade names for these rocks, suggest the range of color and pattern. The Cold Spring Granite Company has grown from its founding in 1889 into one of the largest producers in the United States.

Quarries along the Minnesota and Mississippi river valleys produce dimension stone and crushed rock from the dolomites and limestones of Ordovician age, especially the Oneota and Platteville formations. These softer rocks are easily cut and polished, a factor in their popularity as facings and floors in large buildings. Such names as Winona Travertine and Kasota Dolostone place the point of their origin.

In earlier times, sandstone was a popular building material, both for structures and pavements. Abandoned quarries (opened in the 1870s and 1880s) at Sandstone and elsewhere along the Kettle River have been developed into parks. Banning State Park is an example. A detailed guide to the architecture of downtown St. Paul, written by Sr. Joan Kain (1978), is available from the Ramsey County Historical Society. Called *Rocky Roots*, the booklet provides a geological background on the dimension stones used in the construction of many of the buildings in St. Paul. Hinckley Sandstone has been used in buildings in Illinois, Washington, and Iowa as well as in Minnesota, including the Great Northern depot in Duluth and the Burlington Northern office

Figure 9-1. Gravel pit from which sand and gravel deposited by glacial meltwater are being removed.

Figure 9-2. Granite quarry at St. Cloud.

Figure 9-3. The first 3M plant, an eight-story-high crusher to crush "corundum," was built on the shore of Lake Superior in 1903. (Courtesy of Minnesota Mining and Manufacturing Company.)

building in St. Paul. Appropriately, Pillsbury Hall, which houses the Department of Geology and Geophysics at the University of Minnesota at Minneapolis, is made of this Minnesota stone. Sandstones of the Fond du Lac Formation quarried near the village of Fond du Lac at the west end of Duluth in the 1880s were also widely used. In Duluth, the Board of Trade Building still stands a century later as a monument to this brownstone industry.

Quarries that produce crushed and broken stone for the needs of the construction business accounted for almost $19 million in 1978, compared to $8.6 million for dimension stone. Abrasives, clay, and lime brought in another $10 million (Briggs and Rafn, 1979).

Other Rocky Ventures

In 1902, "corundum" was found on the North Shore of Lake Superior near what is now Illgen City. A new company —Minnesota Mining and Manufacturing (3M)—was formed to mine this rock for use as abrasive (Figure 9-3). The "corundum" proved to be anorthosite, a rock composed largely of plagioclase, which is considerably softer than corundum. This finding hurt the company, and in 1904 3M stock was being traded at "two shares for a shot, and cheap whiskey at that" (Huck, 1955: 23). Eventually the company moved to Duluth and imported garnet for use in abrasive papers. Another move brought them to St. Paul, where the company increased the number of items being manufactured. In spite of its rocky start, the company has survived. In the 1970s the anorthosite was being looked at again, this time as a possible source of aluminum, for it contains 20% to

30% aluminum oxide (Al_2O_3). However, bauxite, the ore of aluminum, contains 55% to 65% Al_2O_3, and the aluminum is much more easily removed from the latter material.)

Another venture based on a rock product was tried a few kilometers west of Ely in the early 1920s. It was decided that the Ely Greenstone, present in almost unlimited quantities, could be crushed and substituted for the crushed green slate that was shipped in with high freight charges from the Appalachians for use as granules on roofing paper and shingles. A quarry was opened, crushing and screening commenced, and several thousand tons were shipped out. Reportedly the development of artificial coloring techniques for use on any type of crushed rock did this new industry in. All that remains is the quarry and a small green "mountain" of the fine-grained waste products near U.S. Highway 169.

Minnesota's only pegmatite mine was located in Minnesota's Northwest Angle. Although a road to that isolated part of Minnesota was not completed until the 1970s, a quarrying operation was tried in the 1930s. Coarse-grained potassium feldspar (some crystals were over 30 cm long) was quarried, moved a short distance to the lakeshore, and hauled by barge about 65 km to the railroad at Warroad. The feldspar was shipped out for use by the ceramics industry.

FUELS

Peat

Minnesota's peatlands, which mark the locations of glacial lakes, are a huge source of potential energy. Of the

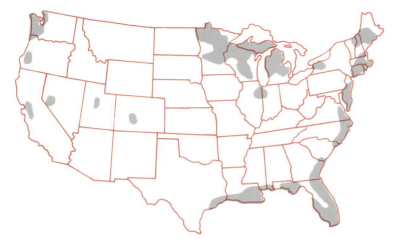

Figure 9-4. Generalized map showing the distribution of major peat deposits in the United States. (After Minnesota Department of Natural Resources, 1978.)

a

Figure 9-5. (a) Peat-mining operation. (b) Close-up view of organic-rich peat after mining and piling.

b

more than 3,000,000 ha (1,000,000 acres) of Minnesota land underlain by peat (Figure 9-4), only about 1,600 ha were under lease from the state in the late 1970s and those were for the mining of peat for horticultural purposes (Figure 9-5). The estimated value of 1978 peat production was about $1.3 million.

Research into peat technology is accelerating, and at least one energy company has developed plans to build a peat gasification plant in northern Minnesota. If it is granted leases in Beltrami, Koochiching, and Lake of the Woods Counties, Minnesota Gas Company (Minnegasco) may construct a full-scale gasification plant at a possible cost of more then $1 billion.

The peatlands of Minnesota are estimated to contain at least 14 billion metric tons of peat. Untreated, peat contains about 2,850 British thermal units (Btu) per kilogram of weight. Partially dried, its energy content rises to 13,250 Btu per kilogram (this figure is for peat with a moisture value of 35%). This is about one-half the value of anthracite coal. The importance of this energy resource is obvious, and its exploitation is inevitable.

Besides having potential value as a fuel, peat is also used for a wide variety of other purposes. Its usefulness is determined by its organic content and its degree of decomposition. Fibric peat, or peat moss, is relatively undecomposed and is formed largely from sphagnum and other mosses. Hemic peat is an accumulation of partially decomposed reeds, sedges, and other wetland plants. A third variety, sapric peat, is so decomposed that the original plant material is unrecognizable. Because organic content per unit volume increases with degree of decomposition, the sapric and hemic peats are best for fuel. Fibric peat, especially, has waterholding characteristics that make it useful as a soil conditioner and oil spill absorbant.

Coal

Minnesota has plenty of cold but almost no coal. Coal forms in swamps, and Minnesota has more swamp area than any state in the union, except for Alaska. Why is there no coal? Most of North America's coal was formed at two times in the past—during the Pennsylvanian and the Late Cretaceous-Early Tertiary. Minnesota has no known rocks of Pennsylvanian age. If they were once present, and that is possible, they were eroded away. Rocks of Late Cretaceous age are present, but they are thin and buried beneath glacial drift. As mentioned in the section on Cretaceous history in Chapter 6, thin beds of the lowest grade of coal, lignite, occur in western and southern Minnesota, and a bed of coal was found in Cretaceous rocks that covered iron ore at Virginia (see Figure 6-37). A 1.5-m-thick coal bed was found in a well at Freeborn, in southern Minnesota, in the 1870s. It is likely that seashore swamps existed on the eastern side of the Cretaceous sea, as they did on the western side in Wyoming, and such swamps may have locally existed during

154

the Early Tertiary. It is at least possible, though improbable, that low-grade Cretaceous coal in economically important quantities could someday be discovered beneath glacial till.

Minnesota's big swamps, actually defined as bogs because of their vegetation, are carryovers from immediate post-glacial times when numerous lakes formed behind glacial moraines and in topographic lows as the glaciers melted. Incompletely decayed plant growth eventually filled them, forming peat. But peat, although it can be a precursor to coal, is not coal. If it were buried beneath hundreds of meters of sand and mud, the water and other volatiles in it might eventually be driven out, leaving coal. For the peat to be covered with sediment would require some special event, such as the uplifting of a mountain range. But there are no mountains on Minnesota's horizon.

Oil and Gas

About 58% of the world's oil and gas comes from sedimentary rocks of Cenozoic age, 27% comes from sedimentary rocks of Mesozoic age, and 15% comes from sedimentary rocks of Paleozoic age. Not much oil and gas is obtained from Precambrian rocks, with which Minnesota is very amply blessed. Oil and gas form by the decay of animal and plant matter, and animals and plants were nearly nonexistent during Precambrian time. (Recall that it was only with the Paleozoic Era that life forms became diverse and abundant.)

Now, you may say, there *are* Lower Paleozoic, Upper Cretaceous, and Pleistocene deposits in Minnesota, so there should be oil and gas. But, in order for an oil or a gas field to be formed, several features must be present including: (1) certain source rocks (usually black muds); (2) favorable reservoir rocks (usually porous sandstone or carbonate) in which the oil or gas can accumulate; (3) a structural trap (such as an upfold or anticline) into which the oil and gas will rise, floating on top of the water in these porous beds; and (4) an impermeable caprock (usually a shale) to keep the oil and gas from rising to the surface of the ground and thus escaping. Even if all of these requirements were satisfied, Minnesota still would not qualify for an oil or a gas field for another reason: the petroleum-yielding source rocks generally must be buried 1,500 m or more in order to attain the heat and pressure necessary for the oil and gas to be driven out of the source rocks. Minnesota's Paleozoic rocks have a maximum total thickness of about 600 m, and the thickest Cretaceous sequence is about 150 m.

Does this mean oil and gas will never be found in Minnesota? Geologists are commonly a somewhat cautious lot, and few would say never in response to this question. As soon as such words were uttered, someone just might find gas or oil. Probably no geologist would invest money in drilling for oil or gas in Minnesota, because he or she would know the odds. Yet, oil and gas can, under very exceptional conditions, migrate into unlikely places. For example, oil has been produced from jointed Precambrian granite in Oklahoma.

Despite the unlikelihood of oil and gas deposits being found in Minnesota, there is a history of search for such deposits. In the late 1880s, natural gas was discovered in Pennsylvania, Ohio, and Indiana. This launched a nationwide "gas rush," and Minnesotans were quite involved, too. N. H. Winchell (1889), state geologist at the time, described the boom. "Experts" from Pennsylvania stated that geologists knew nothing about gas and that it could be found anywhere if properly sought. Gas wells were drilled at many places, including Minneapolis, North St. Paul, South St. Paul, Hastings, Duluth, Mankato, Moorhead, and Freeborn. Earlier shows of gas in water wells further stimulated interest in many localities.

A "gas witcher" from Ohio was brought in to locate gas veins at Mankato. Its supposed presence caused him to jerk and shake so violently that he had to stand on one foot, placing his other foot against his knee in order to break the electric flow. A 300-m-deep well failed to find the gas that shook him up.

Winchell expounded at length on the well at Moorhead. Upon seeing cuttings from the well, already 367 m deep, he realized that the drillhole had entered Precambrian granitic gneiss at a depth of 120 m. Consequently, he urged that the drilling be stopped, especially since it was being paid for by taxpayers. The *Moorhead News* contained this article a few days after Winchell's visit:

Mr. Sam Partridge this morning received a communication dated May 8 [1889], from Prof. N. H. Winchell, state geologist, addressed within to the mayor of Moorhead, stating that he had just examined the samples of drillings taken from the Moorhead artesian well which were sent to him a short time ago, and expressing regret that he had not before been able to obtain samples, "because," said he, "I could at once have told you that there was no earthly use of your going to further expense on the well. You ought to have stopped when the drill struck the rock at the depth of 390 feet, the rock being granitic and of that sort which forbids any hope of obtaining artesian water or other product of value (1889).

More cuttings were soon sent to Winchell along with the appeal that "having gone so far it was heart-rending to give it up now." The drilling continued and the Fargo *Argus* printed the following:

Mayor Hansen, of Moorhead, says they intend to continue sinking the artesian well, in spite of professor Winchell's prognostications. And in this the whole Red River Valley says—"good for Hansen." There is no geological or other prescience that can guess dead sure on Red River Valley matters. Success is what is wanted, and Hansen shows true grit.

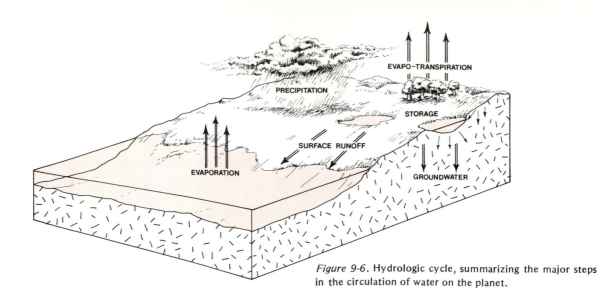

Figure 9-6. Hydrologic cycle, summarizing the major steps in the circulation of water on the planet.

When the Findlay, Ohio, people were first boring for gas, the state geologist delivered a lecture there in which he said it was useless to bore in such a formation, as they would never find gas. But they kept right on and struck a gas well—before the state geologist got away from town. And now Findlay has the biggest gas display in the world. With all due respect to Mr. Winchell, state geologist of Minnesota, why not let Moorhead do the same? (1889).

Chunks of coal, well-preserved seeds, and quicksand mysteriously appeared amid the granite at a depth of 385 m, greatly encouraging the interested parties to drill deeper into the granite. After a time a depth of 434 m was reached. Winchell concluded his report on the Moorhead well with this statement: "According to the latest accounts the mayor of Moorhead is still drilling in this granite. May 26, 1889" (1889: 31).

Moorhead was not alone. The well drilled for gas in Duluth reached a depth of 459 m, mostly in gabbro, before entering the Thomson slaty rocks at a depth of 619 m. Salty water, a common associate of gas and oil, was indeed encountered.

Winchell added pithy comments to the goings-on: "The human eyesight, the sense of smell, the love of lucre, the knowledge of geology, the ignorance of all geologists, have received a sudden and very general popular increment" (1889:01). "The deep well in South St. Paul . . . will also go to posterity with no record except as a monument to the infatuation which can be inspired by the positive assertions of a wanton adventurer in the minds of man ignorant of geology but eager for wealth" (1889: 35). Winchell had concluded that the gas shows were probably from the decay of vegetation in the glacial drift or had moved into the drift from underlying Cretaceous units. However, he regarded the Platteville Limestone, then known as part of the Trenton Formation and equivalent to the gas-producing beds in Ohio, Indiana, and Illinois, as potential source rocks.

The search does go on. In 1979, a deep hole was being drilled in the Middle Precambrian, 1,800-million-year-old Thomson Formation near Kettle River, a village 72 km southwest of Duluth. At the L. Hattenberger farm, a water well drilled in the mid-1970s produced salty water. Whenever a water tap was turned on, it would sputter and release both air and water. The escaping "air" would burn when lit (Canby, 1980). When the flushing of a toilet near which a candle was burning caused an explosion, the gas content of the water was no longer considered trivial. Analyses of the gas by the Minnesota Geological Survey showed that it was not swamp gas, which is common in swamps as a result of the decay of vegetation and which contains methane (CH_4) along with several other aromatics and straight-chain hydrocarbons such as peptane and benzene. The analyses also showed that it was not propane (C_3H_8) from a pipeline that passed within a few kilometers of the farm. Its chemical composition was methane, carbon dioxide and helium. Could it be gas formed in the black muds that now make up the slates of the Thomson Formation? If so, it would be a truly unique discovery, although a bed of coal was found in Michigan in the 1950s in rocks just as old. Certainly plant life in the form of algae existed then, as noted in Chapter 4. Conceivably, the decay of an algal accumulation could produce gas that might now be escaping from the black slates and mixing with waters carried down along joints. Betting against the odds, a local group formed a company and brought in a drill rig. The hole was reportedly drilled to a depth of 2,280 m, but data on the rock units have not been made public. Although neither gas nor oil has been found in any quantity, the drilling has already shown the Thomson Formation to be much thicker than the previously measured maximum thickness of about 1,000 m.

WATER

Earth is called the water planet because it is the only member of the solar system that holds at its surface abundant reservoirs of that wondrous chemical combination of hydrogen and oxygen in the ratio of two parts hydrogen to one part oxygen—H_2O. This planetary fact, decreed by a set of circumstances set into motion at its birth, has been one of the most important controls on the evolution of the surface environment of the earth through time. Water—as liquid in rivers, lakes, oceans, and soil, as solid snow and ice, and as vapor in the atmosphere—is the most valuable of our mineral resources. Without water there could be no life at all.

The hydrologic, or water, cycle summarizes the major steps in the circulation of water at the earth's surface (Figure 9-6). Solar radiation pumps moisture from the oceans by evaporation. Some of the water vapor is transported by wind over the continents, where precipitation results in rainfall and snow. But, since seas cover about 71% of the earth's surface area, most of the water is returned directly to the oceans. Water dropped on the land is eventually returned to the sea, but before it gets there it can have a complicated natural history. Hydrology is the science devoted to studying that history.

Chapter 7 contains a brief summary of the history of Minnesota's rivers and lakes. The pattern of streams that has evolved to carry away surface runoff is a branching system of open channels, each of which joins a larger branch as it flows toward the ocean. The area serviced by a specific network of streams is called a drainage basin, or watershed, and topographic boundaries separating adjoining basins are drainage divides (Figure 9-7). Drainage basins are the basic units in the management of water resources.

Like a giant sponge, the outer part of the earth soaks up part of the water falling on its surface. The result is that the largest and most dependable supplies of fresh water are stored in the void spaces contained in soil and rocks. This water is called *groundwater*. When water has filled all of the available space in a particular part of the underground storage unit, the unit is saturated. The upper boundary of this saturated zone is called the *water table* (Figure 9-8). The intersection of the land surface and the water table produces such common features as springs, swamps, lakes, and streams. The rise and fall of the water table is directly related to the rate of replenishment versus that of withdrawal.

Although almost all earth materials have the capacity for holding groundwater, some are far more porous than others. Natural reservoirs are called *aquifers* when they have the capacity to store and release sufficient volumes of water to satisfy the demands of surface consumers. The best aquifers are coarse-grained, poorly cemented sedimentary rocks and thick, well-sorted sand and gravel deposits associated with glaciation. Thousands of water wells in Minnesota draw from this underground supply. Since water has been legally recognized as a mineral resource, drawing from aquifers with low recharge rates is comparable to mining nonrenewable resources. Some water users are therefore granted depletion allowances from the federal government.

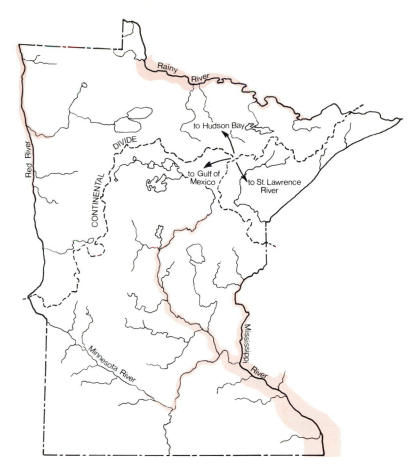

Figure 9-7. Major drainage basins in Minnesota. Note the three-way divide in the northeast. Shading portrays the relative volume of flow. (After Minnesota Conservation Department, Division of Waters, 1959.)

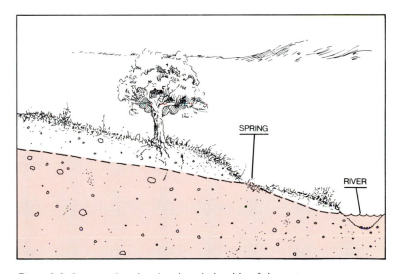

Figure 9-8. Cross section showing the relationship of the water table to the land surface.

Putting a monetary value on the water used in Minnesota is not easy. What is the worth of a scenic lake, a river, or the soil water in 65 ha of a western Minnesota farm? It seems that water is so necessary to our existence that its value is inestimable.

PART IV

Regional Geology

So far in this book, the geological story of Minnesota has been presented in a rather broad way. In this last part, a more detailed approach is taken. The state has been divided, on a geological basis as much as possible, into five regions—northeast, northwest, central, southwest, and southeast. For each region, three introductory review units are presented—one on the lay of its land, another on its Pleistocene geology, and the third on its bedrock geology. Finally, revealing and accessible localities or areas are described in some detail. This part of the book can thus be used by residents of and travelers to each of the five regions to make the geology more meaningful.

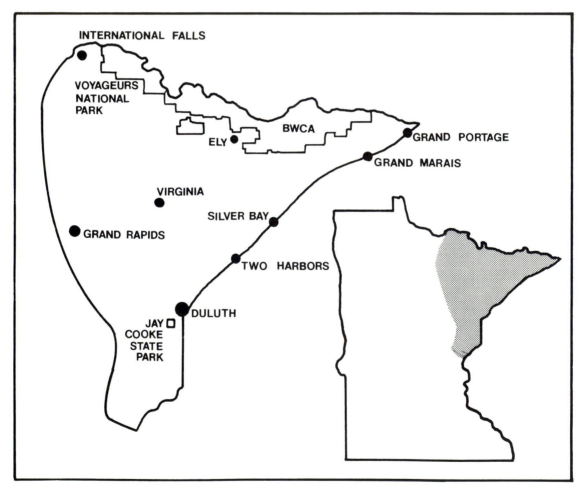

Northeastern Minnesota region.

10

Northeastern Minnesota

THE LAY OF THE LAND

Belying its character as part of an old peneplain surface, northeastern Minnesota holds both the highest and lowest elevations in the state and therefore has the most rugged relief. From a high of 701 m above sea level on Eagle Mountain in central Cook County, the land drops to 183 m at the shoreline of Lake Superior. And were the lake waters to disappear, the land would bottom out about 400 m lower. Much of the landscape owes its character to differential erosion of the bedrock, both by running water and by glaciers. It is surprising to realize that, although ice has been a significant contributor to the modern configuration, many of the present-day details of the landscape are relics of preglacial surface processes.

For the most part, this segment of Minnesota lies far enough north to have been involved in a more complicated glacial history than the areas downstate. Also, it was generally beneath ice that was in a mode of glacial erosion, in contrast to one of deposition. Therefore, drift is relatively thin or even absent over wide areas, being concentrated mainly in belts of moraine. Because of this condition outcrops are abundant (see Figure 2-5).

Drainage style contrasts sharply from place to place because of the diversity of the land surface. Three drainage basins shed the surface waters, each with a distinctive plumbing system. Lake Superior is the sump for North Shore streams, mainly short, steep watercourses with small drainage areas. The exception is the St. Louis River, which has extended its domain by stream piracy (see Figure 7-19b). The Rainy River carries flowage from the border lake district, gathered from hundreds of individual reservoirs, and finally to Lake of the Woods and Hudson Bay. Nibbling along the edges of the Arrowhead region are headward-eroding tributaries of the upper Mississippi and the St. Croix

rivers, which flow southward toward the Gulf of Mexico.

One of the great natural treasures of the world is the complex of lakes that dominates the landscape of northeastern Minnesota. Most of them are ponded in bedrock basins, in contrast to the southern and western lakes, which are contained in depressions developed in glacial drift. Many owe their shapes to the arrangement of weak elements in the bedrock, such as fault zones, patterns of bedding, and formation contacts (Figure 10-1). Along those lines of weakness, preglacial erosion tended to proceed more intensely, thereby effecting an irregular topography. Glacial scour enhanced the initial differences through many cycles of erosion. But, since lakes are among the most fragile of landforms, their present distribution represents a temporary stage in their life history.

GLACIAL GEOLOGY

The products of glacial quarrying and abrasion are clearly displayed in northeastern Minnesota in the form of boulders, streamlined hills, polished and striated bedrock outcrops, and hundreds of lake basins. Glacial deposits, too, especially in the form of moraines, drumlins, and eskers, are important elements of the landscape. All the features are comparatively young. None is older than about 25,000 years and some are as young as 12,000 years. It is this fairly recent glacial past that adds special character to the entire northeastern part of the state.

Three different lobes of ice were active at various times, each leaving behind distinctive kinds of drift (Chapter 7). In the Lake Superior Basin was the Superior Lobe, which advanced from the northeast. The bordering upland was occupied by the Rainy Lobe, moving in the same direction, sometimes synchronously, sometimes not. And, at the close

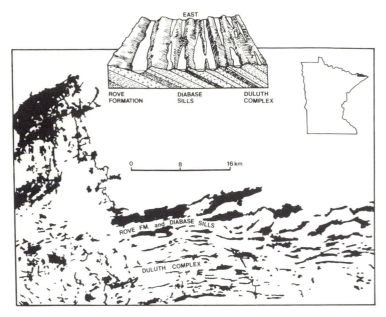

Figure 10-1. The pattern of the border lakes is influenced by the distribution of rock formations of different erodability. In places, they lie in basins preferentially eroded along the strike of diabase sills and formation contacts. (After Zumberge, 1952.)

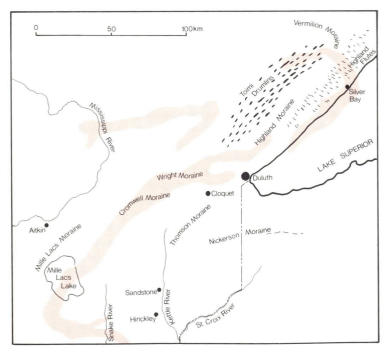

Figure 10-2. Moraines of the Lake Superior region. (After Wright and Watts, 1969.)

of glacial activity, the St. Louis Sublobe flowed from the west, as an offshoot of the Des Moines Lobe.

Both northern lobes carried charges of hard crystalline rock as well as sandstone, both very effective abrasives. The ice itself had a wet, slippery base, a condition necessary for effective scouring. Bouldery, coarse-textured, brown-colored drift, rich in gabbro, granite, and greenstone and including iron-formation and metasediments, was distributed by the Rainy Lobe. At first glance, Superior Lobe drift looks similar because it too is bouldery and reddish brown. A careful look at the stone content reveals important differences, however. Its trajectory along the Upper Precambrian volcanics and sandstones of Lake Superior is reflected in an abundance of those rock types in the till. Furthermore, the last drifts to be deposited by the Superior Lobe were rich in silt and clay. Fine texture is also the major physical property of sediment strewn along the path of the St. Louis Sublobe from the west. Telltale fragments of limestone and Cretaceous shale indicate its relationship to the Des Moines Lobe, for the source of those rock types is the western Dakotas and southern Manitoba.

Important glacial landforms associated with the Superior Lobe are the Highland Moraine, which formed along the western flank of the glacier, and a complicated nest of moraines formed during several fluctuations at its leading edge on the south (Figure 10-2). In retreat the ice mass blocked meltwater outlets long enough to allow Glacial Lake Duluth to form along its margins, as illustrated in Figure 7-18. Its final disappearance uncovered the deep lake basin that the lobe had produced by subglacial erosion; the basin is now occupied by Lake Superior.

In its vigorous advance to the western segment of the long St. Croix Moraine, the Rainy Lobe impressed upon the subglacial surface in northeastern Minnesota an ensemble of streamlined hills and intervening swales, the Toimi Drumlin Field (Figure 10-3). The drumlins, about 1,400 of them, trend southwest, reflecting the flow direction of the ice that formed them. A radiocarbon date from organic sediments in one of the intervening lakes indicates that the drumlins were uncovered by ice at least as early as 15,000 years ago. Farther north, either in retreat or readvance, the Rainy Lobe constructed the bouldery, hilly Vermilion Moraine.

West of the Toimi Drumlin Field, the land surface shows modification resulting from the activity of the St. Louis Sublobe, which moved from west to east. Because drift from this glacier buries Rainy Lobe drumlins and also parts of the Highland Moraine of the Superior Lobe, it is younger than both, about 12,000 years old according to radiocarbon dates. Its advance across old glacial lakebeds gave its drift a clayey consistency, and it is not as stony as the tills from other lobes. Generally, the drift is gray farther west but has acquired a red color from glacial lake sediments left behind by the Superior Lobe and picked up by the St. Louis

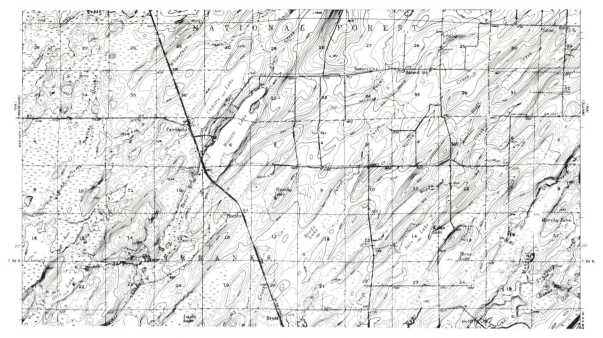

Figure 10-3. Streamlined hills of the Toimi Drumlin Field. They were formed by the Rainy Lobe as it advanced from northeast to southwest to the St. Croix Moraine. (From the Brimson, Minnesota, 7.5' Quadrangle.)

Sublobe. When the St. Louis Sublobe retreated, Glacial Lakes Upham and Aitkin were established in low areas. Their draining via the St. Louis River left a broad, flat plain now occupied largely by bogs and helped to excavate the drift and bedrock along the St. Louis River in Jay Cooke State Park.

BEDROCK GEOLOGY

Northeastern Minnesota has more outcrops of rock than any other part of the state (see Figure 2-5), but even so geological interpretations are commonly hampered by the lack of rock outcrops in key areas. (A "key area" can usually be defined as any place where a geologist really needs an outcrop.) As is to be expected, areas that are being actively eroded—the lakeshores and the river valleys—usually contain the best outcrops (Figure 10-4). These, along with roadcuts, provide bits of a third dimension to what is essentially a flat surface.

Northeastern Minnesota has the best exposures of rocks of Early Precambrian, Middle Precambrian, and Late Precambrian age, as already discussed in Chapters 3, 4, and 5. In addition, some of the best temporary Cretaceous exposures were created during the mining of iron ore on the Mesabi Range, for Cretaceous rocks formed part of the overburden there (see Chapter 6). These, however, have all been covered by slumped glacial deposits. Thus, of the rocks represented in Minnesota's rock column, only the Paleozoic rocks are not present in northeastern Minnesota.

More specifically, northeastern Minnesota includes: (1) the Vermilion district, which is the state's best exposed Lower Precambrian volcanic-sedimentary (greenstone) belt; (2) three well-studied Lower Precambrian batholiths (the Giants Range, the Vermilion, and the Saganaga); (3) part of the Middle Precambrian basin in which the world-famous Biwabik Iron Formation was deposited; (4) well-exposed Upper Precambrian continental lava flows that poured out of North America's largest rift structure; and a complex (5) the Duluth Complex, which is one of the world's largest mafic intrusions as well as a major reservoir of copper and nickel. It is in northeastern Minnesota that much of the Precambrian history of the Lake Superior region and the southern part of the Canadian Shield has been worked out.

The rocks of northeastern Minnesota are valuable not only from a scientific point of view. The iron ores of the Mesabi and Vermilion ranges have long formed an integral part of the economic base of northeastern Minnesota. Now, copper-nickel ore is known to exist in large quantities, and the potential for zinc, copper, uranium, gold, and other metals is better there than in other parts of the state.

PLACES OF INTEREST

Jay Cooke State Park

Jay Cooke State Park, just west of Duluth, is the best place to observe the Middle Precambrian Thomson Formation, as well as maple-clad hills cut from a thick sequence of

Figure 10-4. Glaciated and bare rock outcrops on the shore of a Boundary Waters Canoe Area lake.

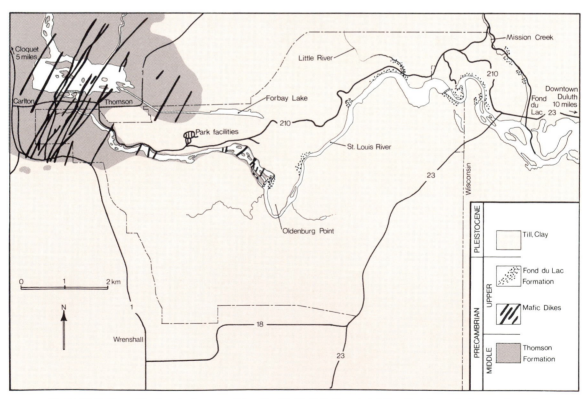

Figure 10-5. Generalized geologic map of Jay Cooke State Park and vicinity. Park boundaries are shown by the dashed line; the eastern boundary is Minnesota Highway 23. (Geology from Wright, Mattson, and Thomas, 1970.)

Figure 10-6. Aerial photograph of a part of Jay Cooke State Park and adjacent areas. Note that the St. Louis River is dammed at the north edge of the park. The east-west trending rock ridges are resistant beds of the Thomson Formation, and the northeast-trending lines are eroded Upper Precambrian mafic dikes. Where the rocks are not visible, they are obscured by glacial drift. The reservoir is about 2.4 km (1.5 mi) long. The village of Thomson is at the center of the photo and Carlton is on the left edge. Compare with the map of Figure 10-5.

red clay, silt, and sand (Figure 10-5). These sediments were deposited in Glacial Lake Duluth when the lake basin was still dammed to the northeast by the retreating Superior Lobe of the last major ice advance (see Figure 7-18). As the ancestral St. Louis River, swollen with glacial meltwater, cut deeply into this easily eroded material, tributary streams and sheet wash (sheets of water flowing downslope) together dissected the surrounding area. The St. Louis carried the sediment into Lake Superior at what is now Duluth with the sand fraction probably contributing to the establishment of Minnesota Point (Park Point) and Wisconsin Point at the mouth of the river.

Finally the river cut through the glacial sediments and into the underlying Middle Precambrian Thomson Formation, the origin of which was described in Chapter 4. Continued erosion along the floor of the river valley has exposed a long (13 km), narrow window of tilted black slates and metasediments (see Figures 4-12 through 4-16) that now make up a series of gorges, falls, and rapids (Figures 10-6 and 10-7). A viewpoint in the park provides a good look at a stretch of river with rapids, and gorges can be observed

Figure 10-7. White water in a gorge of the St. Louis River at the northwest edge of Jay Cooke State Park near Carlton. The rocks are part of the Middle Precambrian Thomson Formation.

Figure 10-8. Duluth's "Point of Rocks," in 1887, from about what is now Sixth Avenue West to Eighth Avenue West. (Courtesy of the Northeast Minnesota Historical Center, University of Minnesota, Duluth.)

from the swinging bridge at the park center or by looking south from the highway bridge over the river at the village of Thomson. By walking about 30 m south of the latter bridge on the east bank of the river and looking back toward the bridge, one can observe an excellent anticline in the Thomson Formation (see Figure 4-15). A thick pod of white quartz just below the northwest corner of the bridge reportedly contains traces of gold. The river's course is commonly structurally controlled; some portions follow eroded Upper Precambrian dikes, eroded fault zones, or eroded softer beds. Some dikes are visible just north of the bridge (see Figure 5-16); recall that these were feeder dikes to Upper Precambrian lava flows that have been removed completely by erosion. Excellent cleavage and bedding relationships are visible on most outcrops (see Figure 4-16). Weathered-out layers of concretions accentuate the original bedding, which is locally difficult to discern (see Figure 4-12a).

This portion of the river constituted a major roadblock—an 11-km portage along the canoe route—for the Indians, voyageurs, and others who traveled from Lake Superior to the border lakes via the St. Louis River, the Embarrass River, the Pike River, Lake Vermilion, the Vermilion River, and Namakan Lake, as well as for those who traveled to the Mississippi River via the Savanna River (from what is now the village of Floodwood) and Big Sandy Lake. This portage, called the Grand Portage of the St. Louis, required three days of hard work for a group with packs and canoes. After another short stretch of navigable rapids, the travelers were faced with Knife Portage, a walk over the upturned Thomson

slates, which terminated at Knife Falls in what is now Cloquet. Several dams between Cloquet and Fond du Lac have, of course, altered the flow since those days of river travel.

Proof that the river once flowed at a higher level during the days of glacial runoff is provided by *potholes* eroded into the Thomson Formation tens of meters above the present river level by rock "tools" swirled in the fast current. A few can be seen on the east bank of the river between the highway bridge and the abandoned railroad trestle adjacent to the village of Thomson, but they are difficult to locate.

At the eastern edge of the park, downstream from the Thomson exposures, the St. Louis has also cut down into the Fond du Lac Formation, which was described in Chapter 5. The red to brown formation, which consists mainly of interbedded sandstones and shales, is exposed in old quarries in the woods just upstream from the Fond du Lac picnic area at the Minnesota Highway 23 bridge over the river. Better exposures occur in the park along the river bank, but they are generally quite inaccessible (see Figure 5-22).

An especially interesting relationship can be seen along Little River upstream from the park highway. There, the gently dipping basal quartz pebble conglomerate of the Fond du Lac is exposed (see Figure 5-22a), and, 30 m or so farther upstream, steeply dipping beds of the underlying Thomson Formation are visible in the stream bottom. This constitutes, of course, a major unconformity between Upper Precambrian and Middle Precambrian rocks, with the erosional surface representing about 1,000 million years of time. In turn, the Fond du Lac in the park is overlain by red Pleistocene lake sediments. This latter unconformity, like the former one, represents a time interval of nearly 1,000 million years, from 900 million years ago to perhaps less than 20,000 years ago.

Duluth

Duluth is on the rocks, geologically speaking. Duluth's 32-km-long (as the seagull flies) hillside, which rises more than 250 m above Lake Superior (elevation, 183 m, or 602 ft) essentially consists of lava flows and gabbroic rocks, both dated as 1,100 million years old, as described in Chapter 5. The flows, the lower portion of the North Shore Volcanic Group, are mainly basalts. Some sills of a similar composition are present within the volcanic sequence. The gabbroic rocks are part of the southwestern end of the large Duluth Complex, which forms a broad arc from Duluth to Ely and back to Lake Superior near Grand Portage (see Figure 5-9). The dividing line between the flows and the gabbros in

Duluth is the "Point of Rocks" (Figure 10-8) at Mesabi Avenue; the flows extend to the northeast up the lakeshore and the gabbro extends to the southwest. Some lava flows, the oldest in the sequence, are exposed near Nopeming southwest of the intrusive gabbro, lying upon a few meters of the Nopeming Quartzite. All of the rock units are delineated on the geologic map of Duluth (Figure 10-9). Not shown on the map are numerous small, slightly younger dikes of granitic rock called "red rock" and black basalt such as that in Figure 5-15.

The hillside posed a problem in railroad construction. Duluth-Superior is the largest iron-ore port in the world (Figure 10-10) as well as a shipping point for wheat from the interior of North America. The railroads wend their paths at as slight an angle as possible down to lake level. Even then several engines are necessary to move trains up and down the slope.

The lava flows of Duluth are best exposed along the shore of Lake Superior and in the three major stream valleys of Duluth—Chester Creek, Congdon Creek, and the Lester River. The rocks are most accessible along the lakeshore in

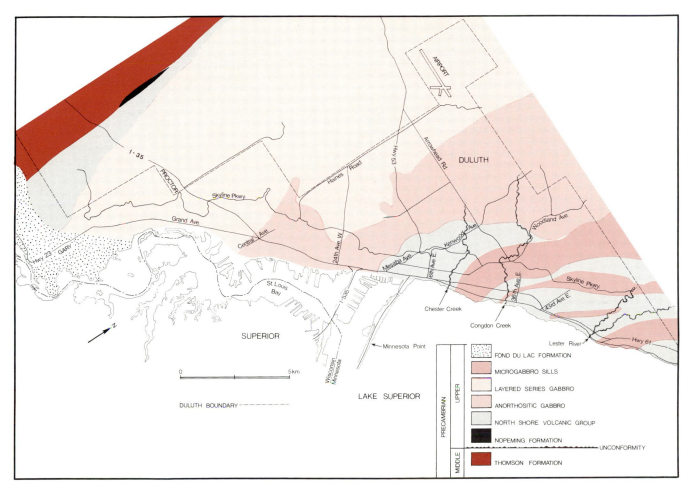

Figure 10-9. Geologic map of Duluth. (After Taylor, 1963 and personal communication, Shoa Motamedi, 1981).

Figure 10-10. Iron-ore docks in Duluth.

Figure 10-11. Lava flows on the lakeshore just southwest of Leif Ericson Park in Duluth. The light spots are amygdules at the top of a flow. The rock across the creek in the background is the massive bottom of the next flow.

and near Leif Ericson Park from about 7th Avenue East to 14th Avenue East (Figure 10-11), along Chester Creek in Chester Park between East 4th Street and Chester Park Drive, and at and near the mouth of the Lester River. At Leif Ericson Park, in addition to flows and dikes, a cross-bedded sandstone unit more than 30 m thick is well exposed (Figure 10-12). It is one of two thick sandstone units exposed between flows on the North Shore of Lake Superior that were deposited during pauses in volcanism by a stream-delta complex on the volcanic surface. Roughly a hundred thinner interflow sedimentary units are present in the North Shore Volcanic Group between Duluth and the Canadian border. The Leif Ericson sandstone unit is faulted against lava flows at the eastern edge of the park.

The gabbroic rocks are easily seen at many places along Skyline Drive. For example, in the vicinity of Twin Ponds and Enger Tower, a variety of gabbroic rocks are dominated by a coarse-grained anorthositic gabbro, a basalt dike in red granite ("red rock") can be observed in the roadcut just north of Twin Ponds, and in high cuts just south of Twin Ponds excellent flow banding is visible in coarse-grained rocks. "Red rock" dikes cut gabbro at the junction of Mesabi Avenue and Superior Street, along skyline Drive at about Seventh Avenue West, and at numerous other places in the city. A slightly older variety of the Duluth Complex, a "layered" gabbro, is present along the western portion of Skyline Drive. The layering, which resembles sedimentary bedding, is the result of crystal settling and movement of the magma prior to final solidification. Layered gabbro is easily seen at the Thomson Hill Information Center and Rest Area at the junction of Interstate Highway 35 and U.S. Highway 2 (Figure 10-13).

The rocks of Duluth were polished, scratched, and grooved by the glaciers that moved southward down what is now the Lake Superior Basin. Abundant evidence of this can be seen on most rock outcrops (Figure 10-14). The ice advanced repeatedly, but most striations are probably the result of the last advance about 14,000 years ago.

As the glaciers melted and retreated northeastward, they formed an ice dam, as discussed in Chapter 7 and shown in Figure 7-18. Thus, water was impounded in the Lake Superior Basin, and this ancestral meltwater lake (Glacial Lake Duluth) became deeper and deeper. Skyline Drive, at an elevation of about 350 m (about 165 m above the surface of Lake Superior), roughly follows the highest beaches of Glacial Lake Duluth. These beaches have been dated by radiocarbon dating as about 12,000 years old, whereas beaches at about 250 m above sea level (about 60 m above

Figure 10-12. Crossbedded sandstone unit at Leif Ericson Park. The unit is more than 30 m thick. It was deposited upon a lava flow and was then covered by another flow. Such interflow sedimentary rock units represent pauses in volcanism, times when streams carried sand over the flow surfaces.

Figure 10-13. Layered gabbro of the Duluth Complex at the Thomson Hill Rest Area at the junction of Interstate 35 and U.S. 2. The layering is the result of the settling of crystals of plagioclase (light colored) and olivine (dark colored) into layers and the movement of magma in the magma chamber before final solidification.

Figure 10-14. Glacially striated bedrock outcrop, typical of most outcrops in Duluth, exposed at the mouth of the Lester River.

the present surface of Lake Superior) are approximately 9,000 years old. Several gravel pits along Skyline Drive in Duluth, such as the one at the intersection of Interstate 35 and U.S. 2, are in delta deposits formed where glacial meltwater streams entered the lake. Out in the lake proper, fine-grained sediment, mostly red clay, was deposited. The abundance of the heavy, sticky, red clay can be confirmed by any gardener or building contractor in the city. Although seemingly worthless, the clay was once used to make bricks at the village of Wrenshall, about 30 km southwest of downtown Duluth. The red clay is prominent in Jay Cooke State Park and along Highway 23 southwest of Duluth. Boulders and pebbles scattered in the clay attest to an abundance of icebergs in Glacial Lake Duluth that dropped these large particles to the lake bottom as they melted. Some near-vertical exposures of gabbro along Skyline Drive are wave-cut cliffs sculpted by waves during the highest stand of Glacial Lake Duluth, formed in the same way that present-day cliffs in East Duluth and along the North Shore are being formed. Glacial Lake Duluth was larger as well as deeper than Lake Superior is at present; its southern shore lay 30 km away in what is now Wisconsin at a ridge of South Shore Volcanic Rock that is visible from Skyline Drive in Duluth on a clear day.

As the glaciers finally melted from the basin now occupied by Lake Superior, uplift due to postglacial rebound began in the northeastern part of the lake basin. This uplift in the northeast caused a tilting of the basin that resulted in the piling up of water at the western end of the basin, much as would occur if one end of a bathtub full of water were elevated. The excess water has drowned the mouth of the St. Louis River, which enters Lake Superior at Duluth,

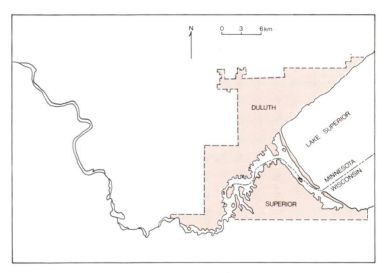

Figure 10-15. Map showing numerous bays of the drowned mouth of the St. Louis River.

helping to make Duluth an excellent port (Figure 10-15; Plate 15). Water levels at this end of the lake are still rising, at the rate of about 15 cm per century. If this process continues, the parking lot of the Duluth arena-auditorium complex will be under water in only a few thousand years. In 1816, near the upstream end of the drowned river, the American Fur Company, owned by John Jacob Astor, took over from the French a 25-year-old trading post at Fond du Lac, the "head of the lake" and long the site of an Indian village.

Sand coming down the St. Louis River and sand moved by long shore drift along the southern and northern shores of Lake Superior was piled up in shallow water by wave

a

b

Figure 10-16. (a) Minnesota Point (Park Point), one of the largest baymouth bars in the world. This bar helps to make the Duluth-Superior harbor an excellent inland port. (b) Sand dunes on Park Point. Duluth's hillside is in the background.

action, thus forming Minnesota Point and Wisconsin Point (Figure 10-16; Plate 15). This is one of the finest baymouth bars in the world, and it is the key factor in making the Duluth-Superior harbor an excellent, safe inland port on a lake that can be truly wicked during storms. The strong winds that blow off Lake Superior have created, and still create, sand dunes on the points; most of these have been modified by human activity. This bar, probably formed about 5,000 years ago, is only the newest in a series of Duluth bars—Rice's and Conner's points, farther up the bay, are older but smaller models much altered by human activity. Minnesota Point was a logical focal point for activity in

the early days of human habitation in the area, for it was a beautiful sandy point adjacent to a rather swampy bay. Indians lived and camped on the point and Daniel de Greysolon Sieur DuLhut, the French explorer for whom Duluth is named, landed there (Figure 10-17) with seven voyageurs and three Indian guides on June 27, 1679, 20 years after the French fur traders and explorers led by Pierre Radisson and Sieur des Groseilliers had passed the area. George Stuntz, who discovered iron ore near Lake Vermilion, established a trading post there in 1852.

The configuration of the Duluth-Superior port is not entirely natural. Much low-lying swampy ground was filled in and built upon or removed by dredging. The natural outlet of the St. Louis River into the lake is near Superior, Wisconsin, at the southern end of Minnesota Point. Duluth's city fathers feared as early as 1857 that this natural passage-way would cause Superior, rather than Duluth, to become the major port at the head of the lake, for some of Duluth's docks were actually on the lakeshore and were subject to the whim of the weather. A plan to dig a canal through Minnesota Point at the Duluth end of the point was conceived but still had not reached fruition by April, 1871 (Figure 10-18). Anxiety over a court injunction sought by Superior (many Superiorites were sure that the presence of the canal would cause the currents to change and the natural river exit in Superior to silt up) to stop the canal stimulated a multitude of Duluthians to join in a volunteer effort to assist the dredging process by pick and shovel. The injunction from Washington, D.C., arrived too late. Boats were already moving through the new canal. A judge had ruled that since federal monies had improved Superior's harbor, the canal should not be built; however, he had added that, if a dike were built between Minnesota Point and Rice's Point, thus isolating the Duluth part of the harbor from the rest and preventing changes in Superior's part of the harbor, the injunction would be set aside. The dike was built with pilings across part of the harbor, thereby preventing the St. Louis River from exiting via the newly dug ship canal. However, Superior then objected to the obstruction in the bay, a public waterway. According to one story, vigilantes dynamited the dike under cover of darkness. Eventually the dike was removed, but the expected dire consequences of the canal never came to pass and Duluth began to grow as a port (Figure 10-19). Today, both Duluth and Superior are major ports, although dredging is necessary to keep the channels deep enough for large ships. And, without the Duluth ship canal, there would have been no need to build one of Duluth's major attractions, the Aerial Bridge, which, by the way, was originally constructed with a suspended car that ferried cars and people across the canal.

A much more detailed guide, *The Geology of the Duluth Metropolitan Area,* was published by G. M. Schwartz (1949) as Bulletin 33 of the Minnesota Geological Survey. A detailed geologic map by R. B. Taylor (1963) is also available.

Figure 10-17. Daniel de Greysolon Sieur DuLhut landing at the "Little Portage" on Park Point near where the Aerial Bridge is now located, on June 27, 1679. The original painting by C. C. Rosenkrantz hangs in the St. Louis County Historical Society at the Duluth Depot. (Courtesy of the St. Louis County Historical Society.)

The North Shore of Lake Superior

The Minnesota portion of the North Shore of Lake Superior is a 280-km avenue of varied beauty. Although its relief is generally not as great as along the Canadian portion, it is considerable and abrupt. The entire length is essentially a random alternation of vertical rocky cliffs, gravelly beaches, and lower rock outcrops (Figure 10-20). Twenty-two major streams have cut steep gorges into solid rock in their plunge to reach the new erosional base level set by Lake Superior, a much lower base level than existed 12,000 years ago during the reign of Glacial Lake Duluth. The Cascade River, for example, descends 275 m in less than 5 km. Most of the streams head within 8 km of the lake and thus

Figure 10-18. Duluth and Minnesota Point during the 1860s. Note that the ship canal had not yet been built. The trees have been removed, perhaps to improve the scenic view of the lake and harbor. (Courtesy of the Northeast Minnesota Historical Center, University of Minnesota, Duluth.)

Figure 10-19. The busy port of Duluth, in 1883. (Courtesy of the Northeast Minnesota Historical Center, University of Minnesota, Duluth.)

Figure 10-20. Beach and rocky point north of Hovland.

Figure 10-21. Glacial striations and small whalebacks on glaciated basalts at the mouth of the Lester River. These features indicate that the ice moved toward the soutwest.

do not drain large areas. The wave-cut cliffs are ever being modified and renewed by nearly constant wave action (see Figure 5-10). Where they encounter the massive lower parts of flows, the waves make slow progress. Where they hit the softer amygdaloidal flow tops, deeper notches and sometimes caves are formed.

Virtually all of the bedrock of the North Shore consists of Upper Precambrian volcanic rocks poured out from the rifts of the midcontinent gravity high 1,100 million years ago, as described in Chapter 5. A few intrusive rocks exist, too, such as the massive jointed sill at Silver Cliff (see Figure

5-19) and irregular intrusive bodies such as the diabase at Silver Bay and Beaver Bay.

Every little bit of the long shoreline offers its surprises, and hidden nooks and bays will greet the more adventurous explorers. The river valleys are especially impressive. For the many people who experience much of the shore in one visit by driving part or all of U.S. Highway 61 between Duluth and Canada, a generalized road log is the most useful way to describe some of the major points of interest. The log commences at the northeastern edge of Duluth; parts of it are based on a more detailed and scientific road log by J. C. Green (1979) available from the Minnesota Geological Survey. Note that Leif Ericson Park, described briefly in the Duluth section of this chapter, might be the first stop on a North Shore trip. The stops are numbered on Figure 5-9.

Stop 1: The Lester River

The Lester River at the east edge of Duluth has cut a long gorge into lava flows. The mouth of the river, a favorite smelting spot during the spring run of the silvery *Osmerus mordax*, has less relief here than it does a short distance upstream, but this is a convenient place to see lava flows of a basaltic composition. Glacial grooves and striations are common right at the lakeshore, and whalebacks with steep ends toward the southwest indicate that the glacier that formed them flowed in that direction (Figure 10-21). A short distance up the lakeshore from the Lester, a dirt road turns off to the right and follows the shoreline for 1.5 km (1 mi) or so. More flows, a coarse-grained sill (the Lester River Sill), and pebble-gravel beaches with a smattering of agates can be found along this strip of lakeshore, which is a popular picnic spot. About half a kilometer from the Lester River bridge, the traveler has a choice of whether to travel to Two Harbors via the expressway or the older and slower but much prettier Scenic North Shore Drive (County Highway 61).

Stop 2: Two Harbors

Two Harbors, the "agate city," has huge iron-ore docks from which millions of metric tons of Mesabi Range taconite pellets are shipped to steel mills via large ore boats. The *Three Spot*, a tiny wood-burning railroad locomotive that was built in 1880 and carried the first iron ore from Tower to Lake Superior on July 31, 1884, still stands at the edge of the harbor at a depot-museum. The *Three Spot* was originally moved by barge on Lake Superior from Duluth to Two Harbors, barely arriving safely in a big storm (Fawcett, 1970). The Two Harbors Tourist Park on Burlington Bay at the eastern edge of town has agates in the beach gravels as well as *in situ* in the lava flows. Recall from Chapter 5 that the origin of agates is gas cavity fillings or amygdules (see Figure 5-13). Note that the agates in the rock are generally white whereas those in the gravels are generally an orange-red color that has come to be widely renowned in Lake Superior agates (See Plate 10). This color change is due to the oxidation of trace amounts of iron during or after removal from the rock by weathering and erosional processes. (Don't even bother trying to remove agates from the flows. Nature has done the hard work for you by leaving an abundance of them in the gravel.)

Note: A few kilometers northeast of Two Harbors near the Encampment River the highway traverses a tract of virgin white pine forest.

Continue northeast on Highway 61 to a point 6.5 km (4 mi) from Two Harbors. Silver Creek Cliff offers a high and scenic view of the lake as the highway winds over this black, coarse-grained mafic sill. Although there is no place to stop here, the lower speed limit allows the observer a quick glance at massive rock columns formed by steeply dipping cooling joints in what was once a 75-m-thick body of magma that intruded the lava flows (see Figure 5-19). Also visible is the lower contact of the sill against a lava flow that was baked and reddened (the iron was oxidized, forming the mineral hematite) by the heat. White calcite veins also mark the contact. The dark rock columns are weathering to form a yellow-brown soil.

Note: Similar columnar jointing and weathering can be observed in several dark gray to black roadcuts along the North Shore. These nearly always signify the presence of a sill.

Stop 3: Gooseberry Falls State Park

Gooseberry Falls State Park, on the Gooseberry River, is an excellent place to observe scenic waterfalls in a steep gorge. At the first falls just downstream from the highway bridge, three lava flows of the Gooseberry River Basalts can be distinguished; the main falls is held up by a thick flow (Plate 14). Note the vertical cooling joints that have created vertical columns in the flows. The broad rock ledge or bench below the falls is a smooth, slightly undulating flow top with an abundance of calcite and zeolite amygdules, some of which have already weathered out, leaving small cavities. A small natural bridge has been formed by differential erosion a short distance below the falls. This falls and the one below it drop a total of 22.5 m. Just above the bridge is a falls with a drop of 8 m, and 1.5 km upstream is still another falls. Additional flows can be distinguished in the roadcuts just west of the bridge. The dark stonework in the park was done during the 1930s as a government project, with the stone coming from a quarry southwest of Beaver Bay.

Stop 4: Split Rock Lighthouse

Split Rock Lighthouse (now a state park), about 10 km (6 mi) farther up the shore, was operational as a lighthouse from 1910 to 1968, warning ships of the rocky and dangerous shoreline. The 30-m-high cliff upon which the lighthouse was built has been cut by wave action out of a diabase sill that also contains some inclusions of anorthosite, a light gray to greenish rock made up almost entirely of plagioclase.

Stop 5: Silver Bay

Reserve Mining Company's taconite plant in Silver Bay is literally on the highway. Stop at the high roadcut in the rock by the plant. This was the world's first taconite plant, built here during the 1950s so as to be close to a large water supply. The iron ore comes from the Reserve Mining Company open pit at Babbitt via a 69-km-long railroad. The plant produces 10 million metric tons of taconite pellets per year. (Chapter 8 covers the taconite process and environmental concerns.) Whereas the tailings were previously dumped into the lake, they have been disposed of since mid-1980 at Milepost 7, an on-land disposal site 8 km from the plant (Figure 8-21). The rock in the roadcut is part of the Beaver Bay Complex, an irregular intrusive composite rock body made up largely of black diabase, which can be thought of as a mafic rock intermediate in grain size between basalt and gabbro. The diabase contains numerous inclusions of light gray to greenish anorthosite, like that at Split Rock Lighthouse (see Figure 5-18c). The source of the anorthosite inclusions, which occur almost exclusively in diabases, is presumably at great depth. The inclusions may represent samples of a more ancient and lower crust, or they may be the result of crystallization and separation of plagioclase in the same magma chamber from which the basalt, diabase, and gabbro originated. Both the diabase and the anorthosite in the roadcut are cut by veins of red granitic rock, part of the Duluth Complex.

Stop 6: Palisade Head Wayside Rest

Continue 4.5 km (2.8 mi) northeast on Highway 61. Turn off to the right on a very steep, narrow, and winding road that ends at the overlook on top. Be careful near the edge of the cliff, because some columns could collapse. This high cliff is cut into a 90-m-thick lava flow of reddish porphyritic rhyolite with small phenocrysts of smoky quartz, plagioclase, and potassium feldspar. Shovel Point, visible 3 km to the northeast, is part of the same rhyolite body, which can be traced for at least 10 km (see Figure 5-10). It is also exposed in the big roadcut at the intersection of Minnesota Route 1 and Highway 61 at Illgen City. To the northeast can be seen the "Sawtooth Mountains" of Cook County, formed by lava flows dipping toward the lake. On a clear day, one can see as far southwest as Split Rock Lighthouse. The islands consist of diabase and anorthosite, as at Silver Bay.

Note: Illgen City, just northeast of Palisade Head, has a unique history, as described in Chapter 9. The very high vertical roadcut just northeast of Illgen City is in a large pinkish anorthosite inclusion in coarse diabase. This relationship is similar to that at Silver Bay.

Figure 10-22. The lowest falls of the Manitou River spill directly into Lake Superior. (Courtesy of J. C. Green.)

Stop 7: The Manitou River

Continue about 18 km (11 mi) northeast on Highway 61. The Manitou River received its name from the Chippewa word for "spirit." The deep, narrow gorge ends at a 30-m-high waterfall a short distance upstream from the bridge; the falls exists because of a massive basalt bed. Three more falls are present between the bridge and the lake, with water spilling directly into the lake from the top of the lowest falls (Figure 10-22). The gorge formed as the upper falls migrated upstream owing to erosion of an underlying amygdaloidal flow top and periodic collapse of the undercut massive flow.

Stop 8: The Temperance River

Continue about 20 km (12 mi) northeast on Highway 61. The Temperance River is so named because it is the only major stream on the North Shore without a gravel bar at its mouth. On the basaltic bedrock floor of the valley are numerous potholes, eroded by water swirling stone tools

around; these can be seen at the bridge. The gorge below the bridge is largely the result of coalescence of several large potholes. The Temperance is one of the longest rivers on the North Shore.

Note: The highest and youngest lava flows in the North Shore Volcanic Group are exposed in the vicinity of the village of Tofte. Carlton Peak, a conspicuous hill 2.5 km inland from Tofte and with its top 278 m above the surface of Lake Superior, consists of two huge and several smaller inclusions of anorthosite in diabase. Rock was quarried here for the breakwater that joins two islands at Taconite Harbor and makes the town a safe harbor. From there, ore boats carry the Erie Mining Company taconite, mined and processed at Hoyt Lakes (see Figure 8-17), to steel mills.

Stop 9: Good Harbor Bay

Continue about 34 km (21 mi) northeast of Tofte on Highway 61. Good Harbor Bay, stretching out beneath a big roadcut and parking area composed of a 50-m-thick basalt flow and a 40-m-thick section of red siltstone, shale, and sandstone (see Figure 5-12), is a famous location for collecting the semiprecious gemstone thomsonite. The black basalt contains the concentrically banded green, white, and pink thomsonite (a zeolite mineral composed mainly of sodium, calcium, aluminum, silicon, oxygen, and water) in amygdules, but the mineral is most easily obtained as weathered out, loose pieces on the beaches of the bay. Native copper has also been found here. The thick basalt flow is one of several that make up the "Sawtooth Mountains" to the west of Grand Marais (Plate 12).

Stop 10: Grand Marais

The village of Grand Marais has one of the best harbors along the North Shore, created by the erosion of softer lava flows behind a resistant massive basalt that is either a flow or a sill (Figure 10-23). Interestingly, this locality was named by the French, and its name means "great marsh." This was probably an apt description in the late 1600s and early 1700s.

Note: A beautiful side trip (176 km, or 110 mi, round trip) is offered by the Gunflint Trail, which leads north (see Weiblen and Davidson, 1972). The trail, no longer a trail but an excellent highway, can be followed to Saganaga and Sea Gull lakes, at the eastern end of the Vermilion district. In numerous roadcuts along the trail, representative examples of rocks from the Duluth Complex and the Saganaga Batholith, as well as a few other rock types, can be seen. The rocks of the Duluth Complex are mostly coarse-grained, dark-colored igneous rocks like those at a point 69 km (43 mi) north of Grand Marais in a roadcut opposite a scenic overlook. The light-colored granitic rocks of the Lower Precambrian Saganaga Batholith (see the section on the Boundary Waters Canoe Area later in this chapter) can be seen from a point 75 km (47 mi) north of Grand Marais and on to the end of the trail, which is 88 km (55 mi) north of Grand Marais. The Gunflint Iron Formation can be seen in roadcuts where the Magnetic Rock and Kekekabic hiking trails leave the Gunflint Trail.

Note: Eastward from the point where the Devil Track River meets the shore of Lake Superior, about 6 km (3.7 mi) northeast of Grand

Figure 10-23. Grand Marais and its small but excellent harbor, viewed from a lookout on the basalt and diabase ridge a few kilometers north of town just off the Gunflint Trail. The island of mafic rock is connected to the mainland by a gravel bar.

Marais, the highway is located on a well-developed 3,500- to 4,000-year-old pebble beach terrace of a higher stand of Lake Superior, the last major stand of the lake before it reached its present level (Figure 10-24). There are wave-cut cliffs and steep slopes just north of the highway. The highest old gravel beach, formed by Glacial Lake Duluth, Lake Superior's ancestor, is 60 m higher in this area than it is at Duluth. This is a measure of the amount of post-glacial rebound since the melting of the glaciers that depressed the crust in the Lake Superior region. A 2.5-km walk up the Brule River Valley will bring you to a deep gorge and spectacular scenery.

Stop 11: Grand Portage

Grand Portage, the site of a 14.5-km portage around the three falls and several kilometers of rapids of the Pigeon River (Figure 10-25) was named by French explorers in the late 1600s. Its name means "great carrying place." The Chippewa Indians called it Kitchi Onigum. It was on the major fur trade route of the 1700s and was the first of more than 40 portages on the route (Figures 10-26, 10-27). The French explorer Medard Chouvart, known as Sieur des Groseilliers, may have reached Lake Superior from Hudson Bay via the Pigeon River about 1659. Daniel de Greysolon Sieur DuLhut passed this spot in 1679 as he headed down

the northern shore of Lake Superior to what is now Duluth. However, this track was undoubtedly used much earlier by the Sioux and the Chippewa Indians. The trail that follows the portage is a difficult one, for it climbs about 225 m above Lake Superior, over resistant diabase dikes and sills.

The original post at Grand Portage, consisting of several buildings at the end of the portage on Lake Superior, was built a few years before the Revolutionary War by the British North West Company, which made it the headquarters for its coast-to-coast fur trade. Britain had wrested the Lake Superior region from the French in 1763 at the close of the French and Indian War, but the real powers of the time, the Sioux and the Chippewa, were still fighting for control of the region. Grand Portage was the first white "settlement" in Minnesota (Figure 10-28). Fort Charlotte was built at the other end of the long portage.

The site of the original trading post on beautiful Grand Portage Bay, part of which is protected from the open lake by Grand Portage Island and Hat Point, is now part of the Grand Portage Indian Reservation and was designated a national monument in 1960 (Plate 13). A walk up Rose Hill at the monument will present an excellent view as well as a close look at the Middle Precambrian Rove Formation.

175

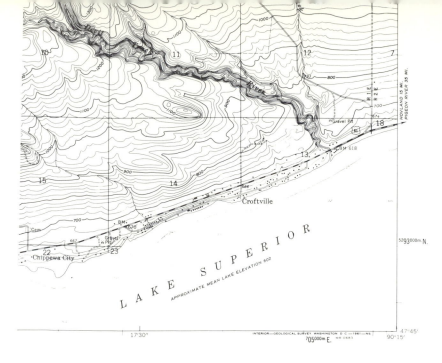

Figure 10-24. Topographic map of a typical segment of Lake Superior shoreline northeast of Grand Marais. Note the deeply incised river cut into lava flows. Croftville is also located on the old lake terrace described in the text. The contour interval is 20 ft. (From the Grand Marais, Minnesota, 7.5' Quadrangle.)

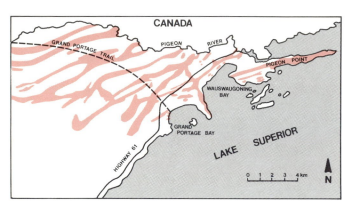

Figure 10-25. Map of the Grand Portage area. The dark areas represent sills and dikes of diabasic (gabbroic) rock that form prominent hills and ridges. Note that one holds up Pigeon Point, the easternmost part of Minnesota.

The 18th century portage was made necessary, specifically, by Horne Falls, Middle Falls, and High Falls (Pigeon Falls) and several kilometers of rough rapids. Middle Falls and High Falls, which is 27 m high (Figure 10-29), owe their existence to vertical diabase dikes intruded into the softer slates of the Rove Formation; the dikes have created local base levels of erosion, preventing the Pigeon River from cutting more deeply into the rock. The Pigeon River forms the international boundary for more than 40 km.

Note: The country between the village of Grand Portage and the Pigeon River is among the most spectacular in Minnesota (Figure 10-30). The dikes responsible for the waterfalls on the river and the thick sills, intrusive into the Middle Precambrian Rove Formation, are parts of the mafic Logan intrusions, which were named after Sir William Logan, the first director of the Geological Survey of Canada, and which may be 1,300 million years old. The areas underlain by the softer slates have been lowered by erosion, whereas those underlain by the dikes and sills stand as high points. One of the resistant sills, 150 m thick and dipping gently southward, holds up Pigeon Point, the easternmost bit of Minnesota (see Figure 10-25). The high relief in the vicinity of Thunder Bay, Ontario, about 60 km east of the Pigeon River, is also due to the Logan intrusions which cap many of the big mesalike hills. The sills have obvious vertical columnar joints whereas the underlying Rove Formation displays horizontal bedding.

Stop 12: The Logan Intrusions

A Logan dike and the Rove Formation into which it was intruded are easily seen in a big roadcut opposite the easternmost of three roadside parking places about 4.5 km (2.7 mi) northeast of Grand Portage (Figure 10-31). From here on a clear day one can see Pigeon Point, Susie and Lucille islands, and even Isle Royale National Park, which lies about 35 km to the southeast in Lake Superior and consists of Upper Precambrian lava flows. Unfortunately, Isle Royale is a part of Michigan and its geology will not be discussed here.

Western Vermilion District (Vermilion Iron Range)

The Vermilion District was discussed in general terms in Chapter 3. It contains the best exposed 2,700-million-year-old volcanic-sedimentary greenstone belt in Minnesota. The eastern half of the district is in the Boundary Waters Canoe Area and is described in the following section. The western and more easily accessible half, between the Lake Vermilion-Tower-Soudan area and the Ely area, will be covered in some detail here.

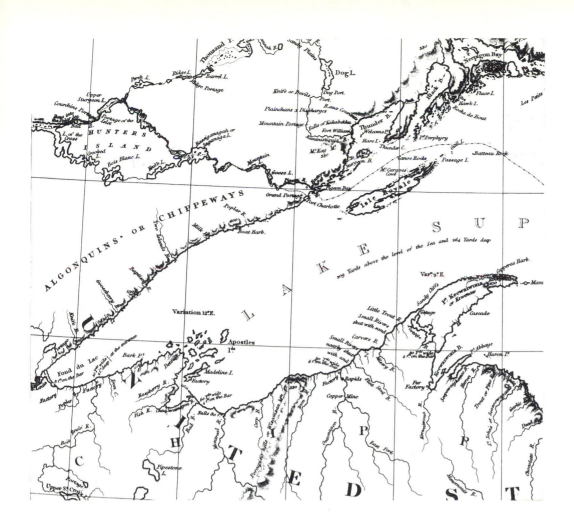

Figure 10-26. An 1816 map of the Lake Superior region. This is from a larger map of Canada and Nova Scotia in *Thomson's New General Atlas* (London). (Courtesy of Archives and Special Collections, University of Minnesota, Duluth.)

Figure 10-27. An 1832 map of Lake Superior published by the Society for the Diffusion of Useful Knowledge. (Courtesy of Archives and Special Collections, University of Minnesota, Duluth.)

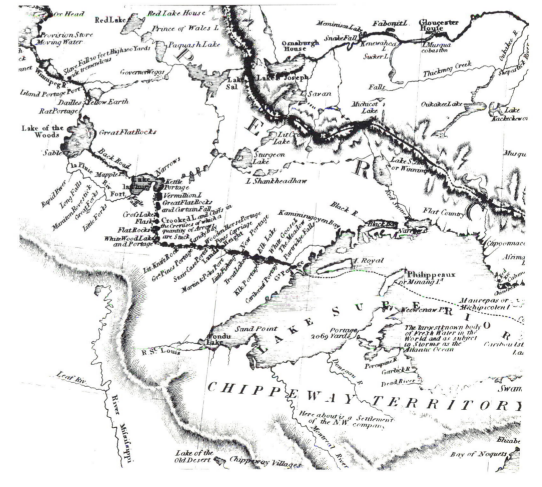

Figure 10-28. Grand Portage.

Figure 10-30. Scenery in northeasternmost Minnesota. Diabase hills tower above Teal Lake. The view is from Highway 61.

Figure 10-31. Roadcut on Highway 61 a few kilometers northeast of Grand Portage. The dark rock exposed at the top of the cut is a ridge-forming vertical mafic dike, one of the Logan intrusions; and the bedded rock in the lower half of the cut is Middle Precambrian Rove Formation.

Figure 10-29. High Falls (Pigeon Falls) on the Pigeon River. (After a sketch by D. D. Owen, 1848, in Winchell et al, 1899.) During logging operations, water and logs were routed around the falls in wooden chutes.

The origin of Minnesota's most ancient iron-formation, the Vermilion Range, and the discovery of iron ore on the Vermilion during the Lake Vermilion gold rush of 1865 were outlined in Chapter 8. The two villages of Tower and Soudan developed side by side. Tower was named for Charlemagne Tower, the Pennsylvanian who built the railroad to Two Harbors and whose fascinating story is told by H. Bridges (1952). Soudan was named for Sudan, Africa. (One version has it that Sudan was in the news at this time. Another version is that the miners who spent their first cold winter in the area named it after the hottest place they could think of.)

The western Vermilion district has been known as the Vermilion Range since at least 1884, when Minnesota's first iron ore was shipped from the Soudan Mine. Soon ore was being shipped from mines in the Ely area as well. The Vermilion Range is pockmarked with iron-ore exploration pits. Numerous lenses and pods of iron-formation are present in the Ely Greenstone, and probably each one was known and transected by a magnetic dip needle survey around the turn of the century. Several mines, such as the La Rue Mine, the McComber Mine, and the North American Mine, contributed much to the rich history and the folklore of the area, but not much to its economic stature.

All that remains of the North American Mine just 2 km south of Tower is a broad dump of rusty rock removed from the mine. Its shaft, now filled with water, was sunk in about 1910 into pyritic black slates and light-colored dike rocks to a depth of 270 m. Its stopes extended as much as 400 m laterally. Although the pyrite may have been a factor in the initial interest in the locality, it was probably largely a stock venture. Reportedly, gold ore from the mine assayed at $20 to $36 per metric ton; this called for a smelter on the site. One story claims that the major promoter boarded an ore train that was hauling high-grade iron ore from the Soudan Mine to the Lake Superior port at Two Harbors and pushed off large ore chunks at the spot where the railroad passed close to the North American. These were then represented to prospective investors as ore from the North American shaft. A fancy resort hotel, complete with palm trees, was to be part of the developing complex on this townsite, which was to be called Walsh, after one of the entrepreneurs.

Because U.S. Highway 169, with its numerous rocky roadcuts, runs the length of the Vermilion Range and because the area has been well studied, a road log with selected stops is an efficient way to portray the geological relationships and to make it possible for the interested person to actually locate and see the rocks. First, however, a general description of the rock units present is essential.

Although the rocks have all been folded by two deformations and faulted three times, geologists have, by detailed mapping (Figure 10-32) and laboratory studies, been able to reconstruct to their satisfaction the conditions at the time each rock unit formed, as well as the original extent of each unit. Because most rock units have been tilted to a vertical or near-vertical position and because the rock units show, on the basis of pillowed lavas (see Figure 3-6) or graded beds (see Figure 3-8), that the tops of the units now face northward, the generalized geologic map of the district (Figure 10-33) is in effect a geologic cross section as well, with the older units on the south and the younger on the north.

A diagrammatic representation of what the eastern Vermilion district may have been like 2,700 million years ago was given in Figures 3-5 and 3-9. Briefly, it was a time of extensive volcanism and related sedimentation. The western Vermilion district was probably quite similar to the eastern part except for more extensive chemical deposition of iron-formation (the Soudan Iron Formation) and a lack of high-level granitic bodies such as the Saganaga Batholith, which intruded the eastern part of the volcanic pile during volcanism. A highly generalized cross-sectional model for the length of the Vermilion district is shown in Figure 10-34. Note its similarity to Figure 10-33.

Now we can proceed with the road log and stop descriptions. In order to cover the rock units in approximate order of geologic age and make it easier to perceive the sequential geologic events that formed the rocks, the log will start at Ely and continue westward to Tower-Soudan and Lake Vermilion. Note that the stops 1 through 8 are keyed to their positions in Figure 10-32. A more extensive field guide for the western Vermilion district by D. L. Southwick and R. W. Ojakangas (1979a) is available from the Minnesota Geologic Survey.

Stop 1: Pillowed Ely Greenstone

The pillowed Ely Greenstone was shown in Figure 3-6. To find the classic exposure in the city of Ely, turn north off Highway 169 onto 13th Avenue East for one block, left on Camp Street for a short distance, right on 12th Avenue East for three blocks, and right on Main Street for one-half block. The exposure is adjacent to the street. The pillows show their original three-dimensional shapes. This exposure is actually a block moved by a glacier from a point about 100 m north of this spot. (Note that no hammering is allowed on this exposure.)

Note: Also on the north side of Ely is a big area of sunken ground that has collapsed into old underground iron mines (see Figure 8-19). It can be reached from the other end of Ely. Ely was an important iron mining center for 76 years (see Figure 8-6).

Note: You can also see pillowed Ely Greenstone in a large roadcut just west of Ely on Highway 169. This metabasalt outcrop shows the classic green color (owing to three metamorphic minerals—chlorite, epidite, and actinolite) from which greenstone belts receive their name. The pillows are best seen by looking upward to the tops of the roadcuts where weathering of their original glassy rinds has accentuated their shapes. A walk to the top of a cut will show how difficult it is to distinguish rock types in lichen-covered, weathered exposures.

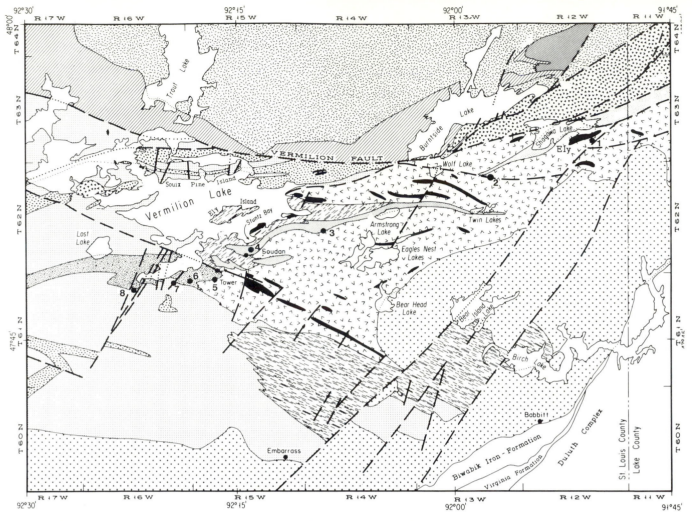

Figure 10-32. Geologic map of the western part of the Vermilion district and adjacent areas. Numbers 1 through 8 are stops for the field trip described in this chapter. (From Ojakangas and Morey, 1972.) See p. 181 for legend and scale.

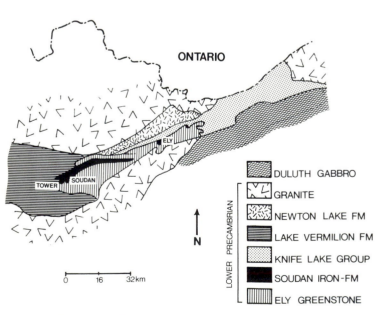

Figure 10-33. Generalized geologic map of the Vermilion district and adjacent areas.

LOWER PRECAMBRIAN

- DULUTH GABBRO
- GRANITE
- NEWTON LAKE FM
- LAKE VERMILION FM
- KNIFE LAKE GROUP
- SOUDAN IRON-FM
- ELY GREENSTONE

0   16   32km

N

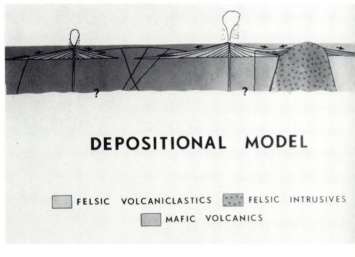

DEPOSITIONAL MODEL

FELSIC VOLCANICLASTICS   FELSIC INTRUSIVES
MAFIC VOLCANICS

Figure 10-34. Depositional model along the length of the Vermilion district showing two main felsic explosive volcanic centers, one near Lake Vermilion in the west and one near Knife Lake in the east. The felsic intrusive represents the Saganaga Batholith, which intruded into the volcanic pile. Arrows represent turbidity currents moving down volcanic edifices into deeper water.

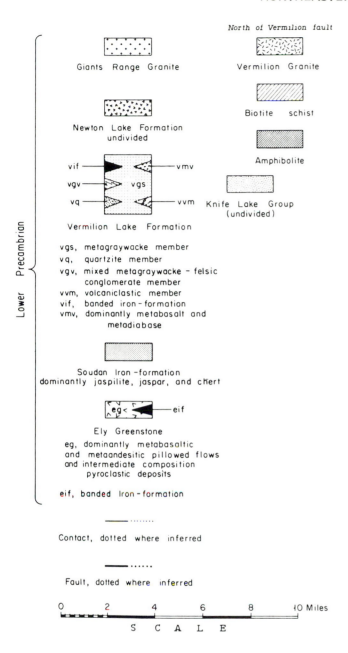

North of Vermilion fault

Giants Range Granite

Vermilion Granite

Newton Lake Formation
undivided

Biotite schist

Amphibolite

vif ——— vmv
vgv ——— vgs
vq ——— vvm

Knife Lake Group
(undivided)

Vermilion Lake Formation

vgs, metagraywacke member
vq, quartzite member
vgv, mixed metagraywacke – felsic
conglomerate member
vvm, volcaniclastic member
vif, banded iron–formation
vmv, dominantly metabasalt and
metadiabase

Soudan Iron–formation
dominantly jaspilite, jaspar, and chert

eg< ——— eif

Ely Greenstone

eg, dominantly metabasaltic
and metaandesitic pillowed flows
and intermediate composition
pyroclastic deposits

eif, banded iron–formation

Contact, dotted where inferred

Fault, dotted where inferred

0   2   4   6   8   10 Miles
S   C   A   L   E

Stop 2: Fault Zone

Pull off to the right of Highway 169 at a point 6.4 km (4 mi) west of the western edge of Ely onto a small grassy road just east of a pine-covered hill along the highway. Walk 20 m to the railroad tracks and then about 50 m to the left (west) on the tracks to a railroad cut. A fault zone, part of the Wolf Lake Fault, a large east-trending fault about 20 km long, is exposed in this cut. The rock here, probably originally a felsic volcaniclastic rock within the Knife Lake Group, has been highly sheared and is now a very shiny phyllite, a rock whose grains are midway in size between those of slate and schist. Scratches called *slickensides*, caused by the pressing of rock into rock during movement

along the fault, are present on every piece of rock, as are small-scale folds or crenulations.

Stop 3: Soudan Iron-Formation

Continue on Highway 169 to a point about 5.3 km (3.3 mi) west of County Highway 408 (Mud Creek Road). Just over the crest of a hill with a large roadcut, turn off to the right onto an abandoned road. Walk back to the cut, but be careful of traffic. The Soudan Iron Formation, a member of the Ely Greenstone, is exposed in this cut, with beds standing in a vertical position. Cherty iron-formation with magnetic, hematitic, and sulfide-rich layers is interbedded here with black muddy beds. (The origin of this type of iron-formation is discussed in Chapter 8.) About 10 m east of the iron-formation, a crosscutting igneous dike of felsic composition (dacite) is exposed. Many similar dikes cut across the iron-formation and the Ely Greenstone between Ely and Soudan. They are interpreted as feeder dikes to explosive felsic volcanic centers that were built upon the largely pillowed greenstone platform (see Figure 10-34).

Stop 4: The Soudan Mine

Continue west along Highway 169 into Soudan. Follow the signs to the entrance of Tower-Soudan State Park. (For information on the park, see the section on the Vermilion Range in Chapter 8 and Figures 8-2 through 8-5. The geology of the park has also been described by P. K. Sims and G. B. Morey [1966].) Park near the headframe of the Soudan Mine. This mine was donated, along with 400 ha of land, to the state by the United States Steel Corporation in 1963. Visitors can don hard hats, ride the skip down a 78° incline to a depth of over 700 m in three minutes, and then ride an underground train about 1 km to the old workings, where high-grade hematite is still visible.

The 14 million metric tons of ore produced by this mine came largely from underground, but the operation started as a number of open pits.

Note: A short walk eastward from the shaft, past the end of the open pit, through a gate, past the first park building on the left, and then a short distance up a gravel road to the left is a classic, much-photographed outcrop of Soudan Iron Formation displaying highly folded beds of alternating white chert, red chert (jasper), and metallic gray hematite (see Figure 3-11a). (No hammering or specimen collecting is allowed!) The folds have been interpreted to be the result of two periods of folding. If the park is closed, this outcrop can be reached from the eastern village limits as follows. Turn right from Highway 169 into Soudan. When the street ends, turn right and go up the hill toward the mine headframe. At the road junction by the buildings, turn right and follow the road about 50 m or so to the crest of Soudan Hill and pull off to the right at the outcrop.

Note: Returning to Highway 169 and driving west into and through the village of Tower, you will note a monument on the right erected in memory of the late President McKinley when he was assassinated in 1901. The mining engineer in charge of the Soudan Mine had been McKinley's roommate at Harvard.

Figure 10-35. Coarse explosive felsic volcanic debris on Lake Vermilion.

Figure 10-36. Mafic Middle or Upper Precambrian dike in Lower Precambrian volcanic sandstone. It has been dated at 1,570 million years, but is more likely about 2,000 million years old.

Stop 5: Lake Vermilion Formation

Just west of Tower on Highway 169 is a large rusty road-cut in a felsic (dacitic) volcaniclastic unit of the Lake Vermillion Formation. Note the stretching of large fragments of dacite and smaller fragments of pyrite that are now highly weathered causing rusty spots and streaks. Better outcrops of this type are present to the north on islands and private land on Lake Vermilion, but they are not easily accessible (Figure 10-35). The rather large size of the fragments, up to 10 cm in diameter, suggests relative proximity to an explosive volcanic center. Detailed mapping of the area has shown such a center to have been located a few kilometers to the northeast in the vicinity of Lake Vermilion. There the dacite fragments are as large as 50 cm. The iron-oxide staining on this outcrop is the type of thing exploration geologists look for as a guide to ore-bearing sulfide minerals. This outcrop, however, contains no such minerals, only pyrite (FeS_2) and pyrrhotite (also FeS_2).

Stop 6: Lake Vermilion Formation

About 2.5 km (1.6 mi) west along Highway 169 is a large roadcut of light gray volcanic sandstone of the Lake Vermilion Formation. It has the appearance of a quartzite, but close field and laboratory study show that it is indeed of volcanic origin, composed of sand-sized grains of volcanic quartz, plagioclase, and finer-grained dacite. Thus, it has the same composition as the felsic dike of stop 3 between Ely and Soudan and the same composition as the fragmental volcaniclastic rock at stop 5. However, it is finer grained, suggesting deposition farther from the explosive volcanic center. At the west end of the south cut, a 6-m-thick dike of gabbro is exposed (Figure 10-36). It has been dated radiometrically at about 1,570 million years, probably a minimum age. Note the "chilled" (finer-grained) borders and the inclusions of volcanic sandstone.

Stop 7: Lake Vermilion Formation

Continue west on Highway 169 for 1 km (0.6 mi) to a big roadcut on the south side of the highway. This roadcut of folded slate (black) and metagraywacke (weathering white) of the Lake Vermilion Formation is an excellent exposure in which to see graded beds like those in Figure 3-8. (Please do not hammer on them.) Two major folding events have been proposed for this outcrop (see Figure 3-11b), but there is also evidence of some deformation before the sediment became hard rock. Although these rocks appear quite different from the felsic rocks at the two previous stops in Lake Vermilion Formation, their chemical and mineralogical compositions are very similar. A likely interpretation of the origin of these beds is that turbidity currents moved down off the volcanic center and deposited the graded beds in the adjacent deeper sea, with the slates originally having been beds of fine mud (probably in part volcanic ash)

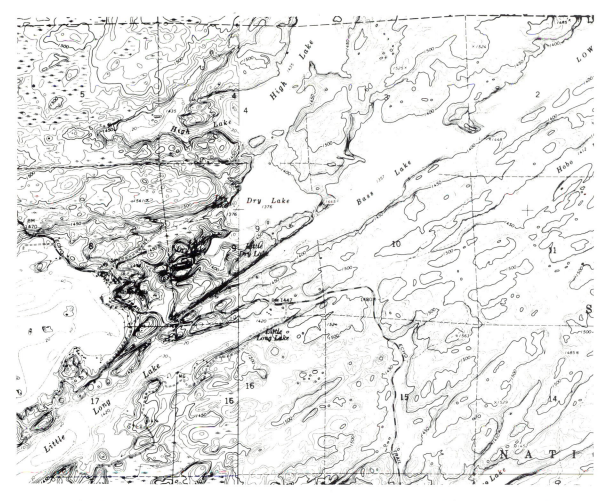

Figure 10-37. Topographic map showing a topographic low along the Burntside Lake Fault, a splay of the Vermilion Fault. (See also Figure 3-16a.) (Shagawa Lake and Ely, Minnesota, 7.5 'quadrangles.)

that slowly rained down onto the sea floor from above between the periodic turbidity currents. Figure 3-9 depicts the suggested environment of deposition away from the volcanic center and perhaps on a submarine fan.

Stop 8: Lake Vermilion Formation

Continue west on Highway 169, for another 3 km (1.8 mi) to the junction of County Highway 77. Turn right and proceed for about 0.8 km (0.5 mi) to the bridge over the Pike River. Park on the left beyond the bridge. Outcrops both east and west of the bridge consist of the same metagraywacke and slate as at the previous stop. The beds are not tightly folded here and can all be seen topping to the south (i.e., the fine-grained tops of these graded beds are to the south). Also visible here on the east side of the river is a northeast-trending fault along which about 280 m of horizontal movement has occurred (see Figure 3-17). The Pike River was part of the major Indian-voyageur route from the tip of Lake Superior to the border lakes country mentioned

in the section on Jay Cooke State Park earlier in this chapter, and many, many canoes must have passed under this bridge.

Three additional sites outside of the western Vermilion district are worth mentioning in order to show the relationships of the rocks of the western Vermilion district to adjacent rock units. The first site is on U.S. Highway 1 about 4.2 km (2.6 mi) south of the junction with Highway 169 in Ely. Here the Giants Range Batholith is exposed in a large roadcut. Various phases of granite, late epidote (which is green), and fluorite (which is purple) can be seen here. The second site is north of Ely, about 5.3 km (3.3 mi) north of the junction on the north side of Shagawa Lake, of County Highway 88 and County Highway 116 (the Echo Trail). The Burntside Lake Fault, part of the Vermilion Fault System, crosses the Echo Trail here at County Highway 752. It is present as an elongate topographic low between rock ridges and is accentuated by the presence of a cleared power line along the fault (Figure 10-37). The Vermilion Fault is the northernmost fault on Figure 3-16a; this location is on

Figure 10-38. Granite and amphibolite along the Echo Trail north of Ely. S. S. Goldich, G. B. Morey, and P. K. Sims, all active in the study of the Lower Precambrian rocks of northern Minnesota during the 1960s and 1970s, are actively discussing the origin of the rocks. As can be seen by a close study of the photograph, geologists are not always in agreement on everything.

the northeast-trending portion of the fault. Rocks south of this fault show low-grade metamorphism, and rocks north of this fault show higher-grade metamorphism that probably took place when the rocks were at least 1.5 km deeper than those on the south side. There has been probably 1.5 km of vertical movement on the Vermilion Fault as well as 20 km of horizontal movement, with the north side moving eastward relative to the south side. Northward along the Echo Trail over a distance of 10 km or more, the rocks pass from gneisses with alternating bands of black metavolcanic-metasedimentary rock—now largely amphibolite—and pink to gray granitic rock (Figure 10-38) to quite clean and homogenous granitic rock of the Vermilion Batholith proper, with few if any inclusions of country rock. These relationships show that the batholith is younger than the volcanic-sedimentary rocks of the Vermilion district and that it intruded into these rocks. The third site, a few kilometers north of Virginia and about 1 km south of the junction of U.S.

Highways 53 and 169, is a point on the Laurentian Divide, from which waters flow north to Hudson Bay and south to the Gulf of Mexico. This site, on the southern margin of the Giants Range Batholith, contains a mixture of metavolcanic-metasedimentary rocks and at least two phases of the younger intrusive granite (see Figure 3-12). A roadside rest on the northbound lane makes this site especially accessible, although the best rock exposure is the knob between the two lanes of the expressway. Geology students affectionately (?) refer to this locality as "Confusion Hill."

Boundary Waters Canoe Area

The Boundary Waters Canoe Area (BWCA) of northern Minnesota fascinates thousands of visitors each year because of its true wilderness aspect of lakes and forest (Figure 10-39). But there is much more here, for the rocks exposed along the lakeshores and on the portages between lakes in most of the BWCA reveal a fascinating story of what this wilderness region, including the adjacent Quetico Provincial Park across the border, was like 2,700 million years ago when explosive volcanoes rose out of a practically lifeless sea to dominate the landscape.

The area that is now the BWCA was established as the Superior Roadless Area in 1938 but did not become the BWCA until 1964. However, it has been known to people as a place of beauty for a long time. Indians traversed it hundreds of years ago (Figure 10-40), and as the fur trade developed in the 1700s, Saganaga Lake, Knife Lake, Basswood Lake, Crooked Lake, and Lac La Croix, plus countless smaller lakes between these large ones, were major links on the route between Lake Superior and the Lake of the Woods area and the wilds of Canada to the north and northwest (Figure 10-41). (See also Figure 10-27.) In the mid-1800s, Canadian geologists entered the border country, and since the 1800s geologists of the Minnesota Geological Survey have intermittently studied the area. J. W. Gruner, of the University of Minnesota, mapped most of the BWCA during the summers between 1928 and 1940, when few canoeists traversed the area, and published a detailed map in 1941.

Since the early 1960s, geologists from the Minnesota Geological Survey and the University of Minnesota have conducted research in the BWCA on various topics. In addition, a number of graduate students, mostly from the University of Minnesota at Duluth, have written theses on several parts of the area.

The popularity of the BWCA has grown tremendously. A quota-reservation system, a can and bottle ban, and overused campsites (Figure 10-42) attest to the increased pressure on this wilderness. Gruner, on his last visit to the BWCA in 1967 by United State Forest Service float plane, was astounded to see myriad tents on Ogishkemuncie Lake. "You could hold a dance here!" he said. He also related that he and his field assistant used to sing opera at their campsites on the shores of the lakes, with the loons providing the

Figure 10-39. The Boundary Waters Canoe Area. (Courtesy of Clark Peterson.)

chorus. "Nowadays," he said, "they'd lock you up for disturbing the peace."

Figure 10-43 is a highly generalized geologic map of the BWCA. It shows that over much of the area volcanic rocks, volcanogenic sedimentary rocks, and granites, all dated at 2,700 million years, predominate. The bedded rocks are folded and faulted and are generally in vertical or near-vertical positions. In the southeast, the gently dipping Middle Precambrian Rove Formation (which is 1,800 million years old) is present, intruded by the 1,100-million-year-old gabbro and related rocks of the Duluth Complex. The Early Precambrian history outlined for greenstone belts and granites in Chapter 3, the Middle Precambrian sedimentational history given in Chapter 4, and the intrusion of gabbro outlined in Chapter 5 provide essential background for the understanding of the initial geological events that molded the rocks of the area.

Weathering and erosion by running water over a period of more than 2 billion years have reduced the entire Canadian Shield, including the present BWCA, to a low-lying generally featureless topography. Then came the glaciers of the Ice Age (see Chapter 7) to gouge out rock that was either softer than adjacent rocks or more thoroughly fractured, thus forming deep, commonly elongate lake basins separated by more resistant ridges, as shown in Figure 10-44 and 10-45 (see also Figure 10-1). The distribution of lakes over most of the BWCA is strongly controlled by shear and fault zones, most of which probably date back to Early Precambrian time. Many of these zones of weakness may have had repeated movements in later times.

Volcanic rocks, either flows or shallow intrusive bodies, are best seen on Jasper Lake and at the western end of Eddy Lake. Pillowed greenstones (lichen-covered) are visible on the cliff that forms the faulted north side of Otter Track (Cypress) Lake (Figure 10-46), along the north side of the south arm of Knife Lake, and on the north side of the east

Figure 10-40. Indian paintings on a granite outcrop at Lac La Croix. (Courtesy of D. L. Southwick.)

Figure 10-42. Sign in the Boundary Waters Canoe Area.

Figure 10-41. *Picture Rocks of Crooked Lake*, by Francis Lee Jaques. (Courtesy of the Minnesota Historical Society.)

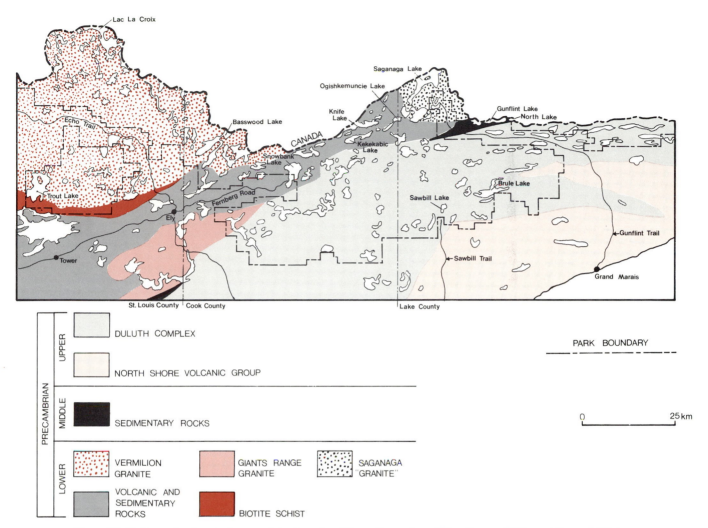

PRECAMBRIAN	UPPER			DULUTH COMPLEX
				NORTH SHORE VOLCANIC GROUP
	MIDDLE			SEDIMENTARY ROCKS
	LOWER			VERMILION GRANITE

VERMILION GRANITE

GIANTS RANGE GRANITE

SAGANAGA "GRANITE"

VOLCANIC AND SEDIMENTARY ROCKS

BIOTITE SCHIST

0   25 km

Figure 10-43. Generalized geologic map of the Boundary Waters Canoe Area. (After Gruner, 1941, and Green, 1981.)

Figure 10-44. Elongate lake in the Boundary Waters Canoe Area.

arm of Jasper Lake. The dark green sandy shores and out-crops on the northern shore of Kekekabic Lake (Figure 10-47) are probably explosive volcanogenic sediments. Much of the BWCA is underlain by steeply dipping interbedded black slates and gray metagraywacke sandstones (Figure 10-48), the products of the metamorphism of sedimentary rocks derived from the erosion of the explosive volcanic islands. The abundant knife-sharp slates long ago provided the names for Knife Lake and Little Knife Lake.

Rock outcrops on several islands and the northeastern shores of Ogishkemuncie Lake provide evidence of a rather unique geologic history. The thick conglomerate and graded sandstone unit there, probably deposited on a submarine fan similar to but smaller than those off the coast of California today, consists largely of pebbles and boulders of volcanic rocks. The conglomerates also contain red jasper pebbles from small "exhaled" iron-formations in the

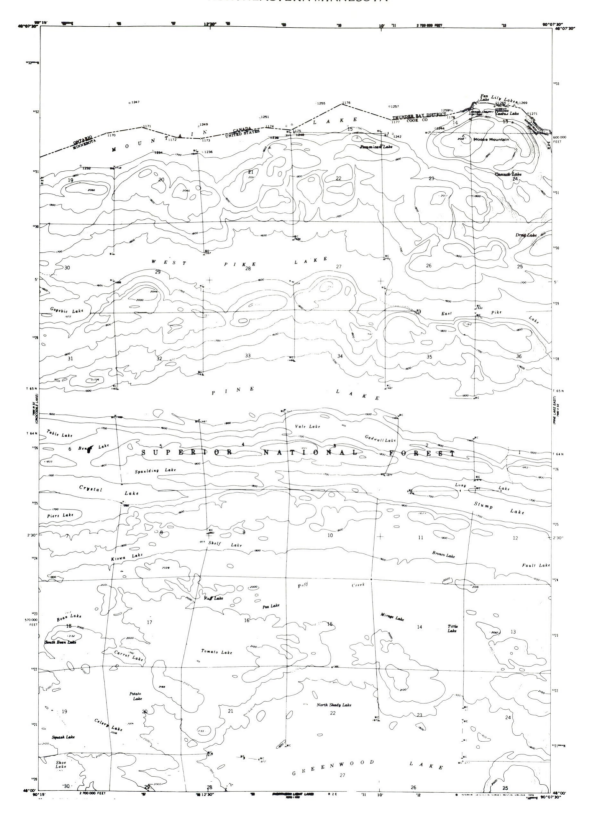

Figure 10-45. Topographic map showing elongate, glacially scoured lakes. The elongate lakes are underlain by soft Rove Formation and the hills between them are formed by diabase sills. The southern one-third of the map area is underlain by gabbro, and there the lakes are more equidimensional. (From the Pine Lake West, Minnesota-Ontario, 7.5′ Quadrangle.)

Figure 10-46. Cliff of pillowed greenstone along the north side of Otter Track Lake.

a

b

Figure 10-47. (a) Coarse explosive volcanic rock on Kekekabic Lake. Large chunks may be volcanic bombs, blebs of magma thrown out of a volcano. (b) Microscopic view of explosive volcanic rock on Kekekabic Lake. Light-colored grains are mostly hornblende. The field of view is 2.5 mm across.

Figure 10-48. Graywackes and slates on Knife Lake. Note the good graded graywacke beds (light gray) and thin dark slates.

volcanic pile (see the section on the Vermilion Range in Chapter 8) and, most importantly, distinctive granitic boulders of Saganaga "granite" clearly derived from the Saganaga Batholith, which lies to the northeast, as shown in Figure 10-49 (see also Figure 10-43). These boulders are tracers that can with certainty be attributed to erosion of the exposed batholith in Early Precambrian time. Furthermore, the interbedding of the conglomerate and associated sedimentary rocks with beds of volcanic tuff shows that the batholith was exposed by erosion in the vicinity of what is now Saganaga Lake while volcanism continued elsewhere in the area. This implies a high-level emplacement of the Saganaga Batholith into a hot crust of volcanic rocks and then rapid erosion and exposure of the batholith, as illustrated in Figure 3-9. Boulders of Saganaga "granite" are found as far away as Ensign Lake, 35 km southwest of Saganaga Lake. In the late 1800s and early 1900s, this granite-bearing conglomerate was interpreted as evidence of the erosion of older mountains formed during a "Laurentian" mountain-building event. Radiometric dating and field relationships now show that this is not true.

189

The Saganaga Batholith itself covers 700 km², with about half of that area in Minnesota. Although mostly a light pinkish gray rock, as in the roadcuts on the last few kilometers of the Gunflint Trail, it contains a darker-colored phase near its borders. A local pegmatitic phase was once prospected for gold on Gold Island in Saganaga Lake. The intrusive nature of the batholith is verified by metamorphism and diking of adjacent rocks (Figure 10-50). What are probably slightly younger granitic bodies are found on Kekekabic and Snowbank Lakes and, in the western BWCA, at Lac La

Croix and Crooked Lake, where the Vermilion Batholith is well exposed.

Gunflint Lake, including Magnetic Bay, is partially underlain by the Middle Precambrian Gunflint Iron-Formation. The cherty and locally magnetic iron-formation can be seen in roadcuts where the Kekekabic and Magnetic Rock hiking trails leave the Gunflint Trail, about 11 km (8 mi) from its north end.

Thick sills of gabbroic rock related to but predating the Duluth Complex have intruded the slates and graywackes

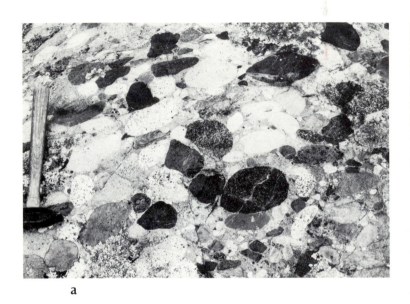

a

b

a

b

Figure 10-49. (a) Outcrop of conglomerate on Ogishkemuncie Lake. White pebbles spotted with black and gray are pebbles of "granite" from the Saganaga Batholith. All other pebbles are various types of volcanic rocks. (b) Close-up view of pebbles of "granite" in the conglomerate of Figure 10-49a. Light gray spots in the pebbles are quartz "eyes," which are very characteristic of the Saganaga "granite."

Figure 10-50. (a) Inclusions of mafic volcanic rock (now amphibolite) in the edge of the Saganaga Batholith at Alpine Lake. (b) Dike of Saganaga granitic rock in metamorphosed mafic volcanic rock (now amphibolite) at Rog Lake.

of the Middle Precambrian Rove Formation in the eastern BWCA and as far east as Thunder Bay, Ontario, and beyond. Differential erosion has left the sills as cliffs, whereas the lakes are underlain by the more easily eroded sedimentary rocks. On some of the cliffs, the Rove Formation can be found exposed beneath the ridge-forming sills (Figure 10-51). (See also Figure 10-31).

The 1,100-million-year-old Duluth Complex in the southeastern BWCA consists of several rock types, but in general it can be divided into two groups—dark gray gabbroic rock and minor "red rock" of granitic composition. The "red rock" is well exposed along the shorelines of several lakes in the southeastern part of the BWCA. The other lakes situated on the Duluth Complex display various phases of the darker-colored gabbroic rocks. It is a darker variety that contains copper and nickel minerals just outside of the BWCA south of Ely (see Figures 8-23 through 8-26 and Plate 7). It is not known whether these minerals are also present in the BWCA.

Voyageurs National Park and Adjacent Areas

The voyageurs who canoed through the border lakes of what is now the Boundary Waters Canoe Area, as well as those who came up from Lake Superior via the alternate St. Louis River-Embarrass River-Pike River-Lake Vermilion-Vermilion River route, eventually reached Namakan and Rainy Lakes. Although the backs of the 8- to 10-man crews were undoubtedly bent to the paddle as the 7.5-m-long birchbark canoes were propelled through the clear waters, these voyageurs probably appreciated the beauty of the large lakes in spite of the dangers and hard work they demanded. Indians undoubtedly had already used the area for hundreds or thousands of years. Kabetogama and Namakan lakes were just two of the many lakes that were centers of Chippewa activity. Chippewa townsites, cemetaries, and grinding holes are present there. In 1971, after decades of preliminary attempts and jockeying, Voyageurs National Park was born, preserving 890 km² of this country so that generations to come may also appreciate its beauties. Its big waters contrast with the more easily paddled routes within the Boundary Waters Canoe Area, for even the voyageurs in their big canoes were often windbound on Rainy Lake. Most of the big trees that the Indians and voyageurs took as commonplace are gone, for some 40 logging camps existed in the area now claimed by the park, but the wilderness feel is nevertheless there among the rocks, waters, and forest.

The geology of the park is relatively simple as far as rock types are concerned (Figure 10-52). Granite, biotite schist, and migmatite (a rock consisting of interlayered granite and biotite schist on various scales) underlie all of the park area, except for the northwest corner on Rainy Lake, where a major volcanic-sedimentary (greenstone) belt projects in from Ontario. There is more granite in the southern part of the park than in the rest of the park, and biotite schist is most abundant in the north. All of these rocks are Early Precambrian, about 2,700 million years old.

The granitic rocks constitute the northern edge of the Vermilion Batholith. Medium- to coarse-grained pink and gray granites are the most abundant granitic rocks, but slightly younger crosscutting pegmatites with potassium feldspar crystals as large as 30 cm across and muscovite crystals up to 10 cm across are conspicuous (Figure 10-53). Although pegmatites can include minerals containing valuable rare elements such as lithium and uranium, the

Figure 10-51. Mafic sill with vertical columns capping a cliff of the Rove Formation with subhorizontal beds on Clearwater Lake. Note the big talus pile at the base of the cliff, the result of frost action.

pegmatites of this area apparently do not; tourmaline and garnet, two rare pegmatite minerals, have been noted at only one locality.

Much of the granitic material is in the migmatites (Figure 10-54). D. L. Southwick, of the Minnesota Geological Survey, who has studied the migmatites of the Vermilion granite-migmatite body for several years, has suggested (1972) that the homogeneous Vermilion Batholith (composed largely of granite) is the deeper part of the tilted and eroded complex. According to this idea, the migmatites would have formed in the "roof" or cover of sedimentary rocks where granitic stringers, whether formed by true magma or by partial melting of the cover rocks during metamorphism, invaded the sedimentary cover. At about the same time, the pressures and temperatures converted the sedimentary cover rocks into biotite schist. Some of the cover rocks may have had a basaltic volcanic origin, for they are now amphibolites. A new, very long and high vertical roadcut on Highway 53 southwest of the park and 49 km (30.7 mi) north of Orr

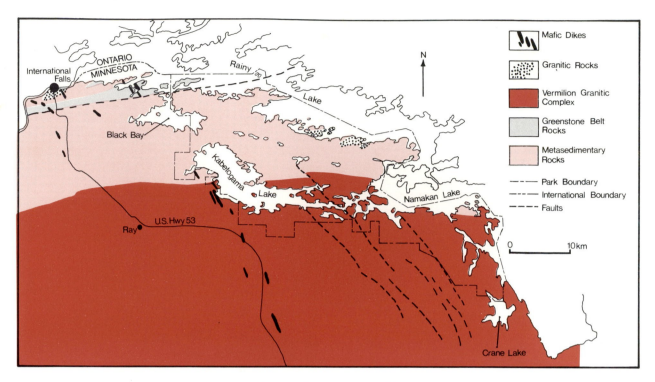

Figure 10-52. Generalized map of Voyageurs National Park. The mafic dikes are Middle Precambrian in age, about 2,100 million years old; all other rocks are about 2,700 million years old. (After the International Falls 1:250,000 geologic map sheet by D. L. Southwick and R. W. Ojakangas, 1979b.)

Figure 10-53. Granitic pegmatite composed of potassium feldspar, quartz, and muscovite in Voyageurs National Park.

provides a spectacular look at a coarsely layered migmatite-granite complex (Figure 10-55).

The northern part of the park contains broad areas of gray to black biotite schist with only a minor granitic fraction. The metamorphic grade in these rocks decreases northward, with more original sedimentary textures and a finer, less recrystallized grain size commonplace near the Rainy Lake Fault. A few kilometers south of the fault (closer to the granites) garnet, sillimanite, and staurolite begin to appear. Well-preserved bedding, including excellent graded beds with coarse sandy bottoms and fine-grained biotite-rich tops, is widespread (Figure 10-56). This rock type extends far to the east in Ontario and far to the west in Minnesota, where it is poorly exposed because of glacial drift. This type of rock is most easily seen outside of the park in the village of Ranier (a few kilometers east of International Falls) in an outcrop right on Main Street; original graded beds are visible in this outcrop.

In the northwestern corner of the park, numerous other rock types are exposed in a narrow Minnesota portion of a major volcanic-sedimentary belt that extends for more than 150 km to the east-northeast in Ontario and 50 km west-southwest beneath International Falls and beyond. The dominant rock type within the belt in Minnesota is cross-bedded feldspathic quartzite, which is exposed west of the park in a roadcut along County Road 109 (1.9 km, or 1.2 mi, south of Minnesota Highway 11), on the mainland south

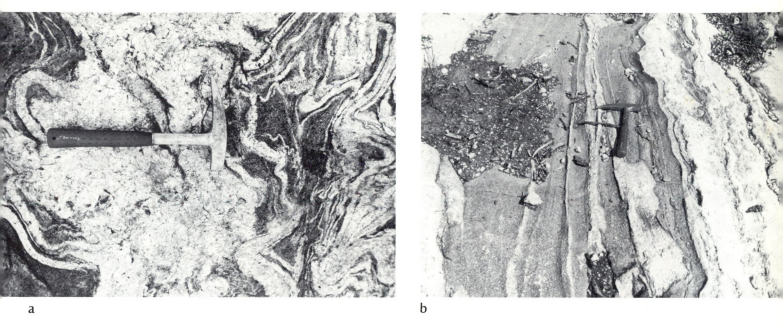

a   b

Figure 10-54. (a) Contorted migmatite in the vicinity of Voyageurs National Park. The light material is granite and the dark material is biotite schist. (b) Migmatite similar to that in Figure 10-54a but not contorted.

Figure 10-55. Granite of the Vermilion Batholith containing large inclusions of black country rock. This is on Highway 53, 49 km (30.7 mi) north of Orr.

of Frank's Bay, on Neil Point, on Grindstone Island, and in the park on Big American and Dryweed islands (Figure 10-57). Conglomerate (Figure 10-58) is associated with the quartzite. A mixed biotite-chlorite schist is common near the Rainy Lake Fault, and some units were probably feldspathic volcanic tuffs. A small island just off the park boundary next to Grindstone Island consists of peridotite, an ultramafic rock containing serpentinized olivine and pyroxene and traces of copper and nickel. A small island in the northernmost part of the park consists of anorthosite, a rock made up almost completely of plagioclase. Other minor rock types are highly stretched pillowed greenstone, greenschist, and various flow or dike rocks. South of the park and Kabetogama Lake on County Road 122 a major mafic dike is exposed in a long roadcut. This 90-m-thick vertical diorite dike probably is one that reappears on islands in Rainy Lake and continues farther into Canada. It has not been dated radiometrically, but its northwest trend suggests that it is part of a major dike swarm more than 2,100 million years old, the result of a major event that fractured the

a

Figure 10-57. Crossbedded feldspathic quartzite between Voyageurs National Park and International Falls. The hammer is parallel to bedding.

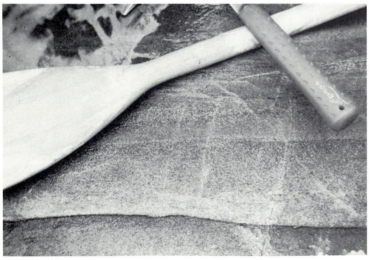

b

Figure 10-56. (a) Microscopic view of totally recrystallized rock, now a garnet-biotite-plagioclase-quartz schist. The large dark grain at the lower right is a garnet crystal, elongate grains are biotite crystals, and the remainder is plagioclase and quartz. (b) Biotite schist beds in Voyageurs National Park. The darker bed with more biotite was once a mud layer, and the lighter bed with less biotite was probably a graywacke sandstone. Before metamorphism, this rock may have resembled that in Figure 3-8.

Figure 10-58. Conglomerate on Neil Point between Voyageurs National Park and International Falls. Pebbles represent various rock types, including quartz diorite (a granitic rock) and numerous volcanic rock types.

Early Precambrian crust from the Lake Superior region to northwestern Canada. Except for glacial drift and modern sediment, the dikes are the only rocks in the region younger than 2,700 million years.

The Rainy Lake Fault, known in Ontario as the Seine River Fault, separates the greenstone belt from the biotite schists (see Figure 10-52). This is truly a major fault, extending over a total distance of 300 km in Ontario and Minnesota. As discussed in Chapter 8, gold was mined in 1894 from quartz along this fault zone on Little American Island of Rainy Lake just west of the park boundary. The fault zone is mostly beneath the waters of Rainy Lake but can be seen west of the park as a topographic low on County Road 138, which runs north from Highway 11 to Neil Point. The fault zone occupies the broad, low swampy zones between the higher rock ridges of the mainland and Neil Point. Adjacent to this zone in a gravel pit on Neil Point, the shearing and faulting in greenschist is obvious, as are quartz and carbonate pods (Figure 10-59). Another fault may underlie the string of lakes on the Kabetogama Peninsula, which includes Locator, War Club, Quill, and Loiten lakes.

Folding is generally hard to discern in the park and adjacent areas since most beds are vertical. However, close field study and plotting of top information from graded biotite schist beds reveals the presence of several important folds with vertical axial planes and vertical sides. Rarely, more gently dipping beds reveal the presence of a fold (Figure 10-60).

In structurally deformed areas such as the Rainy Lake area, different interpretations can be applied to the same rocks, as illustrated below. In 1913, A. C. (Andy) Lawson, of the Canadian Geological Survey, mapped the greenschist, feldspathic quartzite, and conglomerate in the area just west of the northwest corner of the park and showed them folded into a syncline (Figure 10-61, part 1). In 1925, F.F. Grout, of the University of Minnesota, mapped these rocks as an anticline (Figure 10-61, part 2). A third interpretation was put forth by R. W. Ojakangas in 1972 after close study of top indicators such as crossbedding and graded beds; he emphasized the importance of major faults in positioning the rock units (Figure 10-61, part 3). Who is right? Final proof is unobtainable because of the lack of rock exposures in key places.

Figure 10-59. Sheared rock with quartz pods and veins (white in the photo) along the Rainy Lake Fault in a gravel pit on the south edge of Neil Point.

Figure 10-60. Gently dipping beds of biotite schist in Voyageurs National Park on Saginaw Bay, Rainy Lake.

But the International Commission on Geological Nomenclature and the United States Geological Survey deny the existence of such a series of rocks below the Keewatin. Neither of these authorities has, however, so far as their published utterances indicate, visited Rice Bay. The facts which I have recited are open for verification to any geologist and the area is easily accessible. Would it not be well for these eminent authorities before wiping out the Coutchiching series utterly, to examine the Rice Bay section? (1913: 10).

He added:

The facts here recited in regard to this line of contact, particularly near the railway on the shores of Bear passage and the south end of Redgut bay, taken in connexion with the relations of the Coutchiching to the granite, appear to me to prove conclusively the superposition of the Keewatin upon the rocks mapped by me as Coutchiching in the report of 1887. I invite the attention of the International Committee and of the

A much larger controversy over geologic interpretation occurred in the Rainy Lake area. The biotite schists were first studied, especially on the Ontario side of the border, by A. C. Lawson in the late 1880s. (By the way, Lawson's budget for four and a half months of fieldwork in 1887 with a party of five men who studied the rocks in a 2,600-km^2 area, was only $1,608.97!) He named them the Coutchiching Series (after Coutchiching Rapids, now beneath lake level at the point where the Rainy River leaves Rainy Lake). Coutchiching (Koochiching) Falls, with a drop of 7 m, was 4 km down the Rainy River and is now covered by the dam between International Falls and Fort Frances, Ontario. Lawson interpreted them to be older than (i.e., beneath) the green volcanogenic rocks (greenstones) of the Keewatin Series, which he had defined earlier in the Lake of the Woods area and had also mapped on Rainy Lake. If all of these rocks had still been in their original horizontal positions, his statement would have been straightforward. However, the rocks had all been folded and most were steeply tilted or vertical, with their original relative positions obscured.

Several geologists from Minnesota and the United States Geological Survey sharply disagreed with Lawson's (1888) positioning of the Coutchiching Series, claiming these rocks were younger than the Keewatin volcanogenic rocks. (Everyone knew that the Ely Greenstone and correlative rock units were the oldest rocks in the world!) The arguments culminated in a visit to the area in 1904 by an International Stratigraphic Commission, which, in a verdict published in 1905, categorically stated that Lawson was wrong. Lawson returned to the Rainy Lake area in 1911 and in 1913 published a 115-page memoir entitled "The Archaean Geology of Rainy Lake Restudied." The following excerpts present his reassessment of the rock relationships:

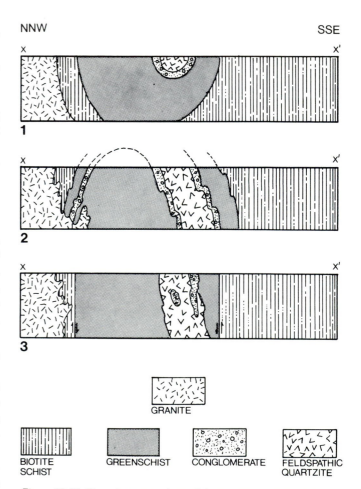

Figure 10-61. Three interpretations of the structure across Rainy Lake: 1, by A. C. Lawson (1913), shows a syncline; 2, by F. F. Grout (1925), shows an anticline; and 3, by R. W. Ojakangas (1972), shows fault relationships. Each interpretation thus assumes different age relationships of the five rock units.

U.S. Geological Survey to this section and challenge them in view of the facts there apparent and easily accessible, to deny the relations of the Keewatin and Coutchiching as I mapped and described them a quarter of a century ago. The fact that these eminent authorities have denied in toto the existence of the Coutchiching series as a constituent member of the Archaean below the Keewatin, without any attempt to verify the very explicit statement of the evidence in regard to this section contained in the report of 1887, places them in a curious light from the point of view of scientific method. The evidence above set forth as to the superposition of the Keewatin upon the Coutchiching is practically the same as that published in 1887. In the course of the work of the past field season this has been supplemented by other observations which support the conclusions then arrived at (1913: 13-14).

Geological work is continuing in the disputed area, especially in Ontario. K. H. Poulsen has mapped complex overturned fold structures (nappes) like those in the Alps, in the Canadian portion of Rainy Lake. This has implications for the "Coutchiching problem," for it indicates that at least locally, the Coutchiching rocks are younger than the volcanic rocks (Poulsen and others, 1980).

This part of Minnesota, along with much of the Canadian Shield, underwent erosion by running water from 2,700 million years ago up to the present, interrupted only by glacial erosion during the Pleistocene (Chapter 7) and perhaps during lower Middle Precambrian time as well (Chapter 4). That this erosion cut deeply into the crust is documented by the presence today, at the earth's surface, of granites that cooled at depth and metamorphic minerals such as staurolite, sillimanite, and garnet that formed under the relatively high temperatures and pressures found only at considerable depths. Probably several kilometers of rock have been slowly weathered and continually removed by erosion over the ages.

During the Plesitocene, glaciers removed the last weathering products and some fresh rock as well. Jointed blocks of rock were lifted out of the positions in which they had rested for 2,700 million years, were ground up, and were carried away to the south, leaving on the Kabetogama Peninsula a rugged terrain of ridges and intervening swamps. Numerous grooved, striated, and rounded rock exposures also attest to the efficacy of the glacial scouring.

Recent erosion and reworking of glacial deposits in the area have produced some delightful secluded sand beaches on the park islands of Rainy and Kabetogama lakes.

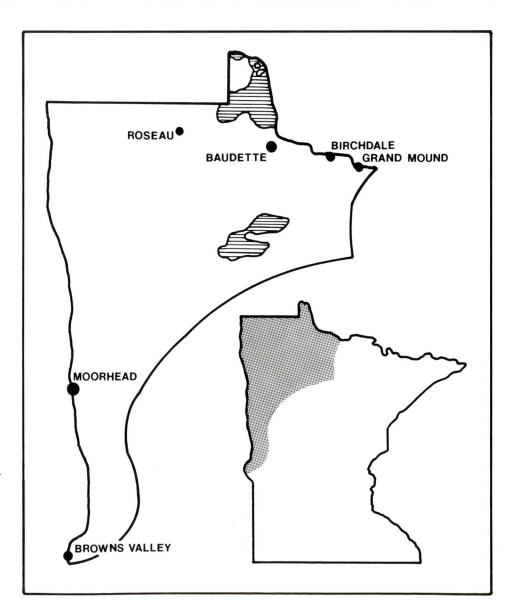

Northwestern Minnesota region.

11

Northwestern Minnesota

THE LAY OF THE LAND

The Red River of the North, which forms the northern half of the western border of Minnesota, flows northward along the axis of a broad topographic basin covered by the silts and clays of Glacial Lake Agassiz. Upper Red Lake, Lower Red Lake, and Lake of the Woods are remnants of that large water body, which drained away sometime before 8,000 years ago. The exposed lakebed, which stretches eastward all the way to International Falls, accounts for some of the flattest terrain to be seen anywhere, interrupted only by the incised valleys of tributary streams and by hundreds of kilometers of drainage ditches (Figure 11-1). From a high of 365 m in the eastern arm, the land surface slopes gently westward to lower than 245 m at St. Vincent, on the Manitoba border. Along the axis of the Red River Valley, elevations decrease from 297 m at Lake Traverse to about 229 m at the international boundary. This gentle gradient encourages aimless wandering of streams, reflected in the tortuous pattern of meanders followed by the Red River and its tributaries (Figure 11-2). The other major watercourse, the Rainy River, follows a straighter path, which may be in part controlled by bedrock structure.

Aside from watercourses, the major relief on the broad plain is furnished by ridges of sand and gravel constructed along former shorelines of Glacial Lake Agassiz. These take the form of linear swells, averaging 4 to 5 m in height and 150 m or more in width and running for tens of kilometers across the countryside. In places, the ridges are splayed into intricate patterns by the vagaries of the currents that deposited them (Figure 11-3). Generally, the elevations of continuous segments rise to the north. The strongest of these ridge systems is traceable in Minnesota from the vicinity of Campbell, in Wilkin County, to Lake of the Woods. Others,

at higher and lower elevations, are not so continuous. The highest level of the lake is represented by the Herman Beach (Figure 11-4). Not only do these old shoreline features contain valuable deposits of sand and gravel (Figure 11-5), but they are also favored as building sites and roadways; one example is Minnesota Highway 11 between Karlstad and Roseau.

The eastern arm of the Lake Agassiz plain is a vast expanse of wetlands and is noted especially for its patterned bogs. The latter consist of tree-covered islands of vegetation, some shaped like teardrops, with intervening water tracks and fens. North of Red Lake is a large boggy area known informally as the "70-mile swamp."

GLACIAL GEOLOGY

The glacial history of northwestern Minnesota is largely the story of Glacial Lake Agassiz. Almost all of the present surface deposits are sediments associated with that lake, as are the landforms. However, deep cuts along some of the rivers and samples taken by drilling indicate a complicated sequence of glacial advances (Figure 11-6), as well as several periods of lake expansion and recession. Finally, the response of the earth's crust to glacial unloading in the form of uplift is superimposed upon the landscape.

A time of warm climate is represented by organic-rich sediments lying beneath glacial till at several exposures along the Red Lake River a few kilometers west of Red Lake Falls. A log enclosed in silt and clay has been radiocarbon dated as greater than 40,000 years old (Figure 11-7); this log apparently came to rest on a lakebed or a floodplain. Next, a glacial advance from the northeast brought till rich in boulders of igneous and metamorphic rock derived from the Canadian Shield. Boulders washed out of this drift furnish

Figure 11-1. High water tables beneath the peat-filled Lake Agassiz plain are troublesome to land use. Drainage ditches are one solution to the problem.

Figure 11-2. The northwest-flowing Red Lake River follows a tortuous route as it approaches its confluence with the Red River near East Grand Forks. Abandoned meander loops are called *oxbows*. (From the Bygland, Minnesota, 7.5' Quadrangle.)

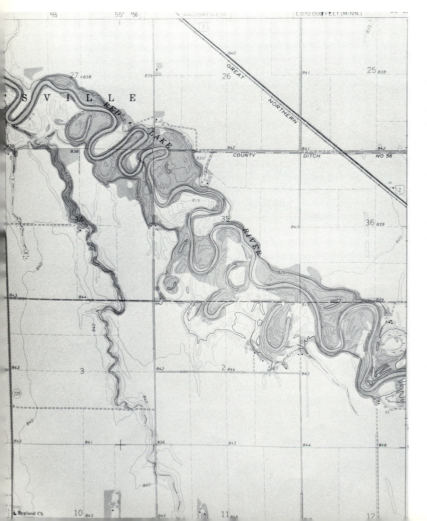

splendid stretches of rapids along the Red Lake River. Burying this drift is a succession of several different tills, all containing limestone fragments derived from the Winnipeg Lowland and some containing the Cretaceous shale of source beds farther west. The last till was deposited by the Des Moines Lobe prior to 12,000 years ago, the time of the instigation of Lake Agassiz. The sequence thus represents multiple glacial advances, first one from the east, then one from the north, and finally a fluctuating ice tongue from the northwest and west. The oldest deposits could predate the Wisconsin glaciation; however, it is possible that only the Wisconsin glaciation is represented.

A general history of Lake Agassiz has already been traced in Chapter 7. Water first began to accumulate when the Des Moines Lobe retreated north of the continental divide in the vicinity of what is today Browns Valley. At first, the ancestral lake was probably a complex of shallow ponds with intervening divides of ice-cored topography, the stagnant remains of the Des Moines Lobe. Eventually, the melting was complete and the waters coalesced to rise and stand for a while at the high stage now marked by the Herman Beach. During this phase of the lake, the margin of the Laurentide Ice Sheet became stabilized on the north, and only one outlet, Glacial River Warren, serviced the spillover. In the southern outlet area, the elevation of the Herman shoreline, named for the town on Minnesota Highway 27 in Grant County, is about 325 m. Evidence for an even higher stage at 335 m, called Lake Milnor, is found in South Dakota.

Three lower beach complexes lie nested in the southern outlet region—the Norcross at 317 m, the Tintah at 311 m,

200

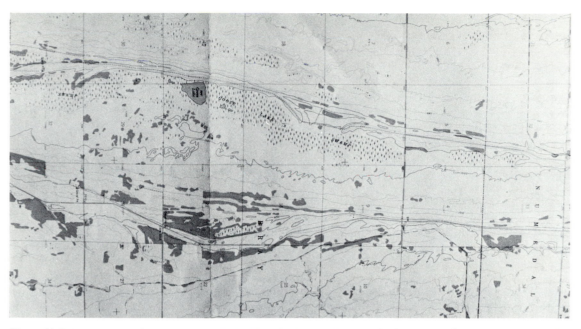

Figure 11-3. Long topographic ridges associated with Lake Agassiz. Goose Lake Swamp is a former lagoon. Note the curved spits. (From the Viking SW, Minnesota, 7.5' Quadrangle.)

and the Campbell at 299 m (Figure 11-8). These features represent periods during which the lake level was stable at least long enough for the shoreline to be marked by strong beach features. The intervals between beaches represent periods when the lake dropped too rapidly for such features to develop. During all of these stages River Warren carried discharge from the lake through several channels north of Browns Valley before cutting the single impressive gorge to what is now Ortonville.

Drops in lake level between stages represent enormous volumes of discharge and floods of catastrophic proportion. Boulder-strewn terraces in the outlet valley between Browns Valley and Ortonville are remnants of River Warren's successively lower channel floors as floodwaters cut deeper. These periods of high discharge and downcutting may relate to episodes of lake expansion triggered by rapid glacial retreat. Lake levels stabilized when the ice margin held steady for a time, causing the outlet river to stop eroding its channel any lower.

An interlude of drying out and subsequent reflooding in the Red River Valley is represented by a sedimentary sequence that contains shallow water and swamp deposits, including wood, pollen, insect parts, and mollusk shells sandwiched between two thick deep-water lake deposits. The wood is about 10,000 years old. From these observations comes the conclusion that two lakes occupied the basin in succession. Lake Agassiz I drained away, probably when melting glacial ice cleared eastern outlets into Lake Superior. Readvance of the ice blocked those outlets, and backflooding to the Campbell level in the Red River Valley formed

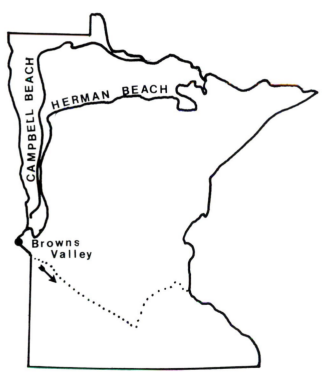

Figure 11-4. The Herman and Campbell beaches are the most continuous of the Lake Agassiz beaches.

201

Figure 11-5. Beach deposits generally contain well-sorted and rounded sand and gravel. They are widely exploited for construction materials.

Figure 11-7. Logs buried beneath glacial drift and exposed along the Red Lake River indicate a warm interval between glaciations sometime before 40,000 years ago, according to radiocarbon dating.

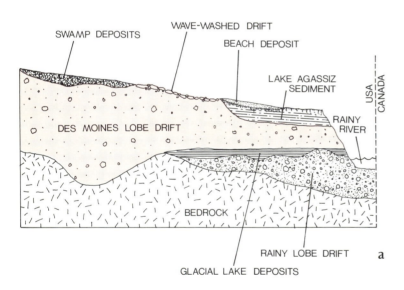

SWAMP DEPOSITS   WAVE-WASHED DRIFT
                    BEACH DEPOSIT
                         LAKE AGASSIZ
                         SEDIMENT
                                    RAINY
                                    RIVER
DES MOINES LOBE DRIFT

BEDROCK

RAINY LOBE DRIFT

GLACIAL LAKE DEPOSITS   a

USA CANADA

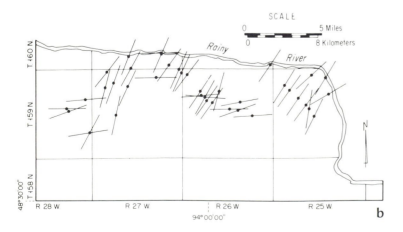

SCALE
0   5 Miles
0   8 Kilometers

Rainy River

N

T-160 N
T-159 N
T-158 N
48°30'00"
R 28 W   R 27 W   R 26 W   R 25 W
94°00'00"   b

Lake Agassiz II, which drowned a forest already established on the exposed lakebed of old Agassiz I. Eventually, all of the ice melted. The lake drained away about 9,000 years ago, and the modern scenery evolved.

Lake Agassiz left behind a vast plain rimmed by gravelly beach deposits but largely underlain by fine-textured silt and clay. Characteristically, these sediments are well bedded. It is not unusual to find pebbles and boulders interspersed with laminated fine-grained material (Figure 11-9); the larger fragments were probably dropped from dirty icebergs afloat in the deeper parts of the lake. Intersecting low-relief linear trenches visible on aerial photographs are thought to be drag marks of grounded icebergs. Beach deposits are sand and well-rounded gravel accumulations, typically stratified and in some places crossbedded. They are generally visible as long, broad ridges.

At Herman, Minnesota, the highest beach lies at an elevation of about 328 m. The same beach ridge is 30 m higher at Fertile, which is 200 km farther north. A similar dramatic rise in the Campbell Beach northward is the best evidence

Figure 11-6. (a) Generalized cross section of northern Koochiching County near the Rainy River at Indus. Two glacial advances are indicated by superposed drift, with subsequent lake stages. (b) Map showing glacial striations on 37 outcrops south of the Rainy River in the Birchdale-Indus area. Two glacial advances are indicated, one from the north-northeast and one from the west. Determining the relative ages of the two sets is difficult, but superposition of limestone-bearing glacial drift upon granite-bearing drift indicates that the west-trending set was formed by the glacier that brought in limestone from the Winnipeg Lowland and is therefore the younger set. (From Ojakangas, Meineke, and Listerud, 1977.)

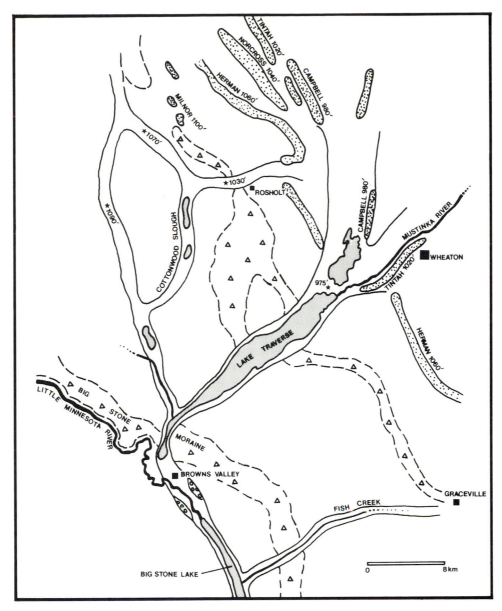

Figure 11-8. Abandoned channels and beaches of Lake Agassiz in the southern outlet area.

that the earth's crust has been elevated vertically since those features formed. Such crustal movements are common to all glaciated regions. They reflect the response of an elastic crust, bent downward under the weight of thick glaciers only to rebound when the load is removed. Greater uplift to the north corresponds to zones of thicker glacial ice.

BEDROCK GEOLOGY

The bedrock geology of northwestern Minnesota is largely buried beneath as much as 150 m of glacial deposits, the thickest glacial cover in the state (see Figure 2-5). Nearly all rock outcrops are in the far north near the Rainy River and Lake of the Woods. Yet surprisingly much is known about the bedrock, which is essentially all of Early Precambrian age and which makes up parts of the volcanic superbelts and gneissic superbelts described in Chapter 3.

The rock types are fairly well known from rock exposures farther east in Minnesota and across the border in Canada, and these rock units can be projected into northwestern Minnesota. Because these rocks, metavolcanics and metasediments of the greenstone belts and younger intrusive granitic batholiths, each have recognizable magnetic signatures, the rock types can be determined from aeromagnetic measurements. The granites generally show as magnetic lows, the greenstones (metabasalts) show as moderately magnetic zones, and iron-formations in the volcanic-sedimentary sequences show as magnetic highs (see Figure 8-28). Similarly, the granites and metasedimentary rocks are less dense than the metavolcanic rocks and register as gravity lows (see Figure 8-29). These differences in magnetism and densities of the rock units are then contoured, much as elevation contours are drawn to show the surface topography of an area. The resulting pattern is a series of northeast-

Figure 11-9. Pebbles in laminated muds were dropped from melting icebergs in Lake Agassiz.

trending zones of higher and lower magnetics and densities. Finally, bedrock samples from water wells and from exploration holes drilled in the search for metals (see Figure 8-32) verify the rock types in many of the individual magnetic or gravity anomalies. The combination of all these data allows the compilation of a geologic map of the bedrock, even though outcrops are rare.

Several large faults are shown on the geologic map of northwestern Minnesota (Figure 11-10). Most of these are projected in from the east and the north. The long northwest-trending Vermilion Fault is exposed in the Vermilion District of northeastern Minnesota and can be followed northwestward nearly to North Dakota as breaks in the magnetic and gravity trends. The northeast-trending Rainy Lake-Seine River Fault is projected in from the Rainy Lake area to the east, and the Quetico Fault, which lies to the north of the Rainy Lake-Seine River Fault, is projected in from Canada. These two faults, which join the Vermilion Fault, are major faults in Ontario, extending for more than 300 km. In the better exposed bedrock of Ontario, these faults are manifested as zones of sheared rock as much as

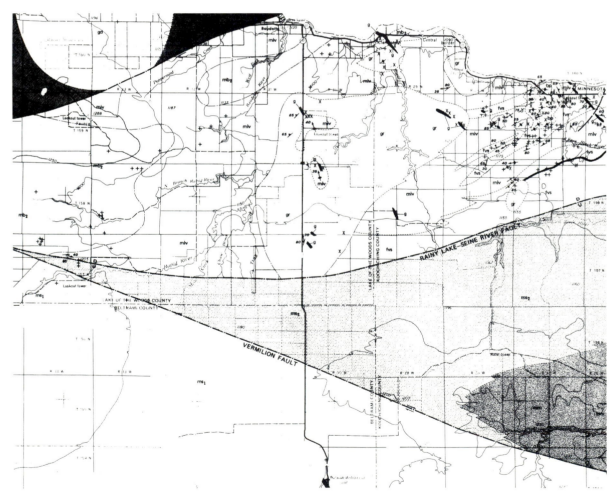

Figure 11-10. Geologic map of the Roseau 1:250,000 sheet. The legend is as follows: gr represents granitic rocks; gd represents grandiorite; miv represents mafic-intermediate volcanic rocks; mb$_2$ represents mafic volcanic rocks; fvs represents felsic-intermediate volcanic rocks; g represents mafic dikes; ms$_2$ represents biotite schist; vsm represents schist-rich migmatite; x indicates a rock outcrop; and + indicates a drillhole in bedrock. The darker areas in the northwest corner of the map are metavolcanic rocks. Compare this with the aeromagnetic map of Figure 8-28. (From Ojakangas, Mossler, and Morey, 1979.)

a b

Figure 11-11. (a) Swarm of narrow mafic dikes in granite on the big peninsula on the south side of Lake of the Woods. (b) View to the northwest along the wide Manitou Rapids mafic dike. Country rock into which the dike intruded is visible at many places along the edge of such dikes. (c) Part of a topographic map showing the topographic expression of the dike shown in Figure 11-11b. The dike here rises as much as 15 m above the adjacent ground. The contour interval is 5 ft. The squares are sections that are 1.6 km (1 mi) square. From the Loman, Minnesota-Ontario, 7.5′ Quadrangle.)

100 m wide, commonly with an abundance of veins and pods of white quartz. (See Figures 10-37 and 10-59). The movement on these faults quite likely was both horizontal and vertical, but because they parallel rather than cut across the northeast-trending rock units, horizontal displacements are difficult to determine. On the Vermilion Fault in northwestern Minnesota, 55 km of horizontal movement has been suggested, with the northern block having moved southeastward relative to the southern block.

Several large northwest-trending vertical mafic dikes from 45 to 60 m wide crosscut all other rocks of the region. Fortunately for geologists, these dikes, in spite of their iron-bearing composition, have not weathered as much as the older country rocks into which they intruded. The large ones stand up as prominent ridges several meters high (Figure 11-11), and the country rocks are commonly exposed along their edges. The rocks of Figure 3-10c, for example, are exposed only because of a resistant mafic dike a few meters away. Such dikes are most easily seen in Minnesota on the Rainy River at Manitou Rapids and Sault Rapids but are intermittently exposed for a total length of nearly 100 km on Ontario. One northwest-trending dike in the Rainy Lake area has been dated at 2,100 million years, and a similar age is likely for these dikes.

Some bedrock exposures exist simply because they are topographic bedrock knobs and stick through the glacial drift. The granite exposed in Highway 11 roadcuts on a hill at Birchdale is an example. In other places, bedrock is exposed due to downcutting by streams. One such accessible spot is at Clementson, about 13 km east of Baudette, in the gorge of the Rapid River adjacent to the bridge. These rocks are metamorphosed sandstones that are better termed volcanogenic graywackes. They dip steeply southward, but a close study of graded textures shows that the tops of the beds are toward the north and thus the beds are somewhat overturned.

c

205

a

b

Figure 11-12. (a) The edge of the Manitou Rapids mafic dike with light-colored volcaniclastic rock in the foreground. The part of the dike shown here is fine grained due to the rapid chilling of the dike edge. (b) Coarse volcaniclastic rock containing large pieces of felsic volcanic rock that have been elongated by deformational stresses in the area.

PLACES OF INTEREST

Manitou Rapids Rest Area

A good spot to observe some of the rare exposed bedrock is at the Manitou Rapids Rest Area, about 6 km west of the village of Indus on Highway 11. (Indus, by the way, was named for the Indus River in India by Marion Smootz, a missionary who homesteaded in the area during the late 1800s.) In the rest area by the northernmost parking places, light gray, coarse-grained, felsic-intermediate volcaniclastic rock (Figure 11-12) is exposed adjacent to a northwest-trending black mafic dike. The dike is exposed all the way to the river, about 100 m to the north.

Mapping of a 500-km² area, of which the rest area is one small portion, has resulted in a geologic map that, when fitted to maps of the Ontario side of the border, yields a map from which further geologic history can be interpreted. The mapping was undertaken by the Minnesota Geologic Survey (Ojakangas, Meineke, and Listerud, 1977) because this region has the potential for containing base metal (copper-zinc) deposits. The three parts of Figure 8-27 illustrate the map, structure sections, and a reconstructed geologic model of the area during volcanism about 2,700 million years ago. Massive volcanogenic sulfide bodies, as discussed in Chapter 8 under Lower Precambrian copper and zinc deposits, occur near such ancient volcanic centers in Canada and elsewhere in the world.

The Northwest Angle

The Northwest Angle, the northernmost piece of land in the 48 contiguous states, has an area of 342 km² and is separated from Minnesota proper by Lake of the Woods. Until the mid-1970s there was not even an all-weather road into the area, but it can now be reached from Canada.

How did the Angle ever become a part of the United States? The Treaty of Paris, which Benjamin Franklin, John Adams, and some of our other forefathers signed in 1783 to end the Revolutionary War, set the international boundary along the water route from Lake Superior to Lake of the Woods and then due west toward the Mississippi River, which was thought to extend north into Canada. Nearly 50 years later, in 1832, Henry Schoolcraft discovered that the source of the Mississippi was Lake Itasca, about 200 km to the south of Lake of the Woods. Obviously, the Treaty of Paris could not be fulfilled as written. Later, when the 49th parallel became the boundary from the Pacific Ocean to Lake of the Woods, the two portions of the international boundary did not meet. The Convention of 1818, which resolved the War of 1812, located the boundary due south from the Northwest Angle to the 49th parallel, but it was not until the Webster-Ashburton Treaty of 1842 that the entire United States-Canadian boundary was set. One "hearsay" reason for the inclusion of the odd-shaped Northwest Angle as part of the United States is that it was compensation for the United States relinquishing its claim to Hunter's Island, a large piece of ground completely surrounded by rivers and lakes in what is now Quetico Provincial Park, that could logically have been construed as being south of a major water route from Lake Superior to Lake of the Woods (see Figure 10-26).

The Angle has a longer history than that already cited. In 1732, Pierre Gaultier de Varennes (Sieur de la Vérendrye), a French army officer, built Fort St. Charles along the northern edge of the Angle after reaching the area via Lake Superior, Grand Portage, and the border canoe route. From here, the French explored the Dakotas, Manitoba, and Saskatchewan and established a flourishing fur trade.

The rocks of the Northwest Angle are all Early Precambrian

in age with the exception of northwest-trending mafic dikes of probable Middle Precambrian age (Figure 11-13). Most of the Early Precambrian rock exposures are along the northern edge of the Angle and on adjacent small islands (Figure 11-14). Most of the rocks are granitic in composition, with granites cut by coarse-grained pegmatites (see Figure 3-14b). However, a thin segment of greenstone belt rocks—metavolcanics—trends into the area from the northeast (Figure 11-15). The major economic interest in the area is for base-metal deposits and gold, for occurrences and deposits are present in the Canadian portion of Lake of the Woods. As mentioned in Chapter 9, a feldspar quarry was once operated in a coarse pegmatite of the Northwest Angle, and there are also minor occurrences of radioactive elements in pegmatites.

Most of the Northwest Angle is very flat, part of the lakebed of Glacial Lake Agassiz. As mentioned earlier, Lake of the Woods itself is a remnant of Glacial Lake Agassiz.

The Grand Mound

The Grand Mound, the largest prehistoric burial mound in the upper Mississippi River Valley region, is situated just off Highway 11, 24 km west of International Falls. It was probably built by Paleo-Indians between about 500 B.C. and A.D. 1000, although no dates have been obtained from the mound itself. It is likely that the site for this mound was chosen because of its proximity to the confluence of the Rainy and Big Fork rivers. The Rainy River, of course, is part of one of the major water routes of northern North America. A journey up the Big Fork River and then into connecting lakes and creeks will bring a canoeist, in the vicinity of Bowstring and Cut Foot Sioux lakes, to within 3 km of lakes and creeks that connect with the Mississippi River. This area is one example of how geological factors affected early inhabitants of Minnesota and no doubt of much of central North America as well.

The mound, now supporting large trees, is about 60 m long by 30 m wide and is 12 m high (Figure 11-16). One of a group of nearby smaller mounds was studied in the 1930s by A. E. Jenks, then the state archaeologist, and yielded more than a hundred "bundle burials" (with the bones of a skeleton disarticulated and buried in a bundle) arranged in four layers, many with evidence of cannibalism (bone marrow and brains had been removed). Pottery, bone, and antler projectile points and copper artifacts were also present. The mounds on the Rainy River belong to the northern forest Laurel culture, a distinctive local Woodland culture. The scarcity of deer bones in the bone refuse is one bit of evidence supporting an adaptation of Laurel peoples in northern Minnesota, Ontario, and Manitoba to heavy forests, for deer survive best where openings are abundant. The Minnesota Historical Society maintains an interpretive center on Highway 11 near the Grand Mound.

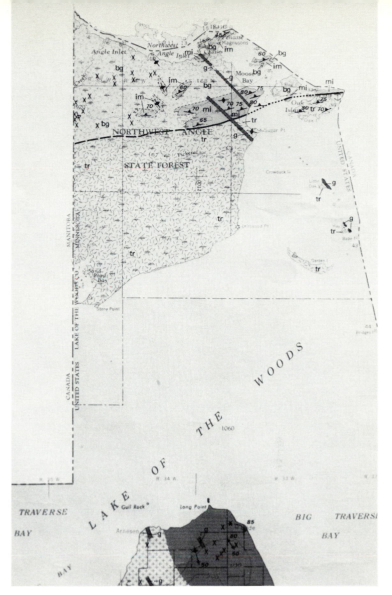

Figure 11-13. Geologic map of the Northwest Angle. Lower Precambrian rocks include predominant granitic rocks (symbolized by tr and bg) and lesser volcanic rocks (symbolized by mi and im). These are cut by northwest-trending mafic dikes (symbolized by g) of Middle Precambrian age. Note a fault cutting across the area. The x's represent outcrops; the symbols with numbers represent outcrops with vertical rock layers. The northern half of the Northwest Angle is about 25 km (15 mi) across. (From the geologic map of the Roseau sheet by Ojakangas, Mossler, and Morey, 1979.)

Roseau Lake

Roseau Lake, on the bed of Glacial Lake Agassiz, near the village of Roseau, has a remarkably circular outline, a shape readily suggestive of a meteoritic impact crater (Figure 11-17). The flat-bottomed and steep-sided lake is 5 km in diameter and 3 m deep. Higher ground just northwest of the lake might be interpreted as material piled up by the impact of a meteorite that hit the earth while moving toward the northwest.

In 1975, geophysicists S. Hammer and C. P. Ervin conducted a gravity survey and analyzed aeromagnetic data and

Figure 11-14. Bedrock island in Lake of the Woods just off the Northwest Angle. The rock is granite cut by coarser pegmatite dikes.

Figure 11-16. Grand Mound, on the bank of the Rainy River 24 km west of International Falls. This photo was taken in 1907; the mound is now covered with trees. (Courtesy of the Minnesota Historical Society.)

Figure 11-15. Pillowed basalts on the northeastern edge of the Northwest Angle. The pillowed unit has been tilted to a vertical position.

concluded that the lake is not a crater. A sediment-filled crater would give low gravity readings, and Roseau Lake does not. Magnetic measurements reflect elongate bedrock structure, including a fault zone, rather than a crater. Furthermore, a meteoritic crater of this size should have a depth of 600 m, and wells show that the glacial drift on the bedrock is only 45 m thick. Hammer and Ervin therefore concluded that the lake is more likely an exceptionally large glacial kettle lake. Yet why is this the only kettle lake in the area, and why is it located on a glacial lakebed rather than in a moraine, the happy home of kettle lakes? Does it indicate the presence of a buried moraine? It is something

of a mystery. And the higher ground northwest of the lake is even more difficult to explain on an otherwise flat glacial lakebed.

Browns Valley

Browns Valley has several claims to fame. Perhaps most important is that it was here that the plug was pulled on Glacial Lake Agassiz, causing it to be drained via Glacial River Warren, as described in Chapter 7.

The Little Minnesota River, which enters Minnesota from South Dakota at Browns Valley, has constructed a sediment dam across the former bed of River Warren (see Figure 11-8). Within this stretch of the valley lies the continental divide, separating Gulf drainage to the south from Hudson Bay drainage to the north. The sediment dam, an alluvial fan, allowed Lake Traverse, the headwater of the Red River of the North, to form. South of the dam, Big Stone Lake can be seen, the result of another natural barrier at Ortonville. The Minnesota River originates there.

A flat, boulder-studded field at an elevation of 319 m just south of Browns Valley along Minnesota Highway 28 is a terrace formed by River Warren. The terrace is a channel-bottom pavement of lagged boulders that established a temporary base level for erosion of the outlet of Glacial Lake Agassiz. The stabilized lake level is marked by the Herman Beach at an elevation of 325 m at the entrance to the gorge about 16 km north of this point. The boulder terrace is traceable downstream in patches for about 32 km as far as the fish hatchery 10 km above Ortonville. Another terrace at 305 m may correspond to River Warren when Lake Agassiz stood at the Tintah shoreline (elevation, 311 m).

To the west, the land surface rises to culminate in the impressive Coteau des Prairies, a high plateau that exists because of a bedrock core and that caused glaciers to be diverted around the area.

Minnesota's best Paleo-Indian site is at Browns Valley. It was discovered in 1933 when gravel was being removed from the municipal gravel pit located at the southern edge

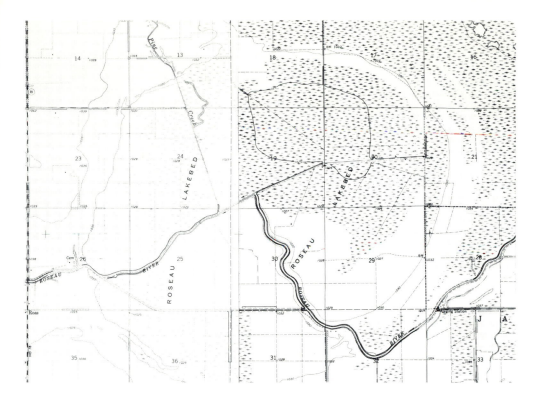

Figure 11-17. Topographic map of the Roseau Lake area showing the circular outline of the lake. Hammer and Ervin (1975) studied this lake to determine whether or not it is the result of meteoritic impact. It apparently is not. (From the Roseau NE and Pinecreek, Minnesota, 7.5 'Quadrangles.)

of the village. There, human bones, including a skull, flaked stone points of the Clovis and Folsom types, and flaked knives were recovered (Figure 11-18). Browns Valley Man, an adult male, was evidently buried in gravels at the outlet of Glacial Lake Agassiz after it finished draining to the south but before soil had developed in the outlet. This may have been about 8,000 years ago.

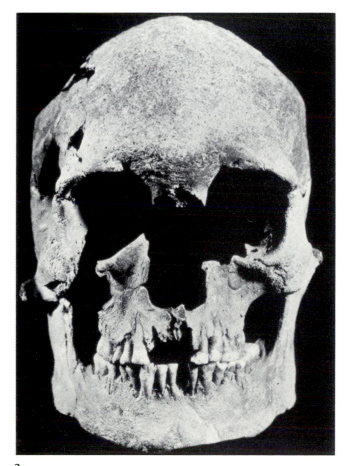

a

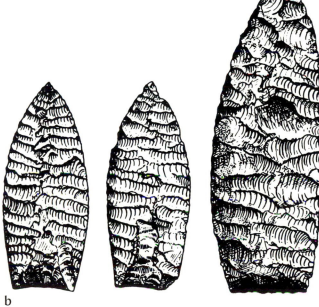

b

Figure 11-18. (a) Skull of Browns Valley Man. (Photo by Eugene Becker and courtesy of the Minnesota Historical Society. (b) Artifacts found with the Browns Valley Man burial. The largest artifact is about 20 cm long. (From Jenks, 1937, and courtesy of the American Anthropological Association and the Minnesota Historical Society.)

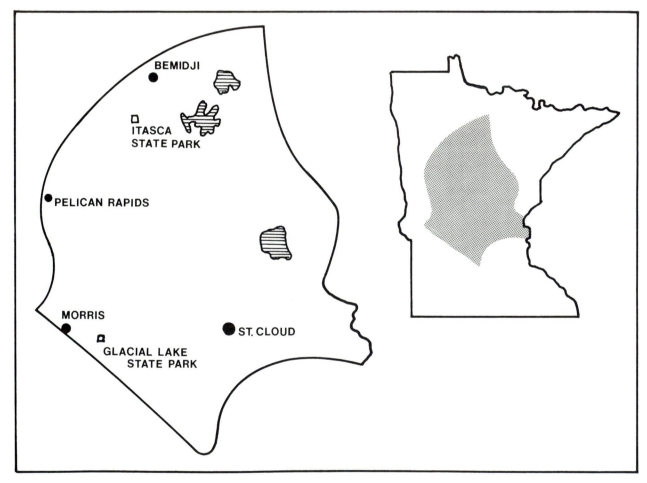

Central Minnesota region.

12
Central Minnesota

THE LAY OF THE LAND

Rugged relief, poor drainage that is reflected in the great number of lakes, and sparse outcrops of bedrock all indicate the importance of glaciation in creating the landscape of central Minnesota. "Moraine terrain" is an apt general name for the area. The upper Mississippi River is the major water-course in the central part of the state, and the St. Croix River forms the eastern boundary of this segment of the state. Mille Lacs Lake is one of the larger lakes in Minnesota. At an elevation of 380 m, its surface is almost 200 m higher than that of Lake Superior.

The strongest topography is contained in moraines constructed by various lobes of ice during the Wisconsin glaciation, notably the St. Croix Moraine, the Alexandria Moraine, and the Itasca Moraine (Figure 12-1). Elevations above 550 m are attained in the latter two; this is about 275 m above the floor of the former Glacial Lake Agassiz. Not only are the glacial sediments piled high in central Minnesota, but they are also among the thickest in the state, more than 150 m thick in some places (see Figure 2-5b). Besides the moraines, other important physiographic areas are the Wadena Drumlin Field, the Eastern Minnesota Tunnel Valley System, and the Anoka Sand Plain. Minor drumlin areas occur in the vicinity of Brainerd and Pierz. North of Pine City are the flat, silt-covered remnants of the bed of Glacial Lake Grantsburg.

GLACIAL GEOLOGY

Central Minnesota contains such a diversity of landforms that it has become an important outdoor laboratory for studying glacial processes. At least four major ice advances are indicated, each having left its own imprint in the strati-graphy and upon the land surface. The Ice Age glaciers left no better record anywhere in the world than here on this magnificent geological stage in Minnesota.

The broad belt of lakes in the west is the Alexandria Moraine. It borders the plain left behind by Glacial Lake Agassiz and encloses the Wadena Drumlin Field. Both features relate to a generally southward incursion of ice, the Wadena Lobe, from the Winnipeg Lowland. The major volume of drift in both the moraine and drumlins is a pale yellow, limestone-rich sandy till. Much of the moraine is capped with an olive-brown silty till, the result of later overriding by the Des Moines Lobe. The drumlins, over a thousand of them, are arranged in a fan-shaped pattern, reflecting the diverging flow of the ice sheet. They are the oldest surface features in the northern half of Minnesota.

On the north and east, the drumlins are truncated and buried by the Itasca Moraine and the St. Croix Moraine. The former represents a readvance of the Wadena Lobe, and the latter is the western margin of the Rainy-Superior Lobe. Drift of contrasting composition is juxtaposed in the inter-lobate area near Walker. When these glaciers melted, about 20,000 years ago, the dirty ice margins collapsed into belts of kettle lakes and kames (Figure 12-2). South of the moraines meltwater deposited a broad fan of sand and gravel. Retreat of the ice eastward unveiled swarms of drumlins in the vicinity of Brainerd and Pierz.

Under a peculiar set of hydrologic conditions, meltwater was drained in certain places through long tunnels beneath the downwasting glaciers. The effect of such subsurface drainage was recognized in eastern Minnesota by H. E. Wright (1973). Driven by the pressure of the enclosing ice, water in the tunnels cut long, straight courses into the drift and bedrock beneath. Eventually, the ice thinned and the tunnels collapsed and, in some, eskers were deposited in

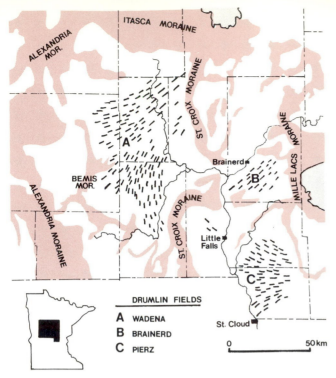

Figure 12-1. Moraines and drumlin fields of Central Minnesota. (After Schneider, 1961.)

response to the diminished energy. Marking the existence of this unique drainage system is a series of tunnel valleys, trending southwest across eastern Minnesota all the way to the Twin Cities (Figure 12-3).

Another advance of the Superior Lobe, much more feeble than the previous glaciation, topped over the Superior Basin to form the Highland Moraine in the north and sent a protrusion southwest to stall and then construct the Mille Lacs Moraine. That pile of sediments dammed the flow of water sufficiently to back up Mille Lacs Lake, a fine example of a moraine-dammed lake (Figure 12-4).

Finally, central Minnesota was affected by the advance of the Des Moines Lobe from the northwest. That glacier overlapped the Alexandria Moraine and the St. Croix Moraine farther east, leaving a swath of shale-rich till derived in part from the eastern Dakotas. An extension of it, the Grantsburg Sublobe, reached into western Wisconsin, blocking drainage to form Glacial Lake Grantsburg north of the Twin Cities (see Figure 7-14). Eventually, that ice, too, melted away, and Lake Grantsburg drained. The Anoka Sand Plain was

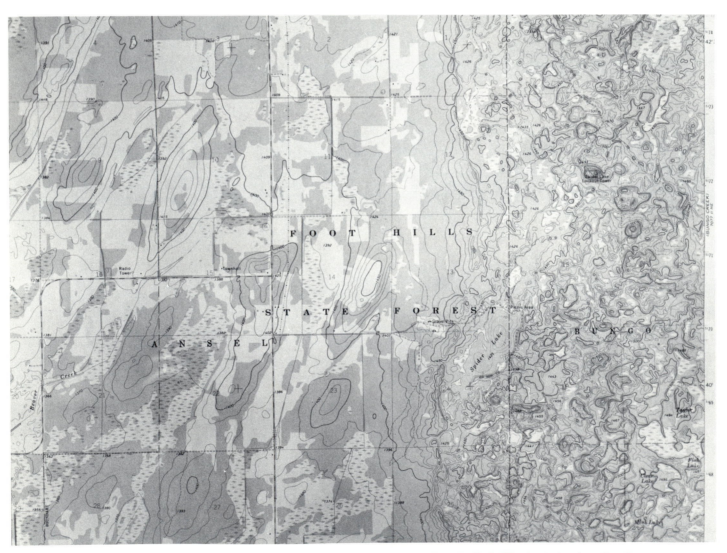

Figure 12-2. The belt of kettle lakes and kames on the east is part of the St. Croix Moraine in Southern Cass County. Here the moraine truncates part of the Wadena Drumlin Field (From the Spider Lake, Minnesota, 7.5′ Quadrangle.)

constructed by shifting meltwater streams that deposited outwash in their wakes. Southwesterly winds subsequently moved some of the sand into patches of dunes as the meltwater streams ran dry. One belt of low sand dunes, up to 3 km wide, stretches from north Minneapolis northwestward to just east of Anoka near Bunker Lake. Others are scattered in patches throughout the sand plain region. Vegetation, mainly grasses, has stopped dune migration, giving the dunes a "fossil" landform status at present.

BEDROCK GEOLOGY

Central Minnesota has few rock outcrops except in certain limited locales, notably the St. Cloud area. However, the limited outcrops when coupled with geophysical data permit interpretations as to the nature of the bedrock. G. B. Morey, of the Minnesota Geological Survey, completed a study of much of this region in 1978 and has classified the rocks into three major groupings—Lower Precambrian rocks (mostly gneisses), Middle Precambrian stratified rocks (an older Mille Lacs Group and a younger Animikie Group, which includes the rocks of the Cuyuna Range), and Middle Precambrian plutonic rocks, which include the granitic rocks of the St. Cloud-Mille Lacs Lake area.

The old gneisses can best be seen at a small abandoned

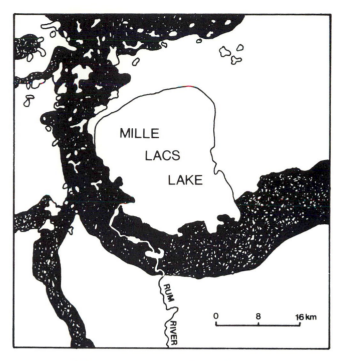

Figure 12-4. Mille Lacs Lake is ponded behind a moraine dam, the Mille Lacs Lake Moraine, constructed by a lobe of ice that advanced southwestward from the Lake Superior Basin about 15,000 years ago. (After Schwartz and Thiel, 1954.)

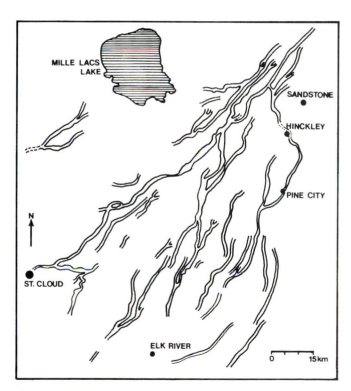

Figure 12-3. The eastern Minnesota Tunnel Valley System is composed of long, linear topographic troughs eroded by meltwater streams enclosed within ice-walled tunnels beneath the wasting Superior Lobe about 16,000 years ago. (After Wright, 1973.)

quarry 3.5 km west of Dad's Corner (Pliney) in Aitkin County, at a point a few kilometers southwest of McGrath in Aitkin County, and south of the village of Denham in Pine County. These outcrops are part of a broad belt of gneiss, named the McGrath Gneiss, that extends northeastward from Mille Lacs Lake and has been dated as at least 2,700 million years old. The old gneisses can also be seen in Stearns County at two abandoned quarries 3 km east of Richmond and at outcrops on Minnesota Highway 23 at the western edge of Rockville. These gneisses appear to be a northern extension of the ancient gneisses of the Minnesota River Valley and, if so, are at least in part 3,600 million years old (see Chapter 3, Figure 3-2, and Plate 2).

The stratified rocks of the Mille Lacs Group are generally poorly exposed. A part of the group, the Denham Formation, is only exposed on private land south of Denham, in Pine County. It consists of metasedimentary and metavolcanic rocks, some of which contain pebbles of McGrath Gneiss, upon which the formation rests. The Little Falls Formation, exposed on the Mississippi River at the center of the city of Little Falls, was a dominantly muddy sedimentary unit that has been metamorphosed to slate, phyllite, and mica schist. Original muddy sandstone beds are now metagraywackes. Several metamorphic minerals are developed in the formation, including garnet and staurolite. Staurolite crystals as large as 1 cm in diameter can be found several kilometers south of Little Falls where the Soo Line

Railroad crosses the Mississippi River (see Figure 4-17). The Little Falls Formation resembles very much the Thomson Formation to the northeast, which is described in Chapters 4 and 10. A metavolcanic unit is exposed in the village of Randall, in Morrison County. Some other units are present only in the subsurface and will not be mentioned here.

The younger Middle Precambrian stratified rocks in central Minnesota, described in a general way in Chapter 4, include, from the bottom up, the Mahnomen Formation (a unit composed of mudstone, siltstone, and quartz sandstone), the Trommald Formation (the iron-formation of the Cuyuna Range), and the Rabbit Lake Formation. All are poorly exposed units that can be correlated, respectively, with the Pokegama Quartzite, the Biwabik Iron-Formation and the Virginia, Thomson, and Rove formations of northeastern Minnesota. The Trommald is discussed in Chapters 4 and 8.

The best-exposed rocks of central Minnesota are the plutonic rocks, which are mainly granites. Granites, of course, crystallize at depth, and hence these probably intruded into the Middle Precambrian sedimentary rock units. This history is also suggested by numerous inclusions of such rocks in the granites. Most of the granites show few or no deformational features, indicating that they were emplaced after the cessation of the Penokean deformational event described in Chapter 4. The increasing grade of metamorphism from north to south in the Thomson Formation is presumably due to the same thermal event that produced the granites. Perhaps the metamorphism is related to the few granitic rock bodies that appear to have been emplaced prior to final folding and metamorphism of the Thomson, or perhaps it is due to heat associated with unknown granitic batholiths at depth.

Abandoned granite quarries are present in and near the village of Warman in Kanabec County, an active quarry is located 8 km south of Isle in Mille Lacs County, and both active and abandoned quarries are abundant in the St. Cloud-Waite Park-Rockville area. From this last area, a number of distinctive types of granite have been quarried, including the "St. Cloud Gray," the "St. Cloud Red," the porphyritic "Pearl Pink," and numerous others with "pink," "gray," or "red" as part of the trade name (see Figure 9-2).

Within the Lower Precambrian bedrock of east-central Minnesota is a major geologic boundary between the greenstone-granite terrane of the northern part of the state and the ancient gneiss terrane of the southern part of the state. G. B. Morey and P. K. Sims (1976) suggested its presence on the basis of geophysical measurements, for the boundary is not exposed, and further interpreted that it is a major fault zone. Independent evidence, from earthquakes, seems to corroborate this. Most of the world's earthquakes are the result of movement along faults. These could be due to deep-seated forces of unknown origin or, as in Minnesota, could be related to vertical adjustments of crust due to post-glacial rebound. Although Minnesota has had fewer earthquakes than most states, it has had some. Ten strong enough to be felt occurred in Minnesota between 1860 and 1979 (Table 12-1), as described by H. M. Mooney (1979) of the University of Minnesota. Of those, five were located along this 50-km-wide boundary zone (Figure 12-5). This structural feature has been named the Great Lakes Tectonic Zone and has been connected with a major fault boundary in Ontario north of Lake Huron (Sims et al., 1980).

The *intensity* of an earthquake, expressed on a scale of I to XII, measures surface effects based on observations of people located near the earthquake *epicenter*, the point on the earth's surface above the earthquake. The rather subjective descriptions of an earthquake's intensity include "hanging objects swing" (III); "standing motor cars rock" or "vibration like passing of heavy trucks" (IV); "sleepers awakened" (V); "many are frightened and run outdoors" or "felt by all" (VI); "everybody runs outdoors" (VII); "damage considerable" (VIII); "general panic" or "damage great" (IX); "most masonry and frame structures destroyed (X); "rails bent greatly" (XI); and "damage nearly total" (XII). The locations of observations of intensity can be

Table 12-1. Earthquakes in Minnesota

Location	Date	Area Affected (in square kilometers)	Maximum Intensity	Magnitude
Long Prairie	1860-1861	. . . . .	V-VII	4.6
Le Sueur	1865-1870	. . . . .	V-VII	4.3
Red Lake	February 6, 1917	. . . . .	. . . .	. . . .
Staples (Motley)	September 8, 1917	48,000	V-VII	4.8
Bowstring	December 23, 1928	. . . . .	III	3.1
Detroit Lakes (Audubon)	January 28, 1939	8,000	IV	3.7
Alexandria	February 15, 1950	. . . . .	V-VI	3.8
Pipestone	February 28, 1964	. . . . .	. . . .	3.4
Morris	July 9, 1975	82,000	VI	4.6
Sauk Centre (West Union)	March 5, 1979	. . . . .	. . . .	2.6

Source: Data from Mooney, 1979.

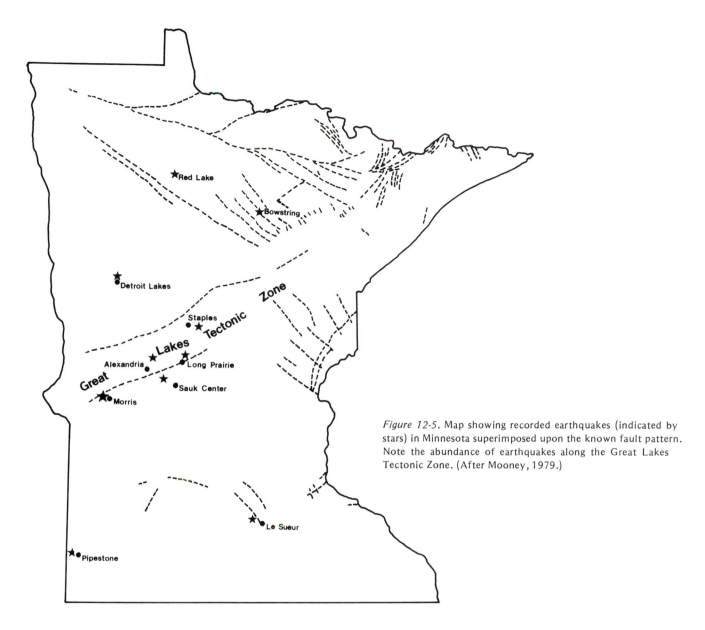

Figure 12-5. Map showing recorded earthquakes (indicated by stars) in Minnesota superimposed upon the known fault pattern. Note the abundance of earthquakes along the Great Lakes Tectonic Zone. (After Mooney, 1979.)

contoured in a "dot-to-dot" manner to produce an isoseismal ("equal seismicity") map such as that of the 1975 Morris earthquake (Figure 12-6).

Magnitude is an indirect measure of the energy released during an earthquake and is based on the measurement of the amplitude of seismic waves. Magnitudes are expressed by a logarithmic scale in which each whole number represents a tenfold increase in the amplitude of the waves and more than 30 times the amount of energy released. Thus, an earthquake with a magnitude of 7 releases about 1,000 times (31 x 31) as much energy as one with a magnitude of 5. An earthquake of magnitude 3 is barely felt, one of magnitude 5 will be felt but will generally not cause much damage, and one of magnitude 7 or more will generally cause major damage. The magnitudes for Minnesota's earlier earthquakes have generally been calculated rather than measured, but the 1975 Morris earthquake, with a magnitude of 4.6 and a depth of 5 to 10 km, was picked up by seismic stations all over North America. For comparison,

the 1906 San Francisco earthquake is estimated at 8.3; the Good Friday, 1964, Alaskan earthquake had a magnitude of 8.5; and two quakes of magnitude 8.6 have been measured in the oceans.

Nevertheless, earthquake risk is very low in Minnesota, even east-central Minnesota, especially when compared to California. You needn't move out of Minnesota because you fear seismic activity.

PLACES OF INTEREST

Interstate State Park and Vicinity

Interstate State Park on the Minnesota-Wisconsin boundary at Taylors Falls, Minnesota and St. Croix Falls, Wisconsin, is a scenic as well as geologically interesting location. Here Upper Precambrian lava flows were covered by Upper Cambrian sedimentary rocks about 575 million years younger. These, in turn, were covered by Pleistocene glacial

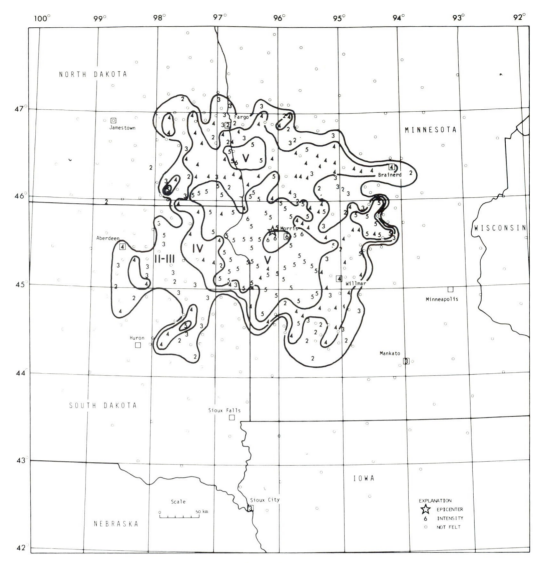

Figure 12-6. Isoseismal map of the July 9, 1975, Morris earthquake. Refer to the text for the meanings of the numbers. (From Stover et al., 1975.)

deposits about 525 million years younger than the sedimentary rocks. This succession is now visible because of erosion.

The lava flows are basalts that poured out into the Mid-continental rift about 1,100 million years ago. They are part of the Chengwatana Volcanics, which are also exposed to the north near Hinckley and Pine City, and are probably contiguous with lava flows of the South Shore of Lake Superior. At least 10 horizontal lava flows are exposed along the 1.5-km-long and 90-m-deep gorge (also known as the dells or dalles) of the St. Croix River (Figure 12-7); each is distinguished by an amygdaloidal top and a more massive base. Sets of vertical and horizontal joints have been instrumental in the development of the present topography.

The contact of the sedimentary rocks upon the lava flows can be seen in the woods behind the miniature railroad at

a private tourist spot just west of County Highway 11 a short distance south of the center of Taylors Falls. Pieces of basalt form a basal conglomerate into which a sandy matrix containing small (6-mm) marine brachiopod shells about 500 to 550 million years old infiltrated from above during sedimentation (Figure 12-8).

Within the park itself, a short hike along a trail west of the highway from the vicinity of park headquarters will bring one to Curtain Falls. The falls is usually quite dry, but erosion over thousands of years has cut an impressive amphitheaterlike bowl beneath the falls in soft sandstones (see Figure 6-13a). These sandstones were formed by sand deposition in a marine sea, as attested to by the presence in certain layers within the bowl of an abundance of poorly preserved internal sandy molds or fillings of trilobite shells that make up a "fossil hash." Another indication of marine deposition is the glauconitic greensand exposed near road level in large roadcuts just north of a broad bend in the highway a few kilometers south of Taylors Falls. (See Chapter 6 for a discussion of glauconite in connection with the Franconia Formation.)

Even though the sandstones rest upon Precambrian volcanics, they are not the oldest Cambrian rocks in the state. The Cambrian sea slowly advanced from south to north over southeastern Minnesota, and the sedimentary rocks in the Interstate State Park area are near the northernmost extent of the sea and were deposited long after many older Cambrian rocks in southeastern Minnesota. The lava flows, which had already been eroded for about 575 million years, stood as irregular seaside bluffs and as islands in the sea but were finally totally inundated.

A younger geologic history is also well illustrated in and near the park. A stop at the roadside scenic viewpoints either just north or just south of Taylors Falls will show that the St. Croix River is considerably smaller than its valley. The broad valley and the deep gorge were cut during the last stages of the Pleistocene glaciation. As the glaciers withdrew northward, leaving behind a layer of glacial till that can be seen capping the present Cambrian bluffs, huge volumes of meltwater flowed out of the region via the ancestral St. Croix River, which was much larger than the present river. Most of this meltwater probably came from Glacial Lake Duluth. As the Superior Lobe melted and water was ponded in the lake basin in front of the ice dam to the northeast and as the basin became full, it overflowed southeastward via the Brule River in Wisconsin, the Kettle River, and the St. Croix River (see Figure 7-18). Proof of a broader and deeper river is provided by numerous potholes found in the park, some with depths as great as 35 m. Such potholes form only in turbulent, high-velocity rivers where rock "tools" (pebbles and boulders) trapped in a depression are swirled around and around acting as an abrasion mill (Figure 12-9).

The gorge has a more recent heritage as well. In 1771, the Chippewa Indians won a bloody and decisive battle against the Sioux (and the Fox) at the portage around the falls, where the two groups accidentally met while canoeing toward each other on the river. It was the site of logjams during the latter half of the 19th century, and a dam constructed at the falls in 1905 has essentially obliterated the "falls," which once consisted of a series of rapids.

Glacial Lakes State Park and Vicinity

Just south of Starbuck on Minnesota Highway 29 is Glacial Lakes State Park, situated in a kame and kettle complex on the southwest margin of the Alexandria Moraine. The park is surrounded by glacial depositional features that originated through the stagnation and collapse of debris-rich

Figure 12-7. The dells of the St. Croix River in Interstate State Park.

Figure 12-8. Rounded basalt boulders in conglomerate at the base of Cambrian sandstones in the village of Taylors Falls. This conglomerate formed at the shoreline of the sea, for small marine brachiopods occur between the boulders.

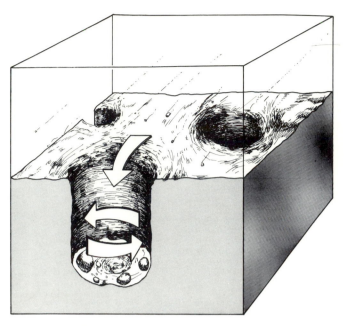

Figure 12-9. The origin of potholes. Pebbles and boulders that acted as abrasive "tools" are often found in potholes. Their width can vary from a few centimeters to a few meters, and the depth can be tens of meters. Those in Interstate State Park are exceptionally large and are locally billed as the "world's largest."

glacial ice (Figure 12-10). The complex makes up a wide belt of high-relief, lake-filled topography that stretches from northwest to southeast for many kilometers.

Lake Minnewaska, to the northeast of the park, is a superb example of a lake resulting from the melting of glacial ice. Its steep eastern shore at Glenwood is an ice-contact slope where a thick glacier margin was buried by outwash deposits. Blue Mound, near the park, is a ridge system composed of bouldery gravel and sand that was a sluiceway carrying outwash and debris across the ice-covered terrain (Figure 12-11). Esker systems, drift plateaus, a meandering tunnel valley, and hundreds of individual kames dot the landscape in the vicinity of Lake Johanna, east of the park (see Figure 7-8).

A closer look at the glacial drift indicates a mixture of till types. The greatest volume of drift was deposited by the Wadena Lobe, which must have stood for a long time at this position. Then, overriding by the Des Moines Lobe left a thinner cap of till derived from the eastern Dakotas. Most gravel pits contain abundant boulders of limestone from the Winnipeg region and crystalline rocks from the Canadian Shield. A dazzling collection of igneous and metamorphic rocks is on display in the beautifully constructed stone walls at the Lake Minnewaska Scenic Overlook at Glenwood. Glacier Ridge Trail, which passes by the park, is a marked

automobile route 300 km long that meanders through some of the most magnificent glacial terrain in Minnesota.

Itasca State Park

Itasca State Park is named for Lake Itasca, the source of the mighty Mississippi River, which here is only a hop and a skip wide (Figure 12-12). The park is located about 50 km southwest of Bemidji and 35 km north of Park Rapids amid glacial moraine country, the Itasca Moraine of Figure 7-13.

The Mississippi River was first seen by the Spaniard Hernando de Soto in 1541 in what is now the state of Mississippi, but it took another 291 years of exploration and misidentifications before its true source was finally reported by Henry Rowe Schoolcraft in 1832. (A fur trader, William Morrison, however, later claimed to have discovered the source in 1804). Schoolcraft, geologist, writer, and Indian agent, discovered the source of the Mississippi while on a trip to visit Indians, vaccinate them against smallpox, and establish missions. The maze of glacial lakes through which the Mississippi passes and the many smaller streams in this portion of Minnesota must have been truly confusing to the mapless early explorers! Schoolcraft, upon climbing a hill with 16 men and their Chippewa guide, Ozawindib, and looking down upon the lake that was the Mississippi's source, "got the first glimpse of the glittering nymph [they] had been pursuing. . . . At a depression of perhaps a hundred feet below, cradled among the hills, the lake spread out . . . presenting a scene of no common picturesqueness and rural beauty" (Holmquist and Brookings, 1963: 136). Schoolcraft asked the missionary who accompanied him on the trip to northern Minnesota, W. T. Bootwell, for a Latin or Greek name expressive of the "head" or "true source." From the last few letters of *veritas* ("truth") and the first two of *caput* ("head"), Schoolcraft coined the name Itasca. Four years later Joseph Nicollet mapped the area around Lake Itasca.

The area is now encompassed by Itasca State Park, Minnesota's largest park. It is a preserve of 13,000 ha, more than a hundred lakes, and abundant virgin Norway pine. Though it is adjacent to the Chippewas' White Earth Indian Reservation, it contains some Sioux burial mounds dating from the times before the Sioux were driven out of northern Minnesota's forests by the Chippewas during the latter half of the 18th century. And people had inhabited the area long before the Sioux arrived. The "Itasca Bison Site," a kill site on Nicollet Creek in the park, was a spot where hunters at some time between 7000 and 5000 B.C. apparently ambushed the now-extinct long-horned bison as they crossed the stream in what was then grassy prairie rather than coniferous forest. Bison bones and tools have been found there.

There is no bedrock exposed in Itasca State Park, for the glacial cover is about 100 m thick. The irregular glacial deposits are part of the Itasca Moraine, which was deposited by the last advance of the Wadena Lobe (see Figure 7-13).

Figure 12-10. The kame and kettle lake complex in Glacial Lakes State Park just south of the Alexandria Moraine Complex, constructed by both the Wadena Lobe and the Des Moines Lobe.

Figure 12-12. The mighty Mississippi at its source at Lake Itasca. Note that both banks show in the photo.

Figure 12-11. This high ridge near the entrance to Glacial Lakes State Park, called Blue Mound, is composed of bouldery sand and gravel deposited in a large sluiceway carrying meltwater and debris across dead and buried glacier ice. (Photo by J. C. Green.)

Figure 12-13. The source of the Mississippi River at Lake Itasca. (From the Lake Itasca, Minnesota, 7.5' Quadrangle.)

The hills near Lake Itasca rise as much as 60 m above the lake surface.

When did the Mississippi River originate? The area could have been deglaciated as early as 20,000 years ago and was certainly deglaciated by 10,000 years ago. The Des Moines Lobe, off to the west, was active at least until 14,000 years ago, and this was probably a time of permafrost and tundra vegetation in the park area. As the climate warmed, ground-water flow was established and precipitation as rain probably increased. Thus, the lake basin would eventually have been filled to overflowing and an outlet would have been established, probably much like the one there today (Figure 12-13). The Mississippi's course was probably determined by the paths of spillover as the various lake basins of the area to the north and east were filled with water. The moraines and other glacial features predetermined the river's course between lakes.

Minnesota Man Site

The site where remains of Minnesota Man were excavated is 5 km north of Pelican Rapids. Here, during road excavations in 1931, the first human skeleton in North American glacial deposits was unearthed about 3 m beneath the ground surface. The material in which the find was made is lakebed clay of Glacial Lake Pelican, which existed about 11,000 years ago as a small glacial lake formed in a depression on stagnant ice of the Des Moines Lobe. Unfortunately, the find had been excavated before a detailed study of the relationships could be made. A radiocarbon date of the skeleton is not entirely satisfactory, partially because of shellac that was used to preserve the skeleton. However, Elden Johnson, state archaeologist, in a report on the prehistoric peoples of Minnesota (1969) concurs with a date of 5000 to 1000 B.C. for the skeleton. A marine shell pendant found with the skeleton supports this conclusion. Also found was a flaking tool made from an elk antler. The great depth of the find, however, argues against an Indian burial. If the skeleton is as old as the glacial lakebeds, it may represent a drowning and it would be the oldest skeleton in North America.

Archaeologists very soon realized that the Minnesota Man skeleton was that of a teen-aged girl, and the remains were dubbed "Minnesota Minnie." Belatedly, in 1976, the Minnesota State Legislature legalized the sexual change of Minnesota Man to Minnesota Woman. "Minnie" now resides at the Historical Society's Museum in St. Paul (Figure 12-14).

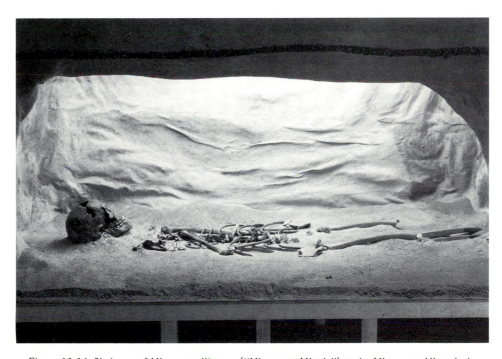

Figure 12-14. Skeleton of Minnesota Woman ("Minnesota Minnie") at the Minnesota Historical Society's Museum in St. Paul. (Photo by Eugene Becker and courtesy of the Minnesota Historical Society.)

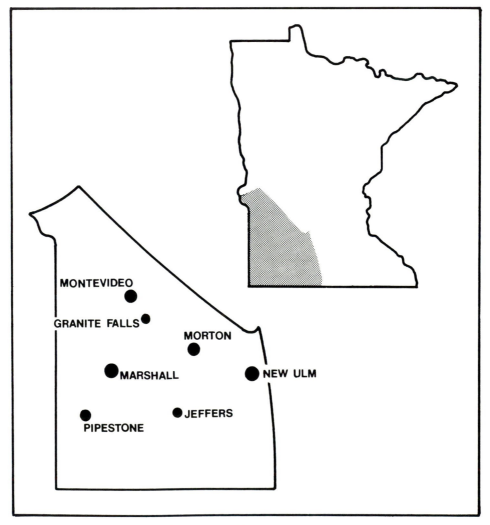

Southwestern Minnesota region.

13

Southwestern Minnesota

THE LAY OF THE LAND

A flatiron-shaped plateau, called the Coteau des Prairies by early French explorers, dominates the regional topography of eastern South Dakota and southwestern Minnesota (Figure 13-1). This distinctive "highland of the prairies" rises dramatically to elevations of more than 640 m in South Dakota. The eastern flank of the plateau descends in a series of broad steps to merge with the Minnesota River Lowland, a southeastern extension of the topographic basin occupied by the Red River of the North. The entire landscape owes its major form to a similar arrangement of the bedrock surface, even though important details of local topography were imprinted by glaciers, running water, and wind.

Two prominent belts of high, hilly terrain on the eastern flank of the plateau are moraines of the Des Moines Lobe, deposited during its last advance southeastward, following the low axis of the Minnesota River Lowland into the Twin Cities (Figure 13-2). The outermost belt, the Bemis Moraine, forms the high crest of the plateau. The other, wider and more lake infested, is the Altamont Moraine, which marks an important standstill during deglaciation. The Bemis Moraine is an impressive topographic barrier, and it has become an important drainage divide. Its well-drained southwestern side sheds water into the Big Sioux River and on to the Missouri. On the other side of the divide, water flows into the upper Mississippi via the Des Moines and Minnesota rivers.

Aside from the fact that streams on both sides of the Bemis Moraine flow into different drainage systems, some of the watercourses have had very different beginnings. Whereas most of the small tributaries that feed into the Minnesota River system flow generally straight down the northeastern slope of the plateau, some of the larger streams, notably the Redwood and the Cottonwood rivers, follow courses that were established by glacial meltwater draining the margins of the Des Moines Lobe. Similarly, large segments of the Des Moines River flow along old meltwater channels. The Bemis Moraine is breached by a series of deep gorges that grade to the southwest, occupied by tiny stream courses that could not be responsible for the capacious valleys through which they flow. Such gaps breach the moraine at Lake Shokatan, at Lake Benton, near Holland, and at Chandler (see Figure 13-2). They were cut by spillover from rising meltwater ponded on the northeast side of the moraine against the receding margin of the Des Moines Lobe.

To the southwest, beyond the Bemis Moraine, forming a cleft in the Coteau des Prairies, is a lake-free, gently rolling surface of older drift that is covered with windblown silt. Here, water is carried to the Big Sioux River by a drainage system that is older and better developed than that of the northeastern slope of the plateau. Much of the arrangement of the stream network seems to be inherited from erosional patterns incised at an earlier time upon the bedrock, now thinly draped with glacial sediments. The exhuming process has been completed in some places, where reddish pink Sioux Quartzite crops out in extensive areas.

By far the most important stream-cut feature is the valley of the Minnesota River, excavated largely by Glacial River Warren when that ancient stream drained Glacial Lake Agassiz (Figure 13-3). H. E. Wright described the valley eloquently as

a narrow sliver of wooded hill slopes in the vast plains to north and south, [holding] within it a diversity of geologic features such as rugged granite knobs, boulder-gravel river bars, broad sandy terraces, gentle colluvial slopes—and a stream along the axis that is almost tiny in the context of these major features (1972a: 575).

223

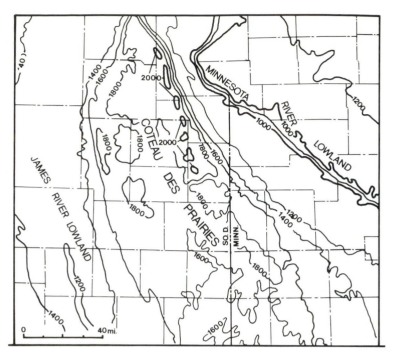

Figure 13-1. Generalized topographic map of the Coteau des Prairies and adjacent Minnesota River Lowland. This "height of land" at over 2,000 ft is the crest of the Bemis Moraine.

As a topographic feature, this incision across the waistline of Minnesota perpetrated by the erosive waters of River Warren furnishes segments of both the western and eastern borders of the state, a testament to the importance of catastrophic events in the shaping of landscapes.

The details of the flood history of River Warren are best displayed in a system of abandoned channels and terrace levels between Rosholt, South Dakota, and Browns Valley (see Chapter 11) and then south to Granite Falls (Figure 13-4). In the latter stretch, River Warren split into a number of channels as it spread across a wide, flat plain, formerly part of a small proglacial lakebed. Eventually all of the channels but one were abandoned, leaving a system of flat-floored, steep-sided valleys, such as Watson Sag. Southeast of what is now Granite Falls, the full force of discharge was enclosed in one valley all the way to the Twin Cities.

Flat, boulder-covered surfaces, called prairies by the early settlers, are remnants of channel bottoms occupied by the glacial river during successive stages in its erosional history. These have been partially buried by aprons of sediment washed down upon them from the steep valley sides (Figure 13-5). One great benefit to geologists is the view afforded in many places of the bedrock that lies beneath the drift. Bedrock islands stud the present valley floor (see Figure

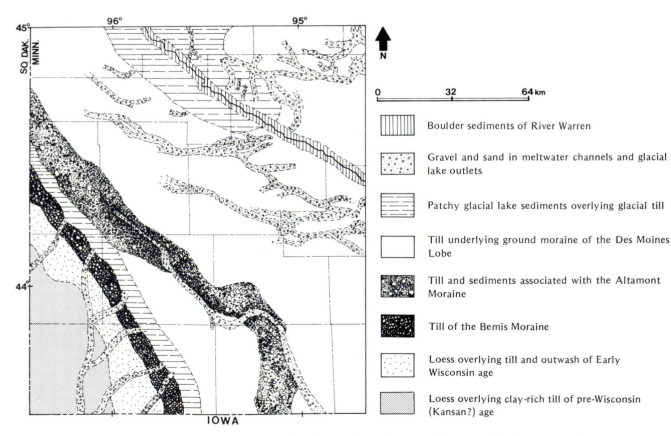

Figure 13-2. Geologic map showing the nature of the surface deposits in southwestern Minnesota. Note the pattern of meltwater channel deposits along the Minnesota River Valley, where the streams probably formed along the retreating margin of the Des Moines Lobe.

Figure 13-3. View of the Minnesota River Valley looking eastward near Granite Falls. The bluffs on the far side are almost 60 m above the valley floor.

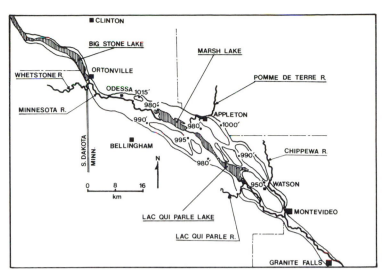

Figure 13-4. Map showing the abandoned channels of Glacial River Warren between Ortonville and Granite Falls. The stretch of branching channels coincides with a flat regional slope across outwash fans and glacial lake plains formed in the wake of the Des Moines Lobe.

1-1) along with bars of bouldery gravel left stranded by the floodwaters. The modern river meanders lazily across the wide valley, depositing silt and fine sand upon its floodplain, quite a contrast to the large caliber of River Warren's bedload.

GLACIAL GEOLOGY

The last glacier to cover southwestern Minnesota was the Des Moines Lobe, a tongue of ice that advanced southward and eastward, first along the present-day Red River Lowland, fed by the Laurentide Ice Sheet in Canada. Eventually, ice penetrated central Iowa and western Wisconsin, reaching a maximum position near Des Moines about 14,000 years

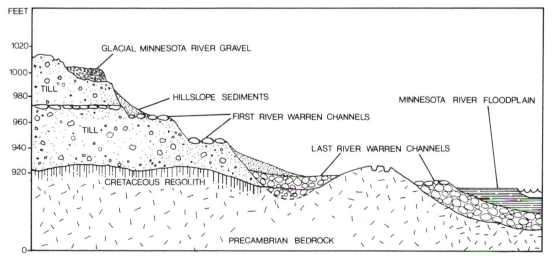

Figure 13-5. Cross section of the Minnesota River Valley near Granite Falls. Bedrock knobs are riddled with potholes formed by the turbulent waters of River Warren. The boulder pavement between tills is faceted and striated (see also Figure 13-6). Boulders on terraces were lagged (left behind) by River Warren.

225

ago (see Figure 7-13). But even earlier—no one knows for sure exactly when—glacial ice from both the Lake Superior region and from the Winnipeg Lowland had invaded southwestern Minnesota. Evidence for these older advances is found as layers of till beneath the sediments left by the Des Moines Lobe and by drift sheets that lie outside the margins of that glacier. Those older deposits of till and outwash, beyond the Bemis Moraine, are buried by a thin veneer of silt called loess that settled from dust-laden winds sweeping the alluvial plains of rivers draining the margin of the Des Moines Lobe.

The color and mineral composition of most of the glacial deposits reflect mixtures of bedrock eroded from sources in northwestern Minnesota, the eastern Dakotas, and southern Manitoba, with contributions from the local substrate. Paleozoic limestone, Cretaceous shale, and Precambrian granite are the most abundant components. The one exception is a buried layer, exposed at only a few places along the Minnesota River Valley and its tributaries, that contains rock fragments carried from the Lake Superior Basin, such as basalt, red sandstone, and gabbro. Even Lake Superior agates can be found. The pink to reddish brown color of this drift, called the Hawk Creek Till for exposures along that stream in northwestern Renville County, contrasts sharply with the buff to dark gray hues of the other layers.

Exposures in many places along the valley indicate that this incursion of ice from the northeast was preceded and followed by several different advances, mainly from the north and northwest (Chapter 7). The last two advances both left limestone-rich tills that at first glance seem to be indistinguishable. But there is an important difference between them. The drift of the last advance, called the New Ulm Till, contains a rich load of shale fragments eroded from the Cretaceous Pierre Formation in the eastern Dakotas. The drift beneath it, the Granite Falls Till, holds almost none. This difference in stone content indicates that the glaciers that deposited the sediment followed slightly different courses.

Another fascinating aspect of the glacial sequence is the occurence of soils, wood fragments, and a boulder layer sandwiched between and developed upon the succession of tills. Most of the wood is older than 40,000 years, according to its carbon-14 content. The boulder layer is commonly found between the last two drift sheets, the Granite Falls Till and the shale-bearing New Ulm Till (Figure 13-6). In places, the boulders are arranged to form a continuous pavement, with the tops planed and striated by glacial abrasion.

The construction of this one-stone-thick layer may be related to the advance of the Des Moines Lobe across the drift left by a predecessor lobe. In this explanation, stones were pressed, one at a time, into the soft, moist mud at the wet base of the sliding glacier. As more and more stones were emplaced, successive rock fragments were dragged across the developing pavement, making striations, until they encountered an unoccupied area large enough to hold them. Eventually, a continuous cover of stones armored the overridden drift sheet, and no more stones could be placed because the resistance was too great. When the Des Moines Lobe finally retreated, a thin blanket of till melted out of the ice and buried the striated boulder pavement. Today, the pavement crops out as a line of boulders on hillsides throughout the Minnesota River Valley area.

The land surface that emerged from beneath the melting Des Moines Lobe owes its major character to its arrangement of end moraines, ground moraine, and meltwater features, including glacial lakes. Without doubt, the Altamont Moraine, with its hundreds of lakes as well as rugged relief, is the most impressive of these landforms. That belt of hummocky terrain is a fine example of glacial deposition associated with stagnant dirty ice. It marks an important phase in the history of the Des Moines Lobe, a time of stalled movement before rapid downwasting destroyed the entire glacier.

Drumlins and eskers, landforms abundant elsewhere in Minnesota, are rare or completely absent in this part of the state. The Des Moines Lobe, for one reason or another, was not capable of originating such forms. But numerous kames, usually expressed as low conical hills with sand and gravel caps, dot the area, along with linear, gravelly ridges unmolded from crevasses in the ice that trapped sediment. Outwash is mainly concentrated in channels cut by streams draining the ice margins, but small patches are randomly distributed throughout the area. Most of the gravel is of poor quality; it contains a large percentage of soft shale fragments, a component that limits its usefulness.

Shallow, short-lived glacial lakes covered large areas during the early stages of glacial retreat. One complex, located south of the Minnesota River Valley in parts of Brown, Watonwan, Blue Earth, Martin, and Faribault counties, has been given the name Glacial Lake Minnesota. Its former presence is recorded by thin deposits of laminated silt, as well as sandy shoreline sediments and a few deltas. In fact, Glacial Lake Minnesota was probably a number of bodies of water trapped in depressions on the surface of the downwasting Des Moines Lobe. The spillover of water trapped behind the Bemis Moraine has already been mentioned. Along the axis of the Des Moines Lobe, other bodies of water were ponded behind ice-cored ridges and quickly drained as the ice melted northward. Therefore, it is not unusual to encounter thin veneers of laminated lake sediments here and there throughout the area.

Outside the Bemis Moraine, in the southwest corner of the state, beneath a cover of loess, lie glacial drifts of several different ages, some of which are probably remnants of the earliest glaciers to cover Minnesota. The older tills are exposed beyond a ramp of younger outwash and till probably deposited by the Des Moines Lobe before it retreated slightly to the position at which it constructed the Bemis Moraine (see Figure 13-2). In South Dakota these younger deposits are termed Early Wisconsin. R. V. Ruhe (1969) ascribed them to the Tazewell substage of the Wisconsin glaciation in Iowa, where he found buried wood fragments that are 20,000 years old.

The older tills are extremely weathered in places, so much that all but the most resistant rock fragments have been altered or completely removed by chemical weathering. Concentrations of pink quartzite boulders and cobbles are encountered at the contact between till and loess. These "stone lines" contain finely polished and faceted ventifacts, the result of natural sandblasting by the winds that eventually deposited the overlying loess (Figure 13-7). Seldom is any rock type other than quartzite found in these layers, indicating that a long period of weathering and soil formation preceded the advance of the last glacier. Such observations lead to the conclusion that the deposits upon which the soil formed are older than the Wisconsin glaciation, perhaps Kansan in age. But that is still a guess.

When ice stood at the Bemis Moraine, the area beyond it experienced conditions of severe cold. Permafrost, a condition of permanently frozen ground that now prevails widely in arctic regions, gripped the soil. Frost-shattered bedrock, as well as glacial erratics and the angular fragments resulting from that time of cold, litter certain areas. Downslope movement of loose sediments during times of thaw resulted in a smoothing of the entire landscape, contributing to the subdued topography of today. And, of course, the extreme antiquity of the entire landscape explains the well-developed drainage system, as well as the relative abundance of bedrock outcrops exposed by erosion of the drift cover.

BEDROCK GEOLOGY

Because of the almost continuous cover of glacial sediments, exposures of bedrock are rare in the southwestern part of the state. The most extensive areas of outcrop are the Minnesota River Valley and parts of the Coteau des Prairies underlain by Sioux Quartzite. Generally, three major rock types underlie the glacial drift: (1) igneous and high-grade metamorphic rocks of Early and Middle Precambrian age (Grant, 1972); (2) the Sioux Quartzite of Late Precambrian age and (3) poorly consolidated marine and continental shales and sandstones of Cretaceous age. The rocks range in age from 3,600 million years for the Early Precambrian rocks, among the oldest on the North American continent, to about 100 million years for the Cretaceous sediments.

Coarse-grained pink or white granite gneiss is probably the major constituent of the older Precambrian complex (see Figures 1-1 and 3-2) with minor rock bodies of more mafic composition. The oldest rocks are mixtures of granitic and mafic components that were subjected to deformation and metamorphism 2,700 to 2,600 million years ago.

Figure 13-6. Boulder pavement sandwiched between two tills in a roadcut near Wilmot, South Dakota. Note the flat tops of the boulders. These facets are also striated. The upper till is shale rich (from the Des Moines Lobe) and the lower till is shale free (from the Wadena Lobe).

Figure 13-7. Wind-polished boulder of Sioux Quartzite lying on top of older till and partly buried by loess. Polishing on some larger boulders and outcrops has been attributed to the buffing action of itchy bison, whose coarse, dirty hides, applied with enough vigor, might have been an effective emery cloth.

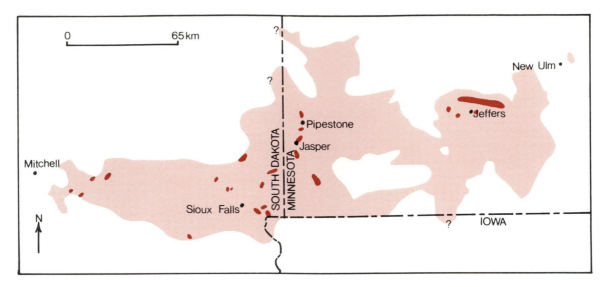

Figure 13-8. Map showing distribution of the Sioux Quartzite in southwestern Minnesota and eastern South Dakota.

Figure 13-9. Exposure of Morton Gneiss in the Cold Spring Granite Company quarry at Morton. Notice the wavy layering of dark minerals and the clots of black-colored mineral material. The vertical marks are drill bit scars. (Courtesy of J. A. Grant.)

Younger granites and basaltic dikes cut through them as intrusions emplaced about a billion years later. At many places, this rock complex has been altered by chemical weathering processes into a soft, clay-rich soil material as much as 30 m thick. Exposures along the Minnesota River Valley near Redwood Falls indicate that the weathering profile developed during the Cretaceous Period under tropical conditions (see Figure 6-39). Rocks of Cretaceous age generally consist of poorly consolidated quartz sand, lignitic clay, and soft, dark gray shale. Within the Cretaceous sediments a distinctive interval that overlies the weathering profile developed on the Precambrian complex is a grayish white to brown, hard nodular clay.

The Sioux Quartzite underlies much of southwesternmost Minnesota, extending westward as far as Mitchell, South Dakota. The Sioux was named for the Sioux River, which has cut a valley into the quartzite at Sioux Falls, South Dakota. The formation is nearly horizontal. It seems to occupy two gentle basins, with conglomerate beds locally outlining the basin forms (Figure 13-8). What is probably almost the base of the formation is exposed near New Ulm, where a conglomerate outcrop lies about 100 m from granite, presumably the basement to the quartzite (see Figure 5-3b). Generally, the Sioux Quartzite is resistant to erosion and therefore lends support to some of the high topography in the Coteau des Prairies and prominent ridges elsewhere in southwestern Minnesota.

PLACES OF INTEREST

Morton

As it approaches Morton from the east, Minnesota Highway 19 descends abruptly into the Minnesota River Valley.

The bare, buff to brown roadcuts expose till of the Des Moines Lobe. Just outside Morton, a conical pile of large angular rock fragments interrupts the flat valley floor. This "quarry kame" contains discarded blocks of granitic gneiss from the Cold Spring Granite Company's active quarry just out of view from the road. Called "Rainbow Rock" in the trade and the Morton Gneiss by geologists, this prized architectural stone is shipped throughout the world. But few appreciate the geological fact that it is one of the oldest known rocks in North America, at least 3,600 million years old.

A good place to stop and explore is the hill behind Morton High School. This large knob of bedrock was too resistant to be eroded by Glacial River Warren. The hill is pitted by round potholes drilled by the turbulent waters of the glacial river that cut the valley. Wavy patterns of dark minerals (mostly hornblende-biotite and pyroxene) enclosed in a pink and gray matrix of quartz and feldspar give an impression of the once plastic nature of the rock (Figure 13-9). The clots and rafts of darker material are frozen in place like great plums in a Christmas pudding. An abandoned quarry on the east side of the hill shows them well in three dimensions. These black inclusions could be samples of an early volcanic crust torn loose by magmas that later crystallized to form the lighter-colored enclosing rock.

Granite Falls

Traveling west on U.S. Highway 212 between Olivia and Granite Falls, the first-time visitor is impressed with the unobstructed view across gently rolling farmlands. Here and there piles of boulders lie along fence lines, stockpiled through years of hauling from the nearby fields. The character of the landscape remains unchanged kilometer after kilometer. One can easily conjure up a vision of the native prairie that flourished here before the farms were established. But then the reverie is interrupted by a steep descent into the valley of Hawk Creek and a subsequent rise to rolling plain, signaling a change in the landscape. A mirage looms up ahead in the form of a huge smokestack rising directly out of the cultivated fields in the distance. The mirage is crystallized into reality when the route suddenly encounters the deep Minnesota River Valley, with Granite Falls and its coal-burning electrical power plant on the valley floor, some 60 m lower than the surrounding upland surface.

The scenic outlook on the valley side offers a sweeping view across the Minnesota River Valley. Several terrace segments are preserved at various heights, and bedrock islands stud the valley floor (see Figure 13-5). Boulder-studded surfaces are remnants of River Warren's bed. On the north side of the highway two till deposits are exposed, separated by a boulder pavement. The upper deposit is the New Ulm Till of the Des Moines Lobe, and the lower is the Granite Falls Till of the earlier Wadena Lobe.

At the municipal park in Granite Falls are excellent exposures of dark-colored gneiss, with fine-grained basaltic dikes cutting across them. The gneiss displays a change in composition along the roadcuts bordering the entrance to the park. One section contains numerous small pink to red garnets. At the north end, the black mica biotite occurs rather than garnet. Here, too, erosion by River Warren cut deep potholes into the outcrops when they were rapids in that river. In fact, to the south, Highway 67 crosses a rugged, rocky terrain composed of ancient crust exposed and sculpted by turbulent glacial river waters.

Pipestone National Monument and Vicinity

Since about A.D. 1600, and perhaps even before that, Indians of many tribes quarried "pipestone" at the sacred Pipestone Quarry in what is now Pipestone National Monument. The "pipestone" is an easily carved, clay-rich, 45-cm-thick rock layer in the Sioux Quartzite (Figure 13-10). It is a red mudstone, colored by disseminated hematite.

Pipes made of this somewhat distinctive rock have been found as far away as Ohio, suggesting that it was widely sought and probably also widely traded, at least in the Midwest, although similar rocks occur elsewhere as well. In the middle of the 19th century, after a period of several hundred years during which Indians of all tribes had free access to the stone, the Yankton-Dakota Indians gained control of the area and thereafter traded the stone with other tribes. Since the spot became a 115-ha national monument in 1937, all Indians have the right, by permit, to quarry the stone.

An Indian legend states that, after the American Indians were created, the Great Spirit summoned them all to the quarry. He broke off a piece of the red stone and made it into a large pipe, which he smoked and pointed successively toward the north, south, east, and west. He told them that the red stone was their flesh and that they should use it for their peace pipes.

Henry Wadsworth Longfellow in his "Song of Hiawatha," published in 1855, added to the fame of the already famous quarry with these words:

> On the Mountains of the Prairie
> On the great Red Pipestone Quarry,
> Gitche Manitu, the mighty,
> He the Master of Life, descending,
> On the red crags of the quarry
> Stood erect, and called to nations,
> Called the tribes of men together.

The pipestone has been called catlinite in honor of George Catlin, who visited the site in 1836 and did much to depict the life of the American Indian (Figure 13-11), but it has no real geological status as a special type of rock. The "Three Maidens" (Figure 13-12) near the quarry are three large granite boulders that, according to an Indian legend,

a

b

Figure 13-10. (a) Pipestone bed at base of quarry (just above water) at Pipestone National Monument. (Courtesy of the National Park Service, Pipestone National Monument.) (b) Pipes carved out of pipestone. (Courtesy of the National Park Service, Pipestone National Monument.)

were eggs of the war eagle. Indian maidens beneath the rocks guarded the quarry, and gifts were left near the boulders as offerings. Some Indians hurled tobacco at the rocks, afraid to move too near. The Indians, even without a knowledge of geology, surely realized that the boulders were not quartzite but were unique in the Pipestone area. George Catlin realized this too. Intrigued by these large boulders and other seemingly anomalous occurrences elsewhere, he wrote in 1836:

I believe that the geologist may take the varieties which he may gather at the base of the Coteau in one hour and travel the continent of North America all over without being able to put them all in place; coming at last to the unavoidable conclusion that numerous chains or beds of primitive rocks have reared their heads on this continent, the summits of which have been swept away by the force of the diluvial currents; and their fragments jostled together and strewed about like foreigners in a strange land, over the great valleys of the Mississippi and the Missouri, where they will ever remain and be gazed upon by the traveler as the only remaining evidence of their native ledges, which have again been submerged or covered with diluvial deposits (quoted in Winchell and Upham, 1884: 65).

In 1836, the concept of continental glaciers had not yet been formulated. We now know that they are frost-split glacial erratics of granite, perhaps from Millbank, in South Dakota, just west of the northwest end of the Minnesota River Valley, 140 km to the north.

The Sioux Quartzite is at or near the surface throughout a wide area in southwestern Minnesota. At Split Rock Creek State Park, the Split Rock River has cut a small gorge into the quartzite, exposing well-preserved bedding features. A large mesa called Blue Mounds is composed of Sioux Quartzite, and within the boundaries of Blue Mounds State Park one can see a buffalo herd as well as native prairie vegetation. At the town of Jasper, misnamed after iron-rich chert instead of the pink quartzite found in the area, quarries have furnished both building stone and road metal. Many outcrops of the Sioux bear the markings of glacial abrasion, preserved because the rock is so resistant to weathering.

Geologists have not been the only people who noticed the Sioux Quartzite. Near Jeffers, a spot on a ripple-marked and glacially striated quartzite ridge is covered with nearly 2,000 rock carvings, or petroglyphs, which apparently date from two periods—3000 B.C. to A.D. 500 and A.D. 900 to 1750 (Figure 13-13). The only other well-preserved petroglyphs in Minnesota, with the exception of a few at Pipestone National Monument that are no longer in place, are on the Nett Lake Indian Reservation in northeastern Minnesota. Another half dozen or so locations of petroglyphs have been vandalized. Native prairie is also present near Jeffers.

Figure 13-11. Diorama in the museum at Pipestone National Monument showing the prairie and quarry area about 1650. This painting is by a National Park Service exhibit team. Catlin's 1836 glamorized painting of the quarry portrayed the Indians in flowing white headgear even though workers in the quarry get covered by red dust. (Courtesy of the National Park Service, Pipestone National Monument.)

Figure 13-12. The "Three Maidens" at Pipestone National Monument. The third maiden was hiding in back and was barely visible when this picture was taken.

Figure 13-13. A few of the many Jeffers petroglyphs on glacially striated and ripple-marked Sioux Quartzite. (Photo by Alan Ominsky and courtesy of the Minnesota Historical Society.)

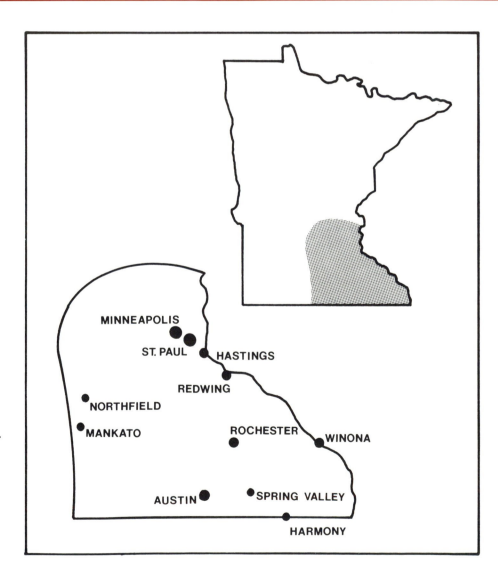

Southeastern Minnesota region.

14

Southeastern Minnesota

THE LAY OF THE LAND

Southeastern Minnesota contains two sharply contrasting landscapes directly related to the glacial history of the area. Bordering the Mississippi River Valley and extending westward to the Bemis-Altamont moraine system is a lake-free terrain with a well-integrated and, in places, deeply entrenched stream system. Inside and to the west of the moraines, the land surface is dotted with lakes, and the streams are not so well developed, a direct reflection of the youthfulness of the surface, ice-free only since the demise of the Des Moines Lobe about 13,000 years ago. The entire region rises topographically from the Minnesota River and Minneapolis lowlands, which are below 305 m, to a strong plateau as high as 427 m. Streams drain into the Minnesota River, into the Des Moines River, and directly into the Mississippi River.

The deep dissection accomplished by rivers flowing into the Mississippi River has resulted in an imposing system of flat-floored valleys with intervening ridges. The steep, thickly wooded valley sides and ridge crests are in sharp contrast to the cultivated fields of the low-relief valley floors. The rivers themselves characteristically meander through tortuous bends, constantly shifting their channels along their way to the Mississippi. Many of the valleys contain terraces, remnants of former river levels left behind by the downcutting streams. Bedrock exposures in the eastern segment are numerous, an indication of both the general thinness of the glacial cover and the severity of stream erosion.

By far the most scenic feature is the Mississippi River Valley itself. Enclosed by steep bedrock walls, the wide valley easily contains the branching system of channels that serves to discharge the waters of the mighty river (Figure 14-1). One long river lake, Lake Pepin, stretches from Red Wing to Wabasha, backed up by the damming effects of the delta of the Chippewa River.

GLACIAL GEOLOGY

Most of southeastern Minnesota was ice free during the Wisconsin glaciation. However, the effects of earlier glaciations are present in the form of glacial erratics and deposits of till and outwash. The earliest history of the Great Ice Age is obscured by the effects of weathering and erosion, which have removed much of the fragile record. In fact, some areas are completely free of glacial sediment. The so-called driftless area, including parts of Minnesota, Wisconsin, Iowa, and Illinois is thought by some to have escaped glacial cover completely. More likely the area was glaciated during one or more of the older glacial stages, with subsequent removal of the evidence by erosion.

Glacial deposits outside the Bemis Moraine have been ascribed to Nebraskan, Illinoian, and Kansan glaciations by various geologists and soil scientists. However, the assignments are tenuous. Multiple layers of till, with soils buried between them, surely indicate a complicated history of multiple glaciation. Study is hampered by a covering of loess, blown from the Mississippi River Valley during the Wisconsin glaciation. And, in places, chemical weathering has decomposed the glacial sediments into a clay-rich gumbo, obscuring their original character.

The last 20,000 years saw the intrusion of two different glaciers (Chapter 7) into the region. First, the Superior Lobe constructed the broad St. Croix Moraine with sediment derived from the Lake Superior region. Then, the Des Moines Lobe advanced from the west, distributing drift rich in limestone and shale across the entire state and into Iowa. In the Twin Cities region and to the north, that glacier overrode

Figure 14-1. Bluffs of Jordan Sandstone on Scenic Drive a few kilometers north of the village of La Crescent. The deeply incised Mississippi River Valley lies to the east enshrouded in fog.

the drift of the Superior Lobe, the double onslaught leaving two contrasting tills, one upon the other. The moraines along the eastern margin of the Des Moines Lobe make up a north-trending belt of lakes, swamps, and hilly terrain. The Twin Cities area is doubly blessed with the topographic effects of both lobes.

Beyond the ice margins, meltwater streams carried sediments and dumped them to form important deposits of sand and gravel. The Mississippi River Valley was choked nearly to the brim with coarse debris washed from the dirty ice of successive glacier lobes (see Chapter 7). One effect of this infilling was to dam the side valleys of the tributary streams, resulting in the formation of temporary lakes, which trapped sediments in them. Later drained and entrenched, the old lakebeds are now exposed along many of the valley sides. Terraces along the Mississippi Valley at various levels are all that remain of the outwash plugs.

BEDROCK GEOLOGY

The Paleozoic marine sandstones, carbonates, and shales of southeastern Minnesota, many containing a rich cache of fossil material, are typical of the rocks that cover a great part of the midcontinent. Three periods of geologic time are represented—Cambrian, Ordovician, and Devonian—as

described in Chapter 6. Scattered deposits of Cretaceous age reflecting both marine and continental environments came to rest at the end of the Mesozoic. On top of everything are the glacial deposits of the Pleistocene Epoch and postglacial sediments of the Holocene, the most recent epoch of geological time. Thus, the total record of exposed rock spans perhaps 500 million years, but by no means does it contain a complete history of that time range. For example, near the Mississippi River downstream from St. Paul, erosion has removed the soft St. Peter Sandstone and any younger units, leaving the Prairie du Chien dolomites as caps on the higher areas.

The designation St. Croixan Series for the rocks of Late Cambrian age recognizes the importance of the exposures along the St. Croix River Valley. For the most part they are poorly cemented sandstones such as the Jordan (see Figure 14-1) and siltier units such as the Eau Claire Formation, which is exposed at the village of Dresbach. A high content of the green micalike mineral glauconite has earned the name greensand for some of the sandstone. The sandstones are sparsely fossiliferous, especially when compared with the overlying Ordovician carbonates. Worm burrows are common, and in places both brachiopods and trilobites are abundant.

Carbonates in the form of dolomite and limestone along

234

with limey siltstone dominate the rock column contributed during the Ordovician Period. One important exception is the St. Peter Sandstone, a pure quartz sand sandwiched between dolomites of the Prairie du Chien Group and the thin Glenwood Formation, and the overlying Platteville Formation, another carbonate. Ordovician rocks hold the richest fossil faunas, and almost any exposure of limestone will yield some specimens (see Figures 6-22 through 6-25). Another sequence of carbonates, the Cedar Valley Formation, was deposited during the Devonian.

Cretaceous rocks in the southeast are generally nonmarine, but they interfinger in the Mankato region with marine sediments. They lie as discontinuous patches on an erosional surface developed upon the Paleozoic rocks (Figure 14-2). Some of the gravels and sands are difficult to distinguish from glacial deposits of younger age. Structurally, the large Hollandale Embayment controlled the major history of marine transgression. However, important smaller structures are superimposed. The Twin Cities Basin was an area of more rapid subsidence. Positive areas rose as either fault-bounded blocks or as uparched folds, such as the Hudson-Afton Anticline and the Red Wing-Rochester Anticline. Although important in a regional way, these broad structures are not tectonically significant.

PLACES OF INTEREST

The Twin Cities

St. Paul and Minneapolis lie atop a pile of Paleozoic sedimentary rocks arranged in the form of a structural basin

(Figure 14-3) that forms an effective trap and reservoir for groundwater. So gentle is the tilting on the rocks that in outcrop they look essentially horizontal. However, from the center of the basin, the strata rise in elevation toward the rim. Because of this, along a level traverse, successively older rocks are encountered as one moves away from the middle of the basin. Sandstone, shale, limestone, and dolomite are the major rock types, all associated with the transgression of seas into the area. Atop the Paleozoic rocks, and completely covering them in many places, are glacial sediments left behind by a succession of glacier tongues during the Pleistocene Epoch. Two advances between 20,000 and 14,000 years ago left rugged moraines with numerous lake basins, giving the area its special topographic character. Finally, the postglacial establishment of the Mississippi and Minnesota river valleys contributed steep-sided gorges and waterfalls to the landscape.

The same stratigraphic sequence is encountered along the Mississippi River below St. Anthony Falls, all the way to Fort Snelling. A geologic section through Minnehaha Falls serves as a reference (Figure 14-4). Beneath glacial drift are three contrasting formations. On the top, forming a hard caprock, is the resistant Platteville Formation, a fossil-bearing dolomitic limestone. Beneath the Platteville is a soft, thinly bedded grayish green shale, the Glenwood, generally just a meter or two thick. Below that is white or light yellow, massive, poorly cemented sand, the St. Peter Sandstone. In a few scattered localities, the Decorah Shale, an olive-gray shale containing thin limestone beds loaded with fossils, is preserved above the Platteville.

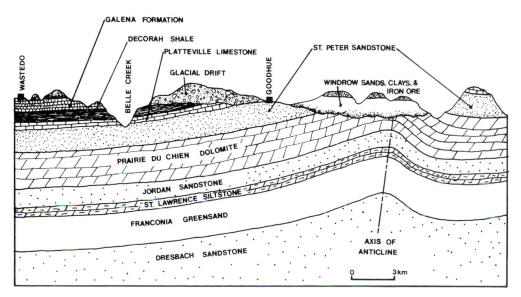

Figure 14-2. Cross section of eastern Goodhue County. The uparched Paleozoic rocks were eroded before the deposition of the Cretaceous Windrow Formation (After Sloan, 1967.)

The bedrock as it is exposed along the bluffs of the river is not continuous beneath the glacial drift. Drilling and excavations have revealed an extensive system of valleys eroded by preglacial rivers, dividing the rock surface into a system of mesas (Figure 14-5). These old valleys are now filled with glacial drift, but in places their location is marked by chains of lakes formed from the collapse of glacial ice that lay thicker along these deeper courses. The most prominent lake chain extends southward from Golden Valley into Minneapolis. It includes Sweeney Lake, Twin Lake, Wirth Lake, Birth Pond, Brownie Lake, Cedar Lake, Lake of the Isles, Lake Calhoun, and Lake Harriet. A longer one, crossing St. Paul and its northern suburbs, contains Snail Lake, Grass Lake, Vadnais Lake, Twin Lake, Savage Lake, Gervais Lake, Kohlman Lake, Keller Lake, Round Lake, and Lake Phalen.

The arrangement of the Mississippi and Minnesota river valleys with their high bluffs was determined by the course of Glacial River Warren and the upstream retreat of St. Anthony Falls from Fort Snelling to downtown Minneapolis (Chapter 7). Fort Snelling, the state's first military outpost, was established in 1819. Its location on top of the 30-m bluff at the junction of the two rivers was selected after the War of 1812 in order to block any British use of the water-

ways to achieve influence with the Indians or to further the British fur trade. Lieutenant Zebulon Pike purchased the site for the fort from the Sioux (Figure 14-6).

Caves and ravines have been excavated in places through the removal of soft St. Peter Sandstone by the seepage of groundwater as springs, especially along the contact with the less permeable Glenwood and Platteville formations. Sand is removed at first at the exposed surfaces along the valley sides. Slowly the opening expands beneath the overlying caprock. If the roof holds, then a cave is formed (Figure 14-7a). Collapse of the roof results in the formation of a ravine. Besides natural processes, the efforts of human excavation have produced more than 65 km of tunnels for storm sewers, pipes, electrical conduits, and other trappings of urban development (Figure 14-7b).

The St. Croix Moraine accounts for the hilly character of St. Paul. That belt of lake-dotted glacial drift stretches eastward to the St. Croix River Valley and south to the big bend in the Mississippi River called Pine Bend. The moraine marks the edge of the Superior Lobe in its last advance; for the most part, it is a stagnant ice complex, with numerous kames and kettle lakes. Broad aprons of outwash extend into the present Mississippi River Valley from the outer margin of the moraine. Its content of gabbro, volcanic rocks,

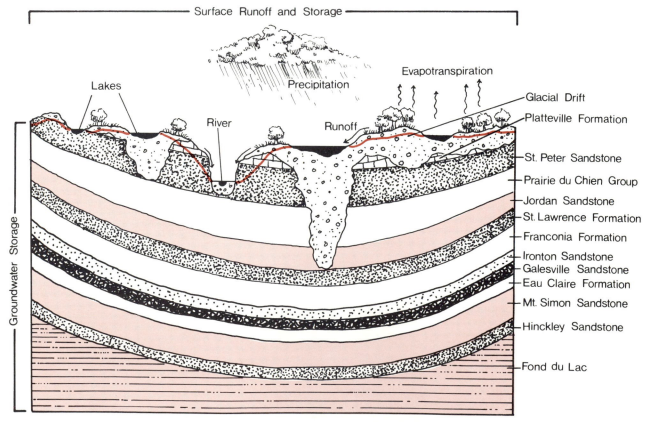

Figure 14-3. Highly generalized geologic east-west cross section of the Twin Cities Basin, showing the water cycle. The geology is much more complex than shown here, based on new work by the Minnesota Geological Survey. The vertical scale is exaggerated. (After Hogbert, 1971.)

red sandstone, and Lake Superior agates was eroded from Precambrian bedrock in northeastern Minnesota.

The much flatter topography of Minneapolis is largely the result of outwash deposited by the melting Des Moines Lobe, which in its advance left fine-textured gray till on top

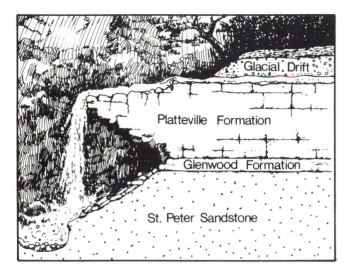

Figure 14-4. The sequence of rocks at Minnehaha Falls is repeated in many outcrops throughout the Twin Cities area. (After a drawing by Ann Cross in Hogberg, 1971.)

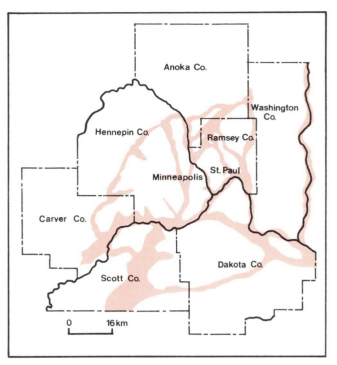

Figure 14-5. Major bedrock valleys in the Twin Cities metropolitan area. All but the Minnesota, Mississippi, and St. Croix river valleys lie buried beneath glacial drift. (After Hogberg, 1971.)

Figure 14-6. Fort Snelling in 1844, as depicted in watercolor and gouache by John Casper Wild. (Courtesy of the Minnesota Historical Society.)

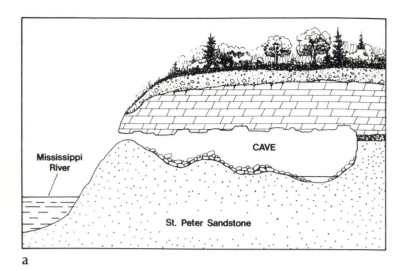

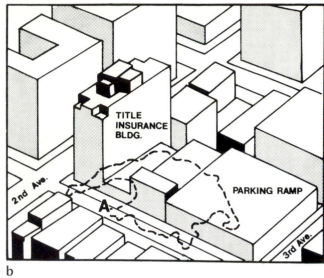

Figure 14-7. (a) Sandstone caves such as this are formed by the removal of sand along the contact between the St. Peter and Platteville formations by groundwater seepage. (b) Surface location of a sandstone cave in downtown Minneapolis. (After Hogberg, 1971.)

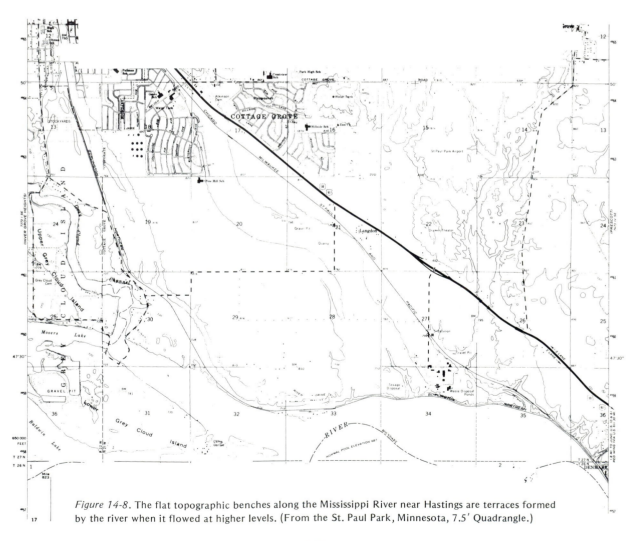

Figure 14-8. The flat topographic benches along the Mississippi River near Hastings are terraces formed by the river when it flowed at higher levels. (From the St. Paul Park, Minnesota, 7.5' Quadrangle.)

of the red sandy drift of the St. Croix Moraine. One large outwash apron grades into the Minnesota River Valley at an elevation of about 250 m near Fort Snelling, making an ideal site, at least topographically speaking, for the Minneapolis-St. Paul International Airport.

Hastings and Red Wing

The bluffs along the Mississippi River near Hastings display a thick section of buff dolomite, part of the Ordovician Prairie du Chien Group, and, at river level near Lock and Dam No. 2, the Upper Cambrian Jordan Sandstone. A gradational contact between the Cambrian and Ordovician systems here consists of alternating dolomite and sandstone layers each a few centimeters thick. Some layers display broken fragments of dolomite in a sandy matrix. These "rip-up conglomerates" are thought to represent the ripping up of cohesive bottom sediments by wave erosion and redeposition before final lithification. Deep gullies in the vicinity of the dam on the north side of the river mark the location of faults along which vertical movements have displaced the rocks as much as 25 m.

Evidence for several periods of alluvial filling and subsequent river downcutting is displayed in the form of terraces (Figure 14-8). On the north side of the river, the Langdon Terrace is nearly 5 km wide, a remnant of the braided meltwater system that drained the Des Moines Lobe (Chapter 7). On the south side of the valley the terrace segments are narrower but still distinct from Hastings to Nininger and along Spring Lake.

The Vermillion River enters the Mississippi River Valley in south Hastings via a 1.5-km-long gorge headed by a waterfall just east of U.S. Highway 61 (Figure 14-9). One of the first flour mills in Minnesota was constructed on this stretch of the Vermillion in 1857 by Alexander Ramsey. Cutting of the gorge began when River Warren had cleared the main valley of its thick alluvial fill, about the same time that the ancestral St. Anthony Falls came into existence at Fort Snelling. Comparing the progress of waterfall retreat points out several controls on rates of erosion. At Hastings the low-volume Vermillion River cut through massive dolomite. The high-volume Mississippi removed the poorly cemented St. Peter Sandstone, thus undercutting the Platteville Formation. It cut an 11-km-long gorge in the time it took the Vermillion to erode 1.5 km.

A few kilometers south of Hastings, vast sand plains grade eastward to the terraces in the Mississippi Valley, from sources in moraines of the Superior and Des Moines lobes to the west. Bordering the outwash is an area of older drift and bedrock, all mantled by loess. In several places, delicate erosional remnants of bedrock in the form of castellated outcrops indicate that the area has not been covered by ice from the last glaciation, for such fragile topographic features would certainly have been leveled (Figure 14-10).

Figure 14-9. The falls of the Vermillion River at Hastings came into existence when River Warren entrenched the Mississippi Valley, into which the Vermillion River flows. (Photo was taken about 1870 by Charles A. Zimmerman. Courtesy of the Minnesota Historical Society.)

Figure 14-10. Chimney Rock, in Dakota County near Hastings, is an erosional remnant of the St. Peter Sandstone standing like a kingpin ready to be bowled over by the next glacial advance.

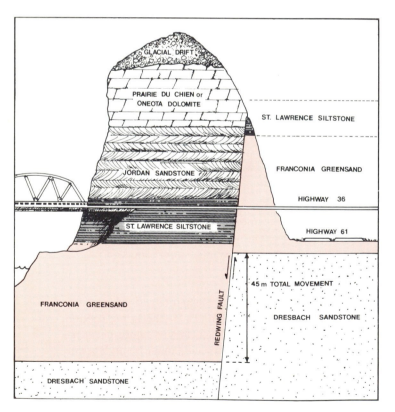

Figure 14-11. Generalized cross section of the Red Wing Fault exposed in Barn Bluff in downtown Red Wing. The view is to the south. (After Sloan, 1967.)

Soldiers' Memorial Park in Red Wing offers a striking view of the Mississippi River floodplain, partially obstructed by Barn Bluff, an outlier of bedrock cut off from the main upland by an earlier course of the ancestral Mississippi River. The bottomland is a complex of sandbars, oxbow lakes, and sloughs, all part of the Mississippi River floodplain. Barn Bluff is capped by resistant dolomite of the Prairie du Chien Group, as is the mesa upon which the park is developed. Hardwood forest covers the sides of the many valleys and gullies cut into the thin drift and bedrock.

At road level near the intersection of U.S. Highways 63 and 61, at the west end of Barn Bluff, is a good view of the Red Wing Fault (Figure 14-11). The exposure shows the yellow Jordan Sandstone in fault contact with the Franconia greensand. The green color of the greensand comes from the mineral glauconite, which occurs here in abundance. Separating the two formations is a 1-m, near-vertical zone of crushed and powdered rock, indicating the actual plane along which the rocks slipped past each other. About 40 m of vertical movement is indicated by matching beds on both sides of the fault plane.

Regional measurements of the elevations of formation contacts have revealed a broad uparching of the rocks, called the Red Wing-Rochester Anticline. Relationships shown on Figure 14-2 indicate that this deformation occurred sometime before the Cretaceous, because sediments of that age are not affected. Such structures in other areas have proved to be good traps for the accumulation of

240

petroleum. So far this anticline has produced nothing but groundwater, indicating that the conditions were just not right for oil (see Chapter 9). What a difference a little geological luck would have made to the economic history of southeastern Minnesota!

Clay deposits south of Red Wing, mainly of Cretaceous age, were responsible for the development of a well-known ceramic industry, Red Wing Potteries.

Cave Country

Wherever limestone and dolomite make up part of the bedrock, caves are likely to be found, because those rock types are susceptible to fairly extensive solution by groundwater. Such is the case in southeastern Minnesota. Fillmore and Olmsted counties contain the majority of known caves in Minnesota, indicating that optimal conditions for their development exist there. Most of these natural cavities are found in limestones of the Ordovician Dubuque and Galena formations. Mention has been made of caves in the St. Peter Sandstone in the Twin Cities area, but those are generally small and few compared to the limestone caves of extreme southeastern Minnesota.

Limestone caves develop best near the water table, the boundary between saturated and nonsaturated rock and soil. Water moving down to the water table becomes more chemically active as it picks up carbon dioxide from decaying organic material in the soil horizon. The resulting mild carbonic acid can dissolve calcite, the mineral that makes up limestone. Such "soil water" quickly becomes saturated as it moves through limey material, and its dissolving capability ends. But at the water table it mixes with water of slightly different acidity, and the result is a chemical phenomenon that is difficult to explain. The net result of the commingling of the waters is renewed chemical energy to consume more limestone. This process of "mixing corrosion" is most effective at the water table, because that is the broadest zone of mixing.

Solution proceeds most rapidly along pathways within the rock that allow the greatest volume of water to flow through. Bedding planes and fracture zones are the major controls, and their location and orientation in the bedrock generally fix the occurrence and geometry of caves. Another important factor in the development of caves is the existence of an outlet for the discharge of saturated water. Deeply incised valleys allow the water to bleed from the rocks surrounding them, thus making room for a fresh supply as well as evacuating the limestone in solution.

As long as a cave remains full of water, it continues to be enlarged by solution. However, when the water table drains and the cave fills with air, another stage in its development begins, that of decoration with cave deposits. The most beautiful of these are the chemical sediments, mainly calcium carbonate, in the form of pillars and pendants, stalagmites and stalactites, and flowstone structures covering the walls,

ceilings, and floor. The floor, of course, also is burdened with fragments of rock and mineral broken loose from above. Even bat colonies, which love the dark, safe quiet environment, contribute deposits to the litter.

The chemical sediments, or *speleothems*, are precipitated when the naturally carbonated mineralized water (bottled and widely sold under, for example, the trade name "Perrier Water") seeps out onto the cave wall and loses some of its carbon dioxide to the air in the cave. When it goes flat, calcium carbonate is deposited as well. Repeated bathing of the ceilings and dripping to the floor build up impressive incrustations.

Exactly when Minnesota's caves were formed is a matter of conjecture. The downcutting of the Mississippi River Valley certainly was an important event in their evolution. Some geologists believe they are as young as the end of the last glaciation, when River Warren cut deeply along the present-day course of the Mississippi. But there is abundant evidence that the main valley is much older and that it has undergone several cycles of downcutting and filling during the Pleistocene. New techniques for determining the ages of cave deposits are being perfected. Early results suggest that some Minnesota cave deposits are older than 30,000 years.

Although roughly 150 caves have been reported in Minnesota, many more no doubt exist. A clue to their presence is the occurrence of small depressions or *sinkholes* on the surface. These natural pits result from solution and collapse. They are commonly recognizable in fields because of clumps of trees growing in them, as well as trash piles. Disappearing streams also indicate subterranean passageways. Terrestrial deposits washed in from above are commonly encountered in cave systems. *Karst topography* is the name given to surfaces developed by solution and subsidence into underground cavities (Figure 14-12).

Two cave systems in southeastern Minnesota are open as commercial ventures, offering guided tours. Mystery Caves, about 10 km southeast of Spring Valley, in Fillmore County, was discovered in 1937 by Joseph Petty (Figure 14-13). Just 1.6 km north of the Minnesota-Iowa border and about 8 km southwest of Harmony, in the same county, is Niagara Cave. Both afford extensive opportunities to view the natural history of Minnesota's caves in comfort and safety.

Rochester and Vicinity

Almost all of southeastern Minnesota east of Interstate Highway 35 lies outside the limits of the last glaciation; that fact is represented strikingly in the landscape: it has no lakes. But, while ice was advancing to the Bemis Moraine, the farthest stand of the Des Moines Lobe, wind deposited an extensive blanket of loess upon the older drift (perhaps Kansan) and bedrock outside the ice border. This blanket of fine-textured, stone-free material makes up much of the parent soil material throughout southeastern Minnesota. Vigorous erosion, by rivers tributary to the deep valleys of

Figure 14-12. Aerial photo of sinkholes in limestone of the
Galena Formation in Fillmore County. (Courtesy of R. E. Sloan.)

the Mississippi River, has stripped much of the soil away, baring the bedrock in many places. Elsewhere, soils have developed directly on rock surfaces, producing a residuum of rusty-colored, sandy-textured sediment with fragments of chert, quartz, and other resistant material. Gravity brings a mixture of everything, including blocks of bedrock, down the valley sides to form thick aprons of slope sediments.

Rochester is situated on the south end of the Red Wing-Rochester Anticline. Erosion along the stretched and fractured crest of that arched structure had cut deeply into the St. Peter Sandstone and younger formations to show its contact with the Prairie du Chien in a few places. Numerous hillsides within the city are underlain by loose white sand derived from and slumped around a core of St. Peter Sandstone. Roadcuts a few kilometers west of Rochester along U.S. Highway 14 show the St. Peter overlain by the greenish shale of the Glenwood Formation and the lower dolomitic limestone of the Platteville. Here, the succession of sandstone, shale, and limestone demonstrates the onlap of the Ordovician sea and sedimentary facies (see Figures 6-17 and 6-19).

In south Rochester, at Golden Hill, a generalized section of rock exposures includes, from bottom to top, a few meters of Glenwood Formation, about 6 m of Platteville Formation, and about 15 m of Decorah Shale, all topped by the lower 11 m of the Galena Formation. In places, these rocks are extremely fossiliferous, and a search of these ex-

posures can be very rewarding to the fossil collector (see Figure 6-23).

Areas underlain by limestone are susceptible to collapse as the result of the development of cavities by solution. The existence of caverns presents certain environmental difficulties, especially to the stability of structures built above them. The Rochester metropolitan areas has such problems. Solution of the Prairie du Chien Group has been followed by the collapse of the overlying formations, promoting instability of the surface above. Active expansion of the caverns appears to be proceeding under the present-day hydrologic conditions, especially along the major river valleys, where groundwater discharges most rapidly.

Another environmental problem concerns the protection of groundwater from contamination by harmful chemical and biological sources on the surface. Karst landscapes are enormous sieves. Surface water has rapid access to underground reservoirs via solution-widened fracture systems and sinkholes. Potentially harmful agricultural chemicals such as pesticides, fertilizers, and weed-killers, as well as the innumerable products of human activity which end up as waste in landfills and sewage systems, all can and do easily trickle down into the groundwater supply. Therefore, extraordinary efforts are necessary to preserve the quality of both surface and groundwater systems, for without a safe water supply, the quality of life itself is drastically diminished.

Figure 14-13. Cave deposits in Niagara Cave, a limestone cave near Harmony. (Courtesy of Rose K. Cremer.)

Bibliography

Bibliography

Ackroyd, E. A., Walton, W. C., and Hills, D. L. 1967. Ground-water contribution to streamflow and its relation to basin characteristics in Minnesota. Minnesota Geological Survey RI 6.

Aguar, C. E. 1971, *Exploring St. Louis County Historical Sites*. St. Louis County Historical Society.

Alcock, F. J. 1947. A century of the history of the Geological Survey of Canada. Geological Survey of Canada Special Contribution No. 47-1.

Austin, G. S. 1972a. Paleozoic lithostratigraphy of southeastern Minnesota. In P. K. Sims and G. B. Morey, eds., *Geology of Minnesota: A Centennial Volume*, pp. 459-73. Minnesota Geological Survey.

Austin, G. S. 1972b. The Sioux Quartzite, southwestern Minnesota. In P. K. Sims and G. B. Morey, eds., *Geology of Minnesota: A Centennial Volume*, pp. 450-55. Minnesota Geological Survey.

Austin, G. S. 1972c. Cretaceous rocks. In P. K. Sims and G. B. Morey, eds., *Geology of Minnesota: A Centennial Volume*, pp. 509-12. Minnesota Geological Survey.

Bath, G. D., Schwartz, G. M., and Gilbert, F. D. 1964, Aeromagnetic and geologic map of northwestern Minneosta, scale 1:250,000. United States Geological Survey Geophysical Investigations Map GP-471.

Berry, E. W. 1939. Fossil plants from Minnesota. *Journal of the Washington Academy of Sciences*, 29:331-38.

Bleifuss, R. L. 1972. The iron ores of southeastern Minnesota. In P. K. Sims and G. B. Morey, eds., *Geology of Minnesota: A Centennial Volume*, pp. 498-505. Minnesota Geological Survey.

Bonnichsen, B. 1972. Southern part of Duluth Complex. In P. K. Sims and G. B. Morey, eds., *Geology of Minnesota: A Centennial Volume*, pp. 361-87. Minnesota Geological Survey.

Bonnichsen, B. 1972b. Sulfide minerals in the Duluth Complex. In P. K. Sims and G. B. Morey, eds., *Geology of Minnesota: A Centennial Volume*, pp. 388-93. Minnesota Geological Survey.

Bonnichsen, B. 1974. Copper and nickel resources in the Duluth Complex, northeastern Minnesota. Minnesota Geological Survey Information Circular 10.

Bray, E. C. 1962. *Billions of Years in Minnesota*. Science Museum of Minnesota.

Bridges, H. 1952. *Iron Millionaire: Life of Charlemagne Tower*. Philadelphia: University of Pennsylvania Press.

Briggs, R. C., and Rafn, E. F. 1979. Minerals in the economy of Minnesota. United States Department of the Interior, Bureau of Mines.

Canby, T. Y. 1980. Water: Our most precious resource. *National Geographic*, August: 145.

Chase, C. A., and Gilmer, T. H. 1973. Precambrian plate tectonics: The midcontinent gravity high. *Earth and Planetary Science Letters*, 21:70-78.

Clark, T. H., and Stearn, C. W. 1960. *The Geological Evolution of North America*. New York: Ronald Press.

Clements, J. M. 1903. The Vermilion iron-bearing district of Minnesota. United States Geological Survey Monograph 45.

Cooper, W. S. 1935. The history of the upper Mississippi River in late Wisconsin and postglacial time. Minnesota Geological Survey Bulletin 26.

Craddock, C. 1972. Late Precambrian geological setting. In P. K. Sims and G. B. Morey, eds. *Geology of Minnesota: A Centennial Volume*, pp. 281-91. Minnesota Geological Survey.

Craddock, C, Mooney, H. M., and Kolehmainen, V. 1969. Simple Bouguer gravity map of Minnesota and northwestern Wisconsin. Minnesota Geological Survey. Miscellaneous Map Series Map M-10.

Darby, D. G. 1972. Evidences of Precambrian life in Minnesota: In P. K. Sims and G. B. Morey, eds., *Geology of Minnesota: A Centennial Volume*, p. 264-71. Minnesota Geological Survey.

Davidson, D. M., Jr. 1972. Eastern part of the Duluth Complex. In P. K. Sims and G. B. Morey, eds. *Geology of Minnesota: A Centennial Volume*, pp. 354-60. Minnesota Geological Survey.

DeKruif, P. 1929. *Seven Iron Men*. New York: Harcourt, Brace.

Dodge, O. 1865. Gold letters no. 2. St. Paul *Pioneer Press*, October 5.

Dott, R. H., Jr., and Batten, R. L. 1981. *Evolution of the Earth*. (3rd ed.) New York: McGraw-Hill.

Driscoll, F. G. 1976. Formation and wastage of Neoglacial surge moraines of the Klutlan Glacier, Yukon Territory, Canada. Ph.D. dissertation, University of Minnesota.

Duluth Board of Education. 1923. *The Story of Duluth*. Duluth Board of Education.

Eames, H. E. 1866. Report of the state geologist, Henry H. Eames, on the metalliferous region bordering on Lake Superior, 1866. St. Paul: Frederick Driscoll, state printer.

Elson, J. A. 1967. Geology of Glacial Lake Agassiz. In W. J. Mayer-Oakes, ed., *Life, Land and Water*, Winnipeg: University of Manitoba Press. pp. 37-96.

Fawcett, G. 1970. *Stories by the Missabe Old Timer*. Duluth, Missabe and Iron Range Railway.

Fritzen, J. 1974. *Historic Sites and Place Names of Minnesota's North Shore*. St. Louis County Historical Society.

Goldich, S. S. 1938. A study in rock weathering. *Journal of Geology*, 46:17-58.

BIBLIOGRAPHY

Goldich, S. S. Nier, A. O., Baadsgaard, H., Hoffman, J. H., and Krueger, H. K. 1961. The Precambrian geology and geochronol-Minnesota. *Geological Society of America Bulletin*, 81:3671-96.

Goldich, S. S., Nier, A. O., Baadsgaard, H., Hoffman, J. H., and Krueger, H. K. 1961. The Precambrian geology and geochronology of Minnesota. Minnesota Geological Survey Bulletin 41.

Grant, J. A. 1972. Minnesota River Valley, southwestern Minnesota. In P. K. Sims and G. B. Morey, eds., *Geology of Minnesota: A Centennial Volume*, pp. 177-96. Minnesota Geological Survey.

Green, J. C. 1972. North Shore Volcanic Group. In P. K. Sims and G. B. Morey, eds., *Geology of Minnesota: A Centennial Volume*, pp. 294-332. Minnesota Geological Survey.

Green, J. C. 1977. Keweenawan plateau volcanism in the Lake Superior region. Geological Association of Canada Special Paper No. 16, pp. 407-22.

Green, J. C. 1978. Why is Lake Superior? *The Minnesota Volunteer*, 41:10-19.

Green, J. C. 1979. *Field Trip Guidebook for the Keweenawan (Upper Precambrian) North Shore Volcanic Group, Minnesota*. Minnesota Geological Survey Guidebook Series No. 11.

Green, J. C. 1982. Geologic map of Minnesota, Two Harbors sheet, scale 1:250,000. Minnesota Geological Survey.

Grout, F. F. 1918. The lopolith: An igneous form exemplified by the Duluth Gabbro. *American Journal of Science* (Fourth Series), 46:516-22.

Grout, F. F. 1923. The magnetite pegmatites of northern Minnesota. *Economic Geology*, 18:253-69.

Grout, F. F. 1925. The Coutchiching problem. *Geological Society of America Bulletin*, 36:351-64.

Grout, F. F., Sharp, R. P., and Schwartz, G. M. 1959. The geology of Cook County, Minnesota. Minnesota Geological Survey Bulletin 39.

Grout, F. F., Stauffer, C. R., Allison, I. S., Gruner, J. W., Schwartz, G. M. Thiel, G. A., and Emmons, W. H. 1932. Geologic map of the state of Minnesota, scale 1:500,000. Minnesota Geological Survey.

Gruner, J. W. 1941. Structural geology of the Knife Lake area of northeastern Minnesota. *Geological Society of America Bulletin*, 52:1577-1642.

Gruner, J. W., 1946. The mineralogy and geology of the taconites and iron ores of the Mesabi Range, Minnesota. Office of the Commission on Iron Range Resources and Rehabilitation in cooperation with the Minnesota Geological Survey.

Hammer, S. and Ervin, C. P. 1975. Crater or kettle? A geophysical study. *Geology*, 3:145-46.

Heinselman, M. L. 1963. Forest sites, bog processes and peatland types in the Glacial Lake Agassiz region, Minnesota. *Ecological Monographs*, 33:327-74.

Hogberg, R. K. 1971. *Environmental Geology of the Twin Cities Metropolitan Area*. Minnesota Geological Survey Educational Series-5.

Hogberg, R. K., and Bayer, T. N. 1967. *Guide to the Caves of Minnesota*. Minnesota Geological Survey Educational Series-4.

Holmquist, J. D., and Brookings, J. A. 1963. *Minnesota's Major Historic Sites*. Minnesota Historical Society.

Huck, V. 1955. *Brand of the Tartan*. New York: Appleton-Century-Crofts.

Jenks, A. E. 1937. Minnesota's Browns Valley Man and associated burial artifacts. American Anthropological Association Memorandum 49.

Jirsa, M. A. 1980. The petrology and tectonic significance of interflow sediments in the Keweenawan North Shore Volcanic Group of Northeastern Minnesota. M.S. thesis, University of Minnesota, Duluth.

Johnson, E. 1969. *The Prehistoric Peoples of Minnesota*. Minnesota Historical Society.

Judson, S., Deffeyes, K. S., and Hargraves, B. 1976. *Physical Geology*. Englewood Cliffs, N. J.: Prentice-Hall.

Kain, J. 1978. *Rocky Roots*. Ramsey County Historical Society.

Kay, M. and Colbert, E. H. 1965. *Stratigraphy and Life History*. New York: John Wiley and Sons.

Klinger, F. L. 1956. Geology of the Soudan Mine and vicinity. In G. M. Schwartz, ed., *Geological Society of America Guidebook Series, Field Trip No. 1, Precambrian of Northeastern Minnesota*, pp. 120-34. Prepared for the Annual Meeting of the Geological Society of America, Minneapolis, Minnesota.

Lawson, A. C. 1888. Report on the geology of the Rainy Lake region. Canada Geological Survey Annual Report No. 3.

Lawson, A. C. 1913. The Archaean geology of Rainy Lake restudied. Canada Geological Survey Memoir 40.

Leith, C. K. 1903. The Mesabi iron-bearing district of Minnesota. United States Geological Survey Monograph 43.

Lesquereux, L., Schuchert, C., Woodward, A., Ulrich, G. O., Thomas, B. W., and Winchell, N. H. 1895. The geology of Minnesota. Volume III, part I, of the Final Report, 1885-1892. Minnesota Geological and Natural History Survey.

Leverett, F. 1932. Quaternary geology of Minnesota and parts of adjacent states. United States Geological Survey Professional Paper 161.

Listerud, W. H., and Meineke, D. G. 1977. Mineral resources of a portion of the Duluth Complex and adjacent rocks in St. Louis and Lake counties, northeastern Minnesota. Minnesota Department of Natural Resources, Division of Minerals, Minerals Exploration Section Report 93.

Lydecker, R. 1976. *The Edge of the Arrowhead*. Minnesota Marine Advisory Service.

Mackin, J. H. 1948. Concept of the graded river. *Geological Society of America Bulletin*, 59:463-512.

Marsden, R. W. 1978. Iron ore reserves of the Mesabi Range, Minnesota. Proceedings of the 51st Annual Meeting of the Minnesota Section of the American Institute of Mining Engineers and the 37th Annual Mining Symposium, pp. 23-1 to 23-57.

Martin, L. 1911. The geology of the Lake Superior district. United States Geological Survey Monograph 52, p. 456.

Matsch, C. L. 1972. Quaternary geology of southwestern Minnesota. In P. K. Sims and G. B. Morey, eds., *Geology of Minnesota: A Centennial Volume*, pp. 548-60. Minnesota Geological Survey.

Matsch, C. L. 1976. *North America and the Great Ice Age*. New York: McGraw-Hill.

Mattis, A. F. 1972. The petrology and sedimentation of the basal Keweenawan sandstones of the north and south shores of Lake Superior. M.S. thesis, University of Minnesota, Duluth.

Mattis, A. F. 1976. Puckwunge Formation of northeastern Minnesota. In P. K. Sims and G. B. Morey, eds., *Geology of Minnesota: A Centennial Volume*, pp. 412-15. Minnesota Geological Survey.

McGinnis, L., Durfee, G. and Ikola, R. J. 1973. Simple Bouguer gravity map of Minnesota, Roseau sheet, scale 1:250,000. Minnesota Geological Survey Miscellaneous Map Series M-12.

Meuschke, J. L., Brooks, K. G., Henderson, J. R., Jr., and Schwartz, G. M. 1957. Aeromagnetic and geologic map of northeastern Koochiching County, Minnesota. United States Geological Survey Geophysical Investigations Map GP-133.

Minnesota Conservation Department, Division of Waters. 1959. Hydrologic atlas of Minnesota. Minnesota Conservation Department Bulletin 10.

Minnesota Department of Economic Development. 1975. Minnesota statistical profile. Minnesota Department of Economic Development, Research Division.

Minnesota Department of Natural Resources. 1978. Minnesota peatlands map. Minnesota Department of Natural Resources.

Minnesota Geological Survey. 1969. The proposed Voyageurs National Park: Its geology and mineral potential. Minnesota Geological Survey.

Minnesota Pollution Control Agency, Minnesota Department of Natural Resources, and the Minnesota Department of Health, 1981, Acid precipitation in Minnesota. (Report to the legislative commission on Minnesota resources, Final Draft)

Mooney, H. M. 1979. Earthquake history of Minnesota. Minnesota Geological Survey, RI 23.

Morey, G. B. 1967. Stratigraphy and petrology of the type Fond du Lac Formation, Duluth Minnesota. Minnesota Geological Survey RI 7.

Morey, G. B. 1972. Middle Precambrian geologic setting. In P. K. Sims and G. B. Morey, eds., *Geology of Minnesota: A centennial Volume*, pp. 199-203. Minnesota Geological Survey.

Morey, G. B. 1974. Cyclic sedimentation of the Solor Church Formation (Upper Precambrian, Keweenawan), southeastern Minnesota. *Journal of Sedimentary Petrology*, 44:872-84.

Morey, G. B. 1976. Geologic map of Minnesota, bedrock geology, scale 1:3,168,000. Minnesota Geological Survey.

Morey, G. B. 1977. Revised Keweenawan subsurface stratigraphy, southeastern Minnesota. Minnesota Geological Survey RI 16.

Morey, G. B. 1978. Lower and middle Precambrian stratigraphic nomenclature for east-central Minnesota. Minnesota Geological Survey RI 21.

Morey, G. B., Green, J. C., Ojakangas, R. W., and Sims, P. K. 1970. Stratigraphy of the lower Precambrian rocks in the Vermilion district, northeastern Minnesota. Minnesota Geological Survey RI 14.

Morey, G. B., Olsen, B. M., and Southwick, D. L. 1981. Geologic map of Minnesota, East-Central Minnesota, scale 1:250,000. Minnesota Geological Survey.

Morey, G. B., and Sims, P. K. 1976. Boundary between two Precambrian terranes in Minnesota and its geologic significance. *Geological Society of America Bulletin*, 87:141-52.

Mossler, J. H. 1978. Cedar Valley Formation (Devonian) of Minnesota and northern Iowa. Minnesota Geological Survey R I 18.

Mudrey, M. G., Jr. 1972. Green prospect, Cook County. In P. K. Sims and G. B. Morey, eds., *Geology of Minnesota: A Centennial Volume*, p. 411. Minnesota Geological Survey.

Nathan, H. D. 1969. The geology of a portion of the Duluth Complex, Cook County. Ph.D. dissertation, University of Minnesota, Minneapolis.

Nute, Grace Lee. 1950. *Rainy River County*. Minnesota Historical Society.

Nute, Grace Lee (editor). 1951, *The Mesabi Pioneer: reminiscences of Edward J. Longyear*. Minnesota Historical Society.

Ojakangas, R. W. 1972. Rainy Lake area. In P. K. Sims and G. B. Morey, eds., *Geology of Minnesota: A Centennial Volume*, pp. 163-71. Minnesota Geological Survey.

Ojakangas, R. W., and Darby, D. A. 1976. *The Earth—Past and Present*. New York: McGraw-Hill.

Ojakangas, R. W., Meineke, D. G., and Listerud, W. H. 1977. Geology, sulfide mineralization and geochemistry of the Birchdale-Indus area, Koochiching County, northwestern Minnesota. Minnesota Geological Survey RI 17.

Ojakangas, R. W., and Morey, G. B., 1972. *Field Trip Guidebook for Lower Precambrian Volcanic-Sedimentary Rocks of the Vermilion District, Minnesota*. Minnesota Geological Survey Guidebook Series No. 2.

Ojakangas, R. W., Mossler, J. H., and Morey, G. B. 1979. Geologic map of Minnesota, Roseau sheet, scale 1:250,000. Minnesota Geological Survey.

Ojakangas, R. W., Sims, P. K., and Hooper, P. R. 1978. Geologic map of the Tower Quadrangle, St. Louis County, Minnesota. United States Geological Survey Map GQ-1457.

Owen, D. D. 1852. Report of a geological survey of Wisconsin, Iowa and Minnesota and incidentally a portion of Nebraska Territory. Philadelphia: Lipcott, Grambo.

Parham, W. E. 1970. Clay mineralogy and geology of Minnesota's kaolin clays. Minnesota Geological Survey Special Publication No. 10.

Parham, W. E. 1972. A possible peneplain of Early Late Cretaceous Age in Minnesota. In C. L. Matsch, *Field Trip Guidebook for Geomorphology and Quaternary Stratigraphy of Western Minnesota and Eastern South Dakota*, pp. 58-68. Minnesota Geological Survey Guidebook Series No. 7.

Peterson, E. C. 1952. Gold mining in northern Minnesota. Unpublished paper, University of Minnesota at Duluth.

Phinney, W. C. 1972. Northwestern part of Duluth Complex. In P. K. Sims and G. B. Morey, eds., *Geology of Minnesota: A Centennial Volume*, pp. 335-45. Minnesota Geological Survey.

Poulsen, K. H., Borradaile, G. J., and Kehlenbeck, M. M. 1980. An inverted Archaean succession at Rainy Lake, Ontario. *Canadian Journal of Earth Sciences*, 17:1358-69.

Rapp, G. R., Jr., and Wallace, D. T. 1966. *Guide to Mineral Collecting in Minnesota*. Minnesota Geological Survey Educational Series-2.

Ruhe, R. V. 1969. *Quaternary Landscapes in Iowa*. Ames, Iowa: Iowa State University Press.

Schneider, A. F. 1961. Pleistocene geology of the Randall region, central Minnesota. Minnesota Geological Survey Bulletin 40.

Schwartz, G. M. 1925. A guidebook to Minnesota trunk highway No. 1. Minnesota Geological Survey Bulletin 26.

Schwartz, G. M. 1936. The geology of the Minneapolis-St. Paul metropolitan area. Minnesota Geological Survey Bulletin 27.

Schwartz, G. M. 1949. The geology of the Duluth metropolitan area. Minnesota Geological Survey Bulletin 33.

Schwartz, G. M. 1964. History of the Minnesota Geological Survey. Minnesota Geological Survey Special Publication No. 1.

Schwartz, G. M., and Thiel, G. A. 1954. *Minnesota's Rocks and Waters*. (Minnesota Geological Survey Bulletin 37.) Minneapolis: University of Minnesota Press.

Silver, L. T., and Green J. C. 1973. Time constraints for Keweenawan igneous activity. *Geological Society of America Abstracts with Programs* 4 (7): 665-66.

Sims, P. K. 1970. Geologic map of Minnesota, scale 1:1,000,000. Minnesota Geological Survey.

Sims, P. K. 1972. Mineral deposits in Lower Precambrian rocks, northern Minnesota. In P. K. Sims and G. B. Morey, eds., *Geology of Minnesota: A Centennial Volume*, pp. 172-76. Minnesota Geological Survey.

Sims, P. K. 1973. Geologic map of western part of Vermilion district, northeastern Minnesota. Minnesota Geological Survey Miscellaneous Map Series Map M-13.

Sims, P. K., Card, K. D., Morey, G. B., and Peterman, Z. E. 1980. The Great Lakes tectonic zone—a major crustal structure in central North America. *Geological Society of America Bulletin*, 91:690-98.

Sims, P. K., and Morey, G. B. 1966. *Geologic Sketch of the Tower-Soudan State Park*. Minnesota Geological Survey Educational Series-3.

Sims, P. K., and Morey, G. B., eds. 1972. *Geology of Minnesota: A Centennial Volume*. Minnesota Geological Survey.

Sims, P. K., Morey, G. B., Ojakangas, R. W., and Viswanathan, S. 1970. Geologic map of Minnesota, Hibbing sheet, scale 1:250,000. Minnesota Geological Survey.

Sloan, R. E. 1967. A teacher's guide for geologic field investigations in southeastern Minnesota. In W. C. Phinney, ed., *A Teacher's Guide for Geologic Field Investigations in Minnesota*, pp. 1-19. Minnesota Department of Education and Minnesota Geological Survey.

Sloan, R. E. 1972. Notes on the Platteville Formation, southeastern Minnesota. In G. F. Webers and G. S. Austin, *Field Trip Guidebook for Paleozoic and Mesozoic Rocks of Southeastern Minnesota*, pp. 43-53. Minnesota Geological Survey Guidebook Series No. 4.

Sloan, R. E., and Austin, G. S. 1966. Geologic map of Minnesota, St. Paul sheet, scale 1:250,000. Minnesota Geological Survey.

Southwick, D. L. 1972. Vermilion granite-migmatite massif. In P. K. Sims and G. B. Morey, eds., *Geology of Minnesota: A Centennial Volume*, pp. 109-19. Minnesota Geological Survey.

Southwick, D. L. 1980. Subsurface research and scientific drilling in western Minnesota. Minnesota Geological Survey Information Circular 18.

Southwick, D. L., and Ojakangas, R. W. 1979a. *Field Trip Guidebook for the Western Vermilion District, Northeastern Minnesota*. Minnesota Geological Survey Guidebook Series No. 10.

Southwick, D. L., and Ojakangas, R. W., 1979b. Geologic map of Minnesota, International Falls sheet, scale 1:250,000. Minnesota Geological Survey.

Stover, C. W., Simon, R. B., Person, W. J., and Minsch, J. H. 1977. Earthquakes of the United States July-September, 1975. United States Geological Survey Circular 749 C.

Taylor, R. B., 1963. Bedrock geology of Duluth and vicinity. Minnesota Geological Survey Map GM-1.

Trethewey, W. D. 1979. Minnesota mining directory. Mineral Resources Research Center, University of Minnesota.

Trevor, R. 1979. *Voyageur Country: A Park in the Wilderness*. Minneapolis: University of Minnesota Press.

Tufford, S. and Hogberg, R. K. 1965. *Guide to Fossil Collecting in Minnesota*. Minnesota Geological Survey Educational Series-1.

Upham, W. 1896. The Glacial Lake Agassiz. United States Geological Survey Monograph 25.

Wallace, A. L. 1876. *The Geographical Distribution of Animals*. Volume 1. London: Macmillan.

Watts, D. R. 1981. Paleomagnetism of the Fond du Lac Formation and the Eileen and Middle River sections with implications for Keweenawan tectonics and the Grenville problem. *Canadian Journal of Earth Sciences*, 18: 829-41.

Webers, G. F. 1972. Paleoecology of the Cambrian and Ordovician strata of Minnesota. In P. K. Sims and G. B. Morey, eds., *Geology of Minnesota: A Centennial Volume*, pp. 474-84. Minnesota Geological Survey.

Webers, G. F., and Austin, G. S. 1972. *Field Trip Guidebook for Paleozoic and Mesozoic Rocks of Southeastern Minnesota*. Minnesota Geological Survey Guidebook Series No. 4.

Weiblen, P. W. 1965. A funnel-shaped gabbro-troctolite intrusion in the Duluth Complex, Lake County, Minnesota. Ph.D. dissertation, University of Minnesota, Minneapolis.

Weiblen, P. W., and Davidson, D. M., Jr. 1972. *Field Trip Guidebook for Precambrian Geology of Northwestern Cook County, Minnesota*. Minnesota Geological Survey Guidebook Series No. 6.

Weiblen, P. W., Mathez, E. A., and Morey, G. B. 1972. Logan intrusions. In P. K. Sims and G. B. Morey, eds., *Geology of Minnesota: A Centennial Volume*, pp. 394-406. Minnesota Geological Survey.

Weiblen, P. W., and Morey, G. B. 1976. Textural and compositional characteristics of sulfide ores from the basal contact zone of the South Kawishiwi Intrusion, Duluth Complex, northeastern Minnesota. Minnesota Geological Survey Reprint Series 32.

Weiblen, P. W., and Morey, G. B. 1980. A summary of the stratigraphy, petrology and structure of the Duluth Complex. *American Journal of Science*, 280-A: 88-133.

White, W. S. 1957. Regional structural setting of the Michigan native-copper district. In A. K. Snelgrove, ed., *Geological Exploration — Institute on Lake Superior Geology*, pp. 3-16. Houghton, Mich.: Michigan College Mining and Technology Press.

Williams, R. 1907. *The Honorable Peter White*. Cleveland: Penter.

Winchell, N. H. 1872. Preliminary geologic map of Minnesota. Minnesota Geological and Natural History Survey.

Winchell, N. H. 1884a. Eleventh annual report for the year 1882. Minnesota Geological and Natural History Survey.

Winchell, N. H. 1884b. Twelfth annual report for the year 1883. Minnesota Geological and Natural History Survey.

Winchell, N. H. 1889. Natural gas in Minnesota. Geological and Natural History Survey of Minnesota, Bulletin No. 5.

Winchell, N. H. 1895. Twenty-third annual report for the year 1893. Minnesota Geological Natural History Survey.

Winchell, N. H., Grant, U. S., Todd, J. E., Upham, W. and Winchell, H. V. 1899. The geology of Minnesota. Volume IV of the Final Report, 1896-1898. Minnesota Geological and Natural History Survey.

Winchell, N. H., and Upham, W. 1884. The geology of Minnesota. Volume I of the Final Report, 1872-1882. Minnesota Geological and Natural History Survey.

Winchell, N. H., and Upham, W. 1888. The geology of Minnesota. Volume II of the Final Report, 1882-1885. Minnesota Geological and Natural History Survey.

Winchell, N. H., and Winchell, H. V. 1891. Iron ores of Minnesota. Geological and Natural History Survey of Minnesota Bulletin No. 6.

Winter, T. C. 1961. A pollen analysis of Kirchner Marsh, Dakota County, Minnesota. M.S. thesis, University of Minnesota, Minneapolis.

Work Projects Administration 1940. The Cuyuna Range: A history of a Minnesota mining district. Minnesota Historical Records Survey Project, Division of Professional and Service Projects.

Wright, H. E., Jr. 1969. Glacial fluctuations and the forest succession in the Lake Superior area. Proceedings of the 12th Conference of the International Association for Great Lakes Research (Ann Arbor), p. 397-405.

Wright, H. E., Jr. 1972. Physiography of Minnesota. In P. K. Sims and G. B. Morey, eds., *Geology of Minnesota: A Centennial Volume*, pp. 561-80. Minnesota Geological Survey.

Wright, H. E., Jr. 1972b. Quaternary history of Minnesota. In P. K. Sims and G. B. Morey, eds., *Geology of Minnesota: A Centennial Volume*, pp. 515-45. Minnesota Geological Survey.

Wright, H. E., Jr. 1973. Tunnel valleys, glacial surges, and subglacial hydrology of the Superior Lobe, Minnesota. In R. F. Black, R. P. Goldthwait, and H. B. Willman, eds., *The Wisconsinan Stage*, pp. 251-76. Geological Society of America Memoir 136.

Wright, H. E., Jr., Mattson, L. A., and Thomas, J. A. 1970. Geology of the Cloquet Quadrangle, colored geologic map, scale 1:24,000. Minnesota Geological Survey Geologic Map Series No. 3.

Wright, H. E., Jr., and Watts, W. A. 1969. Glacial and vegetational history of northeastern Minnesota. Minnesota Geological Survey Special Publication No. SP-11.

Young, G. M. 1973. Tillites and aluminous quartzites as possible time markers for Middle Precambrian (Aphebian) rocks of North America. In G. M. Young, ed., Huronian stratigraphy and sedimentation, pp. 97-128. Geological Association of Canada Special Paper No. 12.

Zappfe, C. 1938. Discovery and early development of the iron ranges. In Lake Superior Iron Ore Association, *Lake Superior Iron Ores*, pp. 13-26. Cleveland: Lake Superior Iron Ore Association.

Zumberge, J. H. 1952. The lakes of Minnesota: Their origin and classification. Minnesota Geological Survey Bulletin 35.

Index

Index

Page numbers in boldface refer to illustrations.

253

Richard W. Ojakangas and **Charles L. Matsch**
are professors of geology
at the University of Minnesota, Duluth.
Ojakangas is coauthor of *The Earth — Past and Present*,
and Matsch is the author
of *North America and the Great Ice Age.*